HAPTIC FEEDBACK CONTROL IN MEDICAL ROBOTICS

HAPTIC FEEDBACK CONTROL IN MEDICAL ROBOTICS

Edited by

PUSHAN KUMAR DUTTA
School of Engineering and Technology, Amity University, Kolkata, West Bengal, India

SHUBHAM GOSWAMI
Institute of Management, Nirma University, Ahmedabad, Gujarat, India

KAMAL KANT HIRAN
Department of Computing & Informatics, Sir Padampat Singhania University, Udaipur, Rajasthan, India

RAUL VILLAMARIN RODRIGUEZ
AI Research Centre, Woxsen University, Hyderabad, Telangana, India

HEMACHANDRAN KANNAN
AI Research Centre, Woxsen University, Hyderabad, Telangana, India

ELSEVIER

ACADEMIC PRESS
An imprint of Elsevier
elsevier.com/books-and-journals

Academic Press is an imprint of Elsevier

125 London Wall, London EC2Y 5AS, United Kingdom
50 Hampshire Street, 5th Floor, Cambridge, MA 02139, United States

ISBN: 978-0-443-27777-1

For Information on all Academic Press publications visit our website at https://www.elsevier.com/books-and-journals

Publisher: Mara Conner
Acquisitions Editor: Sonnini Yura
Editorial Project Manager: Siddharth Khattri
Production Project Manager: Anitha Sivaraj
Cover Designer: Raman Kumar

Working together to grow libraries in developing countries
www.elsevier.com • www.bookaid.org

Typeset by Aptara, New Delhi, India

Contents

Contributors *xiii*

SECTION I Foundatins of haptic medical robotics

1. Introduction to haptic feedback in medical robotics **3**

Panchakarla Aditya Satvik, Valdas Sairam Goud and Manda Sandhya

1.1. Introduction 3
1.2. Fundamentals of haptic feedback 4
1.3. Haptic devices and technologies in medical robotics 7
1.4. Applications of haptic feedback in medical robotics 9
1.5. Advantages of haptic feedback in medical robotics 12
1.6. Challenges and limitations 14
1.7. Emerging technologies in haptic medical robotics 16
1.8. Future research directions 18
1.9. Conclusion 19
AI disclosure 20
References 20

2. Fundamentals of human touch perception **21**

Sanjana S, Raj Jaydeep Patel and Arun Pratap Singh

2.1. Introduction to touch perception 21
2.2. The somatosensory system: An overview 23
2.3. Anatomy and physiology of cutaneous receptors 24
2.4. Afferent pathways and neural coding of touch 28
2.5. Central processing: From thalamus to cortex 30
2.6. Active versus passive touch: The role of haptic exploration 32
2.7. Perceptual phenomena and psychophysics of touch 34
2.8. Affective and social dimensions of touch 36
2.9. Summary and open questions 37
AI disclosure 39
References 39

3. Haptic sensing technologies **41**

Sanskar Agrawal and Anusha Rani

3.1. Introduction 41

3.2. Overview of haptic technologies 42
3.3. Key application areas 46
3.4. Haptic technologies for special needs 49
3.5. Discussion and conclusion 52
AI disclosure 53
References 53

4. Haptic feedback mechanisms 59
Nasrulla K. Khan, Jitha Janardhanan and Simna Shanavas

4.1. Introduction 59
4.2. How sense of touch works 59
4.3. Technical specifications 62
4.4. Critical components and factors to operate and enhance the user experience 64
4.5. Challenges in haptic feedback 65
4.6. Types of haptic device technologies 66
4.7. Applications of haptic feedback 71
4.8. Security concerns 81
4.9. Future of haptics 82
4.10. Conclusion 82
AI disclosure 83
References 83

5. Advanced actuator technologies in medical haptics 85
Boddu Sanjana, Kosinepalli Kumar and Chirikonda Chandana

5.1. Introduction 85
5.2. Fundamentals of haptic actuation 86
5.3. Conventional actuator technologies in medical haptics 87
5.4. Advanced and emerging actuator technologies 91
5.5. Integration of actuators with haptic control systems 94
5.6. Applications of advanced actuator technologies in medical robotics 96
5.7. Design challenges and engineering considerations 98
5.8. Future trends and research directions 99
5.9. Conclusion 100
AI disclosure 101
References 101

6. Haptic interface design for medical applications 103
Polireddy Vasavi Reddy, Lakshita Gandhi and Paramaticona Reddy

6.1. Introduction 103

6.2. Fundamentals of haptic technology 104
6.3. Medical robotics and human–robot interaction 106
6.4. Artificial intelligence-driven haptic interface design 109
6.5. Design principles for medical haptic interfaces 111
6.6. Applications of haptic interfaces in medicine 113
6.7. Governance, ethics, and regulatory considerations 115
6.8. Technical challenges and research gaps 116
6.9. Future directions and emerging trends 117
6.10. Case studies and practical implementations 118
6.11. Discussion 119
6.12. Conclusion 119
AI disclosure 120
References 120

SECTION II Intelligent control and system architecture

7. Haptic feedback control in medical robotics 125

Nasrulla K. Khan, Simna Shanavas and Jitha Janardhanan

7.1. Introduction 125
7.2. Fundamentals of haptic technology 126
7.3. Artificial intelligence-driven applications in haptic medical systems 130
7.4. AI techniques used in haptic medical systems 137
7.5. Challenges and ethical considerations 139
7.6. Future directions and innovations 143
7.7. Conclusion 144
AI disclosure 144
References 144

8. Artificial intelligence frameworks for intelligent haptic control 147

Deepika Roselind Johnson, Srivats Raghavan, Kalyanasundaram Venkataraman, AJ Keerthi, Utkarsh Mishra and Aaryan Gupta

8.1. Fundamentals of haptic medical systems 147
8.2. Limitations in traditional haptic control systems 150
8.3. Artificial intelligence frameworks for intelligent haptic control 154
8.4. Technical and ethical barriers in artificial intelligence-driven haptic systems 158
8.5. Conclusion and future work 163
References 166

9. Design frameworks for multimodal haptic medical robots **169**

Divya Pokar Patel and Nivedita Ravindra Dharne

9.1. Introduction 169
9.2. Literature review 170
9.3. Haptic medical robots 171
9.4. Applications, advantages, and limitations 172
9.5. Proposed multimodal medical robot 172
9.6. Artificial intelligence and machine learning in haptic medical robots 176
9.7. Conclusion and future scope 178
9.8. Future directions 178
AI disclosure 179
References 179

10. Energy efficiency in haptic medical devices **181**

Nikhil Sharma, Pooja Choudhary and Jofia Jose Prakash

10.1. Introduction 181
10.2. Fundamentals of haptic medical devices 182
10.3. Energy consumption in haptic medical systems 185
10.4. Energy-efficient actuation strategies 186
10.5. Control strategies for energy efficiency 188
10.6. Power management and energy harvesting 191
10.7. Case studies in medical haptic systems 192
10.8. Performance evaluation metrics 194
10.9. Challenges and open research directions 195
10.10. Future trends in energy-efficient haptic medical robotics 196
10.11. Conclusion 197
AI disclosure 198
References 198

SECTION III Clinical applications and surgical systems

11. Haptic feedback in minimally invasive surgery: Restoring the sense of touch in modern surgical–practice **203**

Riya and Vishnu M.

11.1. Introduction 203
11.2. Conceptual and theoretical framework 204
11.3. Review of literature 206
11.4. Core Discussion 211
11.5. Case studies and practical applications of haptic feedback in minimally invasive surgery 212

11.6. Implications of haptic feedback in minimally invasive surgery 215
11.7. Future scope and recommendations 217
11.8. Conclusion 218
References 219

12. Tactile feedback systems for telesurgery **221**

Nafisa Aafreen, Debasis Patra, Meghna Dadich and Hemachandran Kannan

12.1. Introduction 221
12.2. Current challenges in telesurgical tactile feedback 223
12.3. Tactile sensing technologies for telesurgery 224
12.4. Tactile feedback mechanisms in telesurgery 226
12.5. Wearable haptic interfaces for telesurgery 228
12.6. Signal processing and control architectures 230
12.7. Integration with virtual reality and training systems 232
12.8. Future directions 233
12.9. Conclusions 234
AI disclosure 236
References 237

13. Haptic interfaces in rehabilitation robotics **239**

Manjunath Devarashetty, Lokesh Venkatraman and Rakesh Ramakrishna Pai

13.1. Introduction 239
13.2. Fundamentals of haptic interaction 240
13.3. Design principles of haptic interfaces for rehabilitation 243
13.4. Classification of rehabilitation haptic devices 245
13.5. Control strategies for haptic rehabilitation robotics 246
13.6. Clinical applications and use cases 250
13.7. Emerging trends and future directions 252
13.8. Challenges and open research problems 254
13.9. Conclusion 256
AI disclosure 257
References 257

14. Virtual reality and augmented reality in medical training **259**

Poojitha Kondapaka and Anand Pindiprolu

14.1. Introduction 259
14.2. Foundational concepts: Virtual reality, augmented reality, and haptic feedback 260
14.3. Applications in clinical and surgical training 262
14.4. Case studies and research findings 264

14.5. Challenges and future directions 266
14.6. Conclusion 267
References 267

15. Brain–computer interfaces and haptic feedback 271

Nihareeka Mohanty and Rudra Pratap Mahali

15.1. Introduction: Why combine braincomputer interfaces with haptics? 271
15.2. Taxonomy of brain–computer interfaces and Their Sensory Roles 273
15.3. Neural recording and stimulation technologies 276
15.4. Haptic feedback using skin sensations, muscle sense, and combined methods 279
15.5. Encoding sensory information for neural and peripheral channels 281
15.6. Signal processing and machine learning in closed-loop brain–computer interfaces 284
15.7. Behavioral and clinical outcomes: Evidence from trials and experiments 286
15.8. System design considerations and constraints 288
15.9. Ethical, legal, and social implications 290
15.10. Technical and scientific challenges 292
15.11. Emerging directions and technologies 294
15.12. Research roadmap and recommendations 295
15.13. Conclusion 297
References 299

SECTION IV Validation, emerging healthcare integration, and future challenges

16. Clinical validation and safety standards of haptic medical robotics 303

Pushan Kumar Dutta and Bhupinder Singh

16.1. Introduction 303
16.2. Clinical validation in haptic medical robotics 306
16.3. Haptic feedback and robot-assisted surgery 309
16.4. Safety standards for haptic medical robotics 310
16.5. Challenges in experimental evaluation 311
16.6. Evaluation metrics and reproducibility 312
16.7. Legal and ethical concerns 313
16.8. Future directions 314
16.9. Conclusion 314
References 314

17. Revolutionizing healthcare: The convergence of medical robotics and digital therapeutics **317**

Mohammad Shahnawaz Shaikh

17.1. Introduction 317
17.2. Literature survey 318
17.3. Significant advances in medical robotics 319
17.4. Robotic-assisted surgery 322
17.5. Methodology 337
17.6. Difficulties and obstacles 341
17.7. Case studies of innovation in action 342
17.8. Conclusion and future work 347
References 348

18. Securing the future: Legal and privacy challenges in haptic medical systems for disabled children **353**

Shashwata Sahu, Navonita Mallick and Hiranmaya Nanda

18.1. Introduction 353
18.2. Research objectives 354
18.3. Research methodology 354
18.4. Conceptual and theoretical framework 354
18.5. Understanding haptic medical systems 354
18.6. International legal instruments 358
18.7. Data collection, storage, and consent issues 361
18.8. Privacy-by-design and ethics-by-design standards 366
18.9. Role of legislators and law reform bodies 370
18.10. Conclusion 373
References 374

Index *377*

Contributors

Nafisa Aafreen
School of Business, Woxsen University, Hyderabad, Telangana, India

Panchakarla Aditya Satvik
School of Business, Woxsen University, Hyderabad, Telangana, India

Sanskar Agrawal
School of Business, Woxsen University, Hyderabad, Telangana, India

Chirikonda Chandana
School of Business Analytics, Woxsen University, Hyderabad, Telangana, India

Pooja Choudhary
School of Business Analytics, Woxsen University, Hyderabad, Telangana, India

Meghna Dadich
School of Business, Woxsen University, Hyderabad, Telangana, India

Manjunath Devarashetty
School of Business Analytics, Woxsen University, Hyderabad, Telangana, India; Data Engineering, Meta Platforms, Menlo Park, CA, United States

Nivedita Ravindra Dharne
School of Business Analytics, Woxsen University, Hyderabad, Telangana, India

Pushan Kumar Dutta
School of Engineering and Technology, Amity University Kolkata, Kolkata, West Bengal, India

Lakshita Gandhi
School of Business, Woxsen University, Hyderabad, Telangana, India

Aaryan Gupta
Department of Artificial Intelligence and Machine Learning, Manipal University Jaipur, Jaipur, Rajasthan, India

Jitha Janardhanan
Department of Computer Applications, Presidency College (Autonomous), Bangalore, Karnataka, India

Deepika Roselind Johnson
School of Computer Science and Engineering, Vellore Institute of Technology, Chennai, Tamil Nadu, India

Hemachandran Kannan
School of Business, Woxsen University, Hyderabad, Telangana, India

AJ Keerthi
School of Computer Science and Engineering, Vellore Institute of Technology, Chennai, Tamil Nadu, India

Nasrulla K. Khan
Department of Computer Applications, Presidency College (Autonomous), Bangalore, Karnataka, India

Poojitha Kondapaka
School of Business, Woxsen University, Hyderabad, Telangana, India

Kosinepalli Kumar
School of Business Analytics, Woxsen University, Hyderabad, Telangana, India

Vishnu M.
School of Business, Woxsen University, Hyderabad, Telangana, India

Rudra Pratap Mahali
School of Masters in Business Administration, Woxsen University, Hyderabad, Telangana, India

Navonita Mallick
School of Law, KIIT Deemed to be University, Bhubaneswar, Odisha, India

Utkarsh Mishra
School of Computer Science and Engineering, Vellore Institute of Technology, Chennai, Tamil Nadu, India

Nihareeka Mohanty
School of Masters in Business Administration, Woxsen University, Hyderabad, Telangana, India

Hiranmaya Nanda
School of Law, KIIT Deemed to be University, Bhubaneswar, Odisha, India

Rakesh Ramakrishna Pai
School of Business Analytics, Woxsen University, Hyderabad, Telangana, India; Data Engineering, Meta Platforms, Menlo Park, CA, United States

Divya Pokar Patel
School of Business Analytics, Woxsen University, Hyderabad, Telangana, India

Raj Jaydeep Patel
Genpact, Hyderabad, Telangana, India

Debasis Patra
School of Business, Woxsen University, Hyderabad, Telangana, India

Anand Pindiprolu
Data Engineer, Deloitte, Hyderabad, Telangana, India

Jofia Jose Prakash
Computer Society, Aldie, VA, United States

Srivats Raghavan
School of Computer Science and Engineering, Vellore Institute of Technology, Chennai, Tamil Nadu, India

Anusha Rani
School of Business, Woxsen University, Hyderabad, Telangana, India

Paramaticona Reddy
School of Business, Woxsen University, Hyderabad, Telangana, India

Riya
School of Business, Woxsen University, Hyderabad, Telangana, India

Sanjana S
Estonian Business School, Tallinn, Estonia

Shashwata Sahu
School of Law, KIIT Deemed to be University, Bhubaneswar, Odisha, India

Valdas Sairam Goud
School of Business, Woxsen University, Hyderabad, Telangana, India

Manda Sandhya
School of Business, Woxsen University, Hyderabad, Telangana, India

Boddu Sanjana
School of Business Analytics, Woxsen University, Hyderabad, Telangana, India

Mohammad Shahnawaz Shaikh
Department of AI and Big Data, Faculty of Engineering and Technology, Marwadi University, Rajkot, Gujarat, India; Lincoln University College, Petaling Jaya, Selangor Darul Ehsan, Malaysia

Simna Shanavas
Department of Computer Applications, Presidency College (Autonomous), Bangalore, Karnataka, India

Nikhil Sharma
Business Analyst, DP World, Hyderabad, Telangana, India

Arun Pratap Singh
Freshworks, California, CA, United States

Bhupinder Singh
Sharda School of Law, Sharda University, Greater Noida, Uttar Pradesh, India

Polireddy Vasavi Reddy
School of Business, Woxsen University, Hyderabad, Telangana, India

Kalyanasundaram Venkataraman
School of Computer Science and Engineering, Vellore Institute of Technology, Chennai, Tamil Nadu, India

Lokesh Venkatraman
School of Business Analytics, Woxsen University, Hyderabad, Telangana, India; Data Engineering, Meta Platforms, Menlo Park, CA, United States

SECTION I

Foundatins of haptic medical robotics

1. Introduction to haptic feedback in medical robotics 3
2. Fundamentals of human touch perception 21
3. Haptic sensing technologies 41
4. Haptic feedback mechanisms 59
5. Advanced actuator technologies in medical haptics 85
6. Haptic interface design for medical applications 103

CHAPTER 1

Introduction to haptic feedback in medical robotics

Panchakarla Aditya Satvik, Valdas Sairam Goud and Manda Sandhya
School of Business, Woxsen University, Hyderabad, Telangana, India

1.1 Introduction

The last 20 years have seen unprecedented technological innovations in healthcare robotic technology, which essentially have changed the nature of contemporary medicine. From the computer-aided surgery platforms to the advanced teleoperation systems, the robotic technologies, in their turn, have gradually entered almost every sphere of clinical practice. These advancements have not only increased the function of the medical practitioners but have also elevated the level of care that the patients can enjoy, especially in the complex surgery procedures that are minimally invasive. The convergence of engineering, computing, and biomedicine still generates innovations that, until recently, belonged to the science fiction world. Surgical robotic-assisted systems have become an indispensable part of the modern surgical setting. They have the capability to provide unprecedented degrees of mechanical accuracy, work in limited anatomical rooms, and remove physiological constraints like hand tremor and fatigue. Robotic platforms like the da Vinci Surgical System have shown that delicate procedures can be carried out, and they can be repeated with accuracy and precision many times better than when performed by a human. Through this, robotic surgery has become widely accepted in fields such as urology, gynecology, cardiac surgery, and orthopedics. Although these accomplishments have been made, a major gap in most traditional robotic surgery systems remains: the lack of meaningful tactile sensing. Conventionally designed robotic systems are largely visual devices that are based on high-definition endoscopic cameras, assisting in guiding surgical movements. Although this is fantastic visual feedback, it cannot be compared to the tactile feedback that surgeons would use when interacting with their hands directly. The loss of the ability to sense tissue stiffness, the forces of interaction between a tool and tissue, and slight changes in the texture can greatly impair the situational awareness of the surgeon and make the tissue more prone to unintentional damage. The human sense of touch is a highly complex sense that combines the mechanoreceptor impulses of the fingertips, palms, and joints to create a multifaceted, multidimensional image of the material world. Surgeons rely significantly on this sense to decide the force required in holding tissues, to identify healthy versus pathological structures, and to identify the

Haptic Feedback Control in Medical Robotics
DOI: https://doi.org/10.1016/B978-0-443-27777-1.00018-4

suture tension of a wound closure. One of the most interesting and difficult directions in medical robotics research is the replication of such a sense of touch in a robot architecture. The haptic feedback technology fills this void by offering a virtual sense of touch and force to the operator of a robotic system. The haptic systems create a feeling of touch that fills the gap between the surgeon and the patient by introducing sensors that can measure the interaction forces and actuators that can feed this information back to the user. It can be said that this technology is central in enhancing surgical accuracy within the scope of force-controlled manipulation of instruments, improving safety, since it can alert to potentially harmful forces during contact, and providing a human–robot interface experience in a wide medical application range.

The field of haptic feedback in medicine has multiple applications. Haptic systems facilitate the feeling of resistance of tissue and tension of suture during a minimally invasive surgery in robotic-assisted surgery. Haptics can be used as a critical force and tactile sensation in teleoperated surgical systems to offset the physical distance between surgeon and patient. Haptic feedback surgical training simulators provide real and immersive training environments where trainees can acquire and practice their procedural skills without being exposed to patients. Haptic feedback is useful in rehabilitation robotics as it allows interactive responsive therapy systems to adjust to the motor recovery progress of the patient. The chapter gives a close-out introduction to the haptic feedback in medical robotics. It starts with an overview of the principal concepts of haptic technology, such as its definition, the major parts, and the most common types of feedback used in the technology. It then proceeds to the analysis of the devices and technologies that make up the modern haptic systems and then to the analysis of their application in the medical robotics field. The chapter also looks at the benefits and the limitations that come with haptic feedback, the surveys that are taking place in the new technology that is transforming the field, identification of future research directions, as well as a conclusion that sums up the main points discussed in the chapter.

1.2 Fundamentals of haptic feedback

1.2.1 Definition of haptic technology

The name haptics is a Greek word, haptesthai, which translates to touch, and is the science of providing tactile feeling and touch feedback in human–computer interface (Better Words, n.d.). Haptic technology, in the context of medical robotics, is a set of hardware and software technologies that allow an individual to perceive and interact with virtual or distant physical worlds, using the sense of touch. This is done by recording physical interaction events, including contact forces between a surgical instrument and a tissue, and recreating similar sensations at a human interface device on which the operator is operating to drive the robotic system. Manual dexterity involves perception of touch and force. By using proprioceptors and tactile information during surgery, the surgeon can determine the mechanical load of the operative area. The perception of force allows detecting the resistance, and the perception of touch provides information on the texture of the surface, temperature fields, and the distribution of pressure locally. The haptic technology attempts to recreate these senses artificially, in such a way that it allows remote or robotic interaction to be more like physical contact. The correlation that exists between tactile sensing and robotic systems is symbiotic. By applying tactile and force sensors, robotic platforms also acquire perceptual behavior by constantly sensing the mechanical surroundings. This sensory information is computed, conveyed, and displayed by actuators in the user interface, forming a closed-loop process in which human motor inputs make robots move and sensory feedback recovers and lessens further input. This two-way communication is the basis of successful haptic interaction as applied in the medical field (Kapoor et al., 2014).

1.2.2 Components of haptic systems

A complete haptic system consists of four major parts, which are sensors, actuators, control algorithms, and user interface devices. The components have a specific and key role to play in ensuring the fidelity and responsiveness of the haptic experience. The perceptual devices of the haptic system are referred to as sensors, and they detect physical quantities like force, pressure, displacement, and torque at the point of contact. In medical robotics, sensors are placed on the tool on the end-effector of the robot, which measure the forces of contact and the properties of surfaces as the tool and the tissue interrelate. These sensors determine the quality of the haptic feedback provided to the user directly in relation to their accuracy, sensitivity, and bandwidth. The effectors that process the sensory information into a physical feeling that can be perceived by the human operator are the actuators. They can generate forces that are counter to hand movement, localized skin deformation, or vibration patterns that convey certain tactile events. Actuation technology, which can be electric motors, pneumatic systems, or piezoelectric elements, is determined by the application needs, which are the force range required, spatial resolution, and size limits. Haptic systems use control algorithms as the computational basis to take sensor data, simulate the dynamics of the interaction, and compute the correct feedback stimuli to be used. Such algorithms have to work in real time to ensure the illusion of physical contact, which is being perpetuated. They also introduce stability mechanisms to prevent instability in the force-feedback loop, which would otherwise result in runaway oscillations or divergent forces. The physical tools used to interact with the haptic system are user interface devices, including pen-based styluses and joysticks, as well as advanced wearable gloves and exoskeleton robots. The interface device design makes it possible to determine which degrees of freedom of motion can be measured, what kinds of feedback can be provided, and how well the device will fit into the clinical workflow.

Table 1.1 Types of haptic feedback used in medical robotics.

Types of feedback	Description	Example medical application
Force feedback	Provides resistance force to simulate interaction with objects	Robotic surgery
Tactile feedback	Simulates surface texture and pressure sensation	Tissue palpation
Vibrotactile feedback	Uses vibration signals to indicate contact or alerts	Surgical training simulators
Kinesthetic feedback	Provides movement-based feedback related to joint motion	Rehabilitation robotics

1.2.3 Types of haptic feedback

Haptic feedback is divided into a number of different types, each of which is associated with various dimensions of tactile and kinesthetic perception. Knowledge of these classes is critical for choosing the right feedback modalities in particular medical cases. Table 1.1 summarizes the key forms of haptic feedback applied in medical robots.

The most popular modality in medical robotics is possibly force feedback, which is particularly used in robotic-assisted surgery. It entails the creation of opposing forces at the user interface that replicate the resistance experienced by the robotic instrument when they come in contact with the tissue or other items in the body. Tasks that demand fine control of force, e.g., suturing, tumor resection, and navigation of catheters, are important to this kind of feedback. Tactile feedback is directed toward the simulation of surface-level sensory data, such as texture, compliance, and topography of pressure. It has found specific application in the field of tissue palpation, especially in diagnostic and therapeutic decision-making, because the possibility to differentiate normal and pathological tissues on the basis of stiffness and texture can be applied. Tactile feedback is normally provided on the fingertips or palm of the user interface device in terms of arrays of small actuators. Vibrotactile feedback is a controlled vibration-based method of communication that is used to convey particular events or states to the user. Vibrotactile signals may be used to indicate the contact between a tool and tissue in the surgical training scenario, as well as to signal tool–tissue contact threshold breakage or warn of a possible danger. In navigation systems, this modality is also used with haptic cues that are used to indicate the instrument placement in relation to important anatomical features. Kinesthetic feedback deals with the sense of position of the body, limb motions, and joint forces. It is best applied directly in rehabilitation robotics, where the haptic exoskeleton machines control and oppose movements of patient limbs in motor recovery training. Kinesthetic feedback systems can be used to dislearn and relearn motor patterns after neurological injury by offering desired resistance or assistance on the therapeutic movements (Laga Boul-Atarass et al., 2025).

1.3 Haptic devices and technologies in medical robotics

1.3.1 Haptic sensors

A wide range of different measurement technologies make up the sensory layer of a haptic system, all of which are appropriate to various facets of force and tactile sensation. The simplest form is the force sensors, which are usually done on a strain gauge design, load cells, or optical sensing principles. The devices can measure both the magnitude and direction of the contact forces with very high precision and are typically installed on robotic end-effectors to enable real-time feedback of the forces during surgical manipulation. Multiaxis force/torque sensors find special application in surgery, where complete three-dimensional information about the forces and torques of the tool–tissue interface is required. Tactile sensors do not just measure scalar forces but provide a measure of the spatial distribution of pressure over a contact surface. These sensors are meant to mimic the work of the mechanoreceptors present in the human body in the skin, such as giving information about the geometry as well as texture of the surface, as well as compliance. These are organized in a matrix format that can be used to map profiles of contact pressure, which can be used in such tasks as tumor detection by palpation of tissues and evaluating the distribution of contact tension in a suture. Pressure sensors are a special type of tactile sense device which is highly sensitive to normal force per unit area. Pressure sensors are also used in medical robotics and are used in areas where gripping forces must be monitored to avoid damage to tissues, including laparoscopic graspers or catheters going through vascular anatomy. The pressure sensor has been miniaturized such that it can now be integrated into very small-diameter surgical instruments, and it can be used in procedures that have extremely limited access (Humanetics Group, n.d.). Microelectromechanical systems (MEMS)-based sensors have become a revolutionary technology in haptic sensing of medical robots. MEMS fabrication methods make it feasible to fabricate sensors of microscale dimensions with outstanding sensory properties, quick response, and the capacity to be more firmly structured into elastic materials. MEMS force and tactile devices may be directly installed on the surfaces of robotic tools, which can sense spatially distributed information with a resolution comparable to human fingertip mechanoreceptors. They are also compatible with conventional semiconductor fabrication processes, which makes them cost-effective in mass fabrication (Guo et al., 2024).

1.3.2 Haptic actuators

Haptic actuators are those devices that transform electrical control signals into mechanical stimuli perceived by the human operator as tactile or kinesthetic stimuli. The most widely used technology in haptic interfaces is based on electric motors, which provide a highly defined mapping between electrical input and mechanical output, are widely available, and can provide controlled forces with a large dynamic range. Use of direct-current (DC) motors and brushless versions is typical in force-feedback joysticks and robotic exoskeleton joints, in which they can produce large torques to provide realistic sensations of resistance (Iakovidis & Vartholomeos, 2025). Pneumatic actuators are based on the concept in which the controlled air pressure is used to generate forces or displacements. The reason they are appreciated is that they can produce enormous forces in comparison with their mass and size, and they are intrinsically compliant with their biocompatibility when made out of the right materials. Pneumatic actuators commonly used as inflatable bladders or bellows in haptic gloves and wearable interfaces create the sense of touch by deforming the skin locally. The pneumatic systems are also intrinsically safe in human interaction because they cannot generate the damaging impact forces in case of control failure. Piezoelectric actuators take advantage of the inverse piezoelectric effect, in which the piezoelectric material receives an electric voltage that results in a mechanical deformation of the material. These actuators find especially favorable applications in vibrotactile feedback applications because of their ability to work at high frequencies, generate very accurate controlled vibration amplitudes, and their ability to be made in very small form factors. Piezoelectric devices are commonly available in tactile displays, where they provide texture on the surfaces, and in warning devices embedded in the handles of surgical instruments. They are very low bandwidth and fast response, which makes them suitable to be used in applications where it is necessary to simulate the tactile events dynamically.

1.3.3 Haptic interfaces

One of the most developed and commonly adopted haptic interfaces in medical robotics is the joystick-based interface. These systems record the hand and wrist movements of the operator and provide resistive forces using the motor-driven systems that mimic the forces on the remote robotic tool. The Phantom Omni and Delta haptic devices are commercial haptic joystick systems broadly used in robotic surgery research and surgical simulation systems. They have a fairly small, desktop-capable form factor that can easily be integrated into existing robotic control architectures. Wearable haptic gloves are a more body-conforming design of a haptic interface, in which the sensing and actuation of the glove are spread over the entire hand, instead of being localized at the point of grip. To achieve the above, haptic gloves generally include a mix of bending sensors to detect finger flexion, fingertip force sensors, and arrays of actuators located on the finger pads and palm. The haptic devices allow very dextrous control in the virtual operating room and offer very rich tactile feedback over a wider spatial range, which is closer to the experience of using a real hand. They are especially useful in surgical simulation systems where learners are supposed to be trained in complicated bimanual operations. Exoskeleton interfaces expand the haptic interaction surface to include whole limb sections, which include all the motion of the limb, not just the distal hand interaction. Medical exoskeletons are mostly used in rehabilitation robotics, whereby they are used to support or oppose the movements of the affected limbs during therapeutic activities. When the resistive or assistive torques applied at each joint are modulated, the exoskeleton systems can be used to follow preset motion paths, deliver progressive resistance workouts, and adjust the strength of the therapy to the improving or varying motor activity in the patient (Verma et al., 2024; Yoon et al., 2024).

1.3.4 System integration

The inclusion of sensors, actuators, and control systems into a consistent haptic feedback design is a major engineering issue. To maintain the perceptually transparent haptic interaction required in real time, sensor–actuator communication should be of sufficiently high data rates to be able to sustain the real-time update frequencies demanded (which are usually over 1 kHz). This requires communication buses with low latency and data processing pipelines that are capable of acquiring, processing, and transferring sensory information within the stringent time constraints of the haptic control loop. Haptic medical robotics control systems have to deal with the complex nature of the dynamics of the robotic mechanism itself and the biological environment that the robot is interacting with. The accuracy and the stability of the force rendering in the model-based control strategies can be enhanced with the help of the representations of tissue mechanics, instrument dynamics, and human operator behavior. Control systems of impedance and admittance are especially common, which are the systems that allow the natural means of controlling the mechanical response of the haptic interface to operator inputs and environmental contacts. Live feedback processing can only be performed by specific computing hardware that is capable of operating the control algorithms at the necessary rate, as well as assuming the overhead of transmitting data to peripheral devices and managing higher-level task management systems. Haptic control platforms built on top of real-time operating systems, field-programmable gate arrays, and application-specific integrated circuits have allowed fulfilling the demanding timing characteristics of medical applications with the flexibility required to support research and clinical adaptation. Table 1.2 shows the popular haptic devices and applications of the device in medical robotics (Castro et al., 2025; Siewert et al., 2025).

1.4 Applications of haptic feedback in medical robotics

1.4.1 Robotic-assisted surgery

Minimally invasive surgery is one of the most revolutionary solutions in surgery, as it allows performing complex surgical operations with the help of long instruments

Table 1.2 Common haptic devices and their applications in medical robotics.

Devices	Technology used	Application area
Haptic joystick	Force feedback motors	Robotic-assisted surgery
Haptic glove	Tactile sensors and actuators	Surgical simulation and training
Exoskeleton device	Kinesthetic actuators	Rehabilitation robotics
Tactile Sensor arrays	MEMS sensors	Tissue stiffness detection

and endoscopic visualization using a small incision. Although such a procedure greatly alleviates the trauma of patients, loss of blood, and recovery time in contrast to open surgery, it restricts the surgeon strongly perceptually and ergonomically. The transmission of the force of touch to the hands of the surgeon is reduced by the rigid and long-shafted character of the instruments, and the two-dimensional character of endoscopic images that are taken conventionally would remove the depth information. All these restrictions decrease the capability of the surgeon to estimate interaction forces and determine the properties of tissues. In robotic-assisted surgery, force sensing is of special interest, as a lack of direct tactile feedback has been found to be one of the major safety concerns. Over-gripping may lead to tissue necrosis, unwanted puncture of the vessel, or suture rupture, whereas under-gripping may lead to poor tissue handling or slipping of the instrument. Haptic feedback systems that indicate real-time forces at the tip of the robotic instrument to the surgeon's hand controller have provided a very important safety net that allows the manipulation of the instrument with a force that is precise and safe in a controlled manner. Haptic feedback is also significant in enhancing surgical precision when it is incorporated in the control loop of robotic surgical systems. Studies have proven that the rates of tissue damage and suture breakages are remarkably lower in surgeons who work in force feedback conditions than in force-blind conditions. The perception and control of the forces of interaction allows the manipulation of delicate tissues, a more precise tying of knots, and a smoother division of the anatomical planes, which in the sum produces better results of the surgical operation (Rae-Dupree, 2025).

1.4.2 Teleoperated surgical systems

Surgical systems are teleoperated to promote the delivery of expert surgical services to previously inaccessible environments such as geographically isolated locations, hospitals in military fields, and in space. The robotic systems have the operating surgeon at a physical distance from the patient and interact with the robot platform through a master-slave architecture, where instructions sent at the master console are relayed to the slave robot at the operating site. The reliability and quality of haptic feedback in this system are important determinants of clinical effectiveness and the safety of surgery. The interaction between the surgeon and the robotic system in teleoperated conditions is not limited to providing kinematic commands but also to the exchange of force information and sensory information in both directions. The faithfulness of this communication defines the accuracy with which the surgeon can perceive and react to the events at the place of surgery. The haptic communication is also of high quality, such that the remote surgeon is able to notice the resistance of tissues, anatomical boundaries, and react to unforeseen contact returns, similar to how they would react in a direct-access surgery case. Safety in teleoperated systems feedback includes the force-based and event-driven haptic cues (Takano et al., 2025). Constant force feedback keeps the surgeon informed of the mechanical condition of the operating environment, and threshold-based alerts warn the surgeon of interaction forces approaching or exceeding safe limits of operation. Unneeded safety features, such as automatic force limiting and emergency stop procedures when haptic overload parameters are met, are important elements of safe teleoperated surgical system design. The incorporation of effective haptic feedback significantly decreases the mental and oversight workload of the distance surgeon.

1.4.3 Surgical training simulators

The use of surgical training simulators with haptic feedback as virtual reality (VR) provides an interesting alternative to the conventional apprenticeship system of surgical training. These systems simulate the three-dimensional virtual environments where the trainees can practice the procedures of these systems using real anatomical models without harming real patients. The educational usefulness of these simulators relies on the fact that such simulators include haptic feedback, with the help of which trainees can acquire the force control skills and impressiveness of tactile patterns, which are the cornerstones of surgical competence. Surgeon simulation training involves cognitive and psychomotor aspects of skill development. Haptic simulators aid the learning of the procedure through practice, where the surgeon can repeat and master the specific operation step by step. This instantaneous feedback of the haptic system in the form of resistive feedback during tissue manipulation, vibrotactile feedback during threshold violation, or akin to kinesthetic feedback during complicated maneuvers facilitates the acquisition of skills by supplying informative and timely feedback to the performance during training otherwise not accessible in conventional training settings. The interaction between tissues can only be realistically simulated by computational models of biological tissue mechanics capable of producing physically feasible responses to forces in real time. The elastic, viscoelastic, and nonlinear mechanical properties of different types of tissues are simulated using finite element models, mass-spring networks, and data-driven tissue models. With high-fidelity haptic rendering methods, these models allow simulation of tissue cutting, suturing, grasping, and palpation to be simulated with the realistic appearance necessary to facilitate significant transfer of skills to the real operating room.

1.4.4 Rehabilitation robotics

One of the most effective areas of the haptic feedback application in medical robots is stroke rehabilitation. Motor deficits are usually common in patients who have suffered a stroke, manifested through weakness, spasticity, and a deficiency of fine motor coordination, especially in the upper limbs. Haptic feedback-controlled robotic therapy systems can be used to provide repetitive and structured motor training that utilizes neuroplastic underpinnings of motor recovery. These systems allow relearning of motor programs that could have been impaired by neurological injury by the provision of controlled assistance or resistance during therapy movements. Haptic feedback provides sensorimotor integration, which facilitates motor skill recovery. By providing the patient with the correct level of kinesthetic cues in the therapies with the haptic system, this stimulates proprioceptive pathways to aid in motor learning and cortical reorganization. The assist-as-needed strategies of using a robotic system to provide the minimal level of assistance needed to achieve a given therapeutic movement are especially effective in fostering active engagement of the patients and achieving the maximum neuroplastic gain. Haptic-based interactive therapy systems may be programmed to adjust dynamically based on the real-time performance of the patient. Kinematic and force measurements, like movement smoothness, time to complete the task, and the tracking error of a force, can be used to determine the objective measures of patient improvement, which will adjust the therapy parameters. This informational flexibility allows specific rehabilitation plans to change with the progress of the patient, keeping a therapeutic challenge optimal in the rehabilitation process.

1.5 Advantages of haptic feedback in medical robotics

All the benefits that haptic feedback integration in medical robotic systems brings to the system and health care facility overall include a wide range of benefits that positively contribute to health care outcomes, increase the potential of medical practitioners, and widen the usability of the robotic systems in health care facilities. These benefits cut across the surgical performance, safety, training, and cognitive ergonomics. One of the most reliable gains of haptic feedback in robotic surgery is enhanced precision in surgery. The haptic systems allow the surgeon to have real-time force and touch information of the operative area at the point where the instrument is touching the patient, allowing an accuracy that is near the skillful handling of the instrument using the hands. Comparative studies of the performance of surgical tasks with haptic devices and controls have reported that, with the extra information of force feedback, the error in force application, tissue injury, and the time taken to complete a particular surgical task are all significantly reduced. This has been a valuable development in terms of process improvement that involves fine structures, including blood vessels, nerves, and thin tissue membranes.

The ability to regulate the force enables direct control of tissue damage, which is directly associated with reduced risk of tissue damage. In the absence of the sense of touch, the surgeons at the robotic-controlled system are forced to resort to visual indications to determine the forces of interaction at all times, which is both cognitively challenging and inherently inaccurate. Haptic feedback bridges this perceptual gap to enable surgeons not only to apply forces within safe limits of operation, but also to sense and react to any unwanted contact before it leads to tissue damage. The direct consequences of this risk minimization are patient safety and surgical morbidity. Better surgeon control will involve the accuracy of each instrument movement and the quality of the whole operating decision-making. Haptic feedback adds value to the informational content that can be presented to the surgeon, which makes the intraoperative decision-making more informed and confident. The capacity to differentiate the tissues based on their mechanical characteristics, the determination of the knot security through the perception of tension, and hemostasis through the perception of vessel resistance are some of the advantages of enhanced surgical control, whose benefits are reflected in improved patient outcomes. The benefit of realism in training simulators is one of the important issues in surgical education. Haptic feedback converts the visual plausibility of virtual surgical environments into a physically empty simulation to perceptual rewarding and physically significant learning platform. Haptic-enabled simulators allow trainees to build more realistic mental models of tissue mechanics and instrument behavior, and transfer this to real operative performance better. The interaction provided by haptic feedback also gives the trainee a sense of realism, which has been known to increase their engagement and motivation, which in turn increases learning. One of the practical advantages of using robotic systems in the work of surgeons is reduced cognitive workload during the work with long and complex procedures. In the absence of haptic feedback, the surgeons have to devote much mental effort to monitor the visual proxies of the force information and simulate the probable force state of the operating environment in their minds. Cognitive resources can be allocated to the higher-level surgical planning and decision-making when this information is conveyed in a direct and intuitive manner via the haptic channel, which may lead to the reduction of error occurrence and the quality of performance in general. The final indicator through which the worth of the haptic feedback technology should be judged is improved patient safety. The presence of increased accuracy, lesser risk of tissue damage, better surgical control, and reduced cognitive burden are all factors that make surgical episodes safer, less complicated, and more successful. Many consequences of haptic technologies have yet to be fully realized as they are more generally incorporated into clinical robotic systems, but overall, their effect on patient safety should be significant (Hou et al., 2024).

1.6 Challenges and limitations

1.6.1 Latency issues

In the haptic feedback system, the latency is defined as the delay between the physical phenomenon that takes place in the robotic instrument and the sensory signal that is delivered to the human operator. Even small delays can seriously deteriorate the quality and safety of haptic interaction. Latencies longer than a few milliseconds may interfere with the force control of the surgeon, resulting in tool oscillation or incorrect tissue characterization in robotic surgery. The human haptic system is highly attuned to temporal perturbation; vibrotactile timing perceptual thresholds are on a scale of several milliseconds, with low-latency processing being an essential factor in high-fidelity haptic feedback. Communication delays pose significant challenges in teleoperation systems, where propagation delays arise from the distance between the master and slave and are further influenced by the medium of communication. Tens and hundreds of milliseconds of round-trip delay, which are many times greater than the stability margins inherent to traditional force feedback control loops, can be added by transcontinental teleoperation using public internet infrastructure. Partial attempts to counter the impact of the communication delay are compensation schemes such as predictive display algorithms, wave variable transmission, and model-based haptic rendering, and none has so far managed to counteract the performance loss of large latencies.

1.6.2 System calibration

The haptic feedback requires precise and reliable feedback that is based on precise calibration of all the sensing and actuation elements in the signal chain. The accurate distance accuracy of sensors in the robotic end-effector is a prerequisite: should the actual interaction forces not be correctly measured by the force sensors in the robotic end-effector, the feedback sent to the surgeon will be distorted in a systematic manner, which may result in erroneous force perceptions and reduced safety. Calibration methods have to consider both time-dependent sensor drift and effects of temperature changes on sensor sensitivity and cross-axis artifacts due to force coupling that may corrupt force measurements in multiaxis sensing systems. The issue of force scaling is based on the necessity to scale the forces measured at the scale of robotic instrument-tissue interface to the forces that are perceptually suitable and comfortable at the interface between the operator and the machine. In minimally invasive surgery, the force desired between instrument and tissue is usually in the range of fractions of a newton to multiple newtons, and the scaling of such forces to the master interface is determined by the impedance properties of the interface device, the feel of interaction desired, and the range of scalable force of the actuators. Inappropriate scaling might lead to insensitive feedback, which does not give meaningful information, or overforces, which makes an operator suffer and fatigue.

1.6.3 Modeling biological tissue

Haptic simulations of surgical processes should be accurately realized by computational models of the mechanical properties of the tissue that are faithful. It is even more complicated by the fact that biological tissues demonstrate very nonlinear mechanical behavior, which is hard to describe fully. Soft tissues are viscoelastic, i.e., the mechanical response is not only determined by the deformation state of the present but is also sensitive to the deformation history and rate. Moreover, tissue characteristics differ significantly across people, across tissue types, and physiological conditions, and it is exceptionally difficult to create a universal tissue model. The dynamic tissue properties are additional modeling challenges. Tissues used in surgery can be actively tense due to muscle activity, pressure gradients caused by effects of fluids, or can be altered in a mechanical manner by the action of the surgery itself, such as cutting, cauterization, or retraction. Simulation algorithms of real-time tissue simulations should be able to support these dynamic changes, and at the same time, be computationally efficient enough to support haptic update rates. There is still a notable research gap between physically realistic tissue simulation and computationally feasible real-time rendering in the area of haptic medical robots.

1.6.4 Stability and control

Haptic feedback stability is the basic issue behind the design of medical robotic systems. Interaction of a haptic system with a stiff physical environment, e.g., a rigid surgical instrument contacting hard tissue or bone, can cause the force feedback loop to become unstable, resulting in the oscillatory or divergent force outputs that are not only perceptually disturbing but may be dangerous. This is because of the delays in the sampling and computation of the digital control systems, which constrain the bandwidth of the possible feedback responses and provide some fundamental limits on the rigidity of the virtual surfaces that can be rendered in a stable fashion. The complexity of control algorithms is another related issue, since the algorithms needed to produce a stable, transparent, and accurate haptic feedback must be capable of multiple and conflicting goals. They should be able to render high-fidelity force feedback over a broad selection of interaction conditions, be stable against the effects of changing environmental stiffness, counteract system nonlinearities and sensor noise, and react to operator perturbations in an appropriate fashion. The creation of an actual, adaptive control algorithm capable of fulfilling all these requirements in real-time operating situations is a long-term aspect of active research in haptic systems engineering.

1.6.5 Cost and implementation challenges

The high costs of high-quality sensing, actuation, and control hardware have limited the commercial application of haptic feedback technology in clinical medical robotics.

Accurate force/torque sensors, high bandwidth servo actuators, and real-time control processors are major capital expenditures that increase the already high cost of robotic surgical platforms. To most healthcare facilities, the incremental cost of haptic capability has to be paid off with proven clinical benefits, which can be difficult to achieve without solid clinical evidence that is yet to be fully established across all applications. The issues of implementation go past the cost of hardware to include the complexity of the engineering efforts involved in adapting haptic systems to the existing robots. To add the force sensing and feedback features to robotic systems originally not intended to support haptics, mechanical redesign, creation of new sensor interfaces, and testing of new control architectures are all necessary. The regulatory approval processes of haptic-enabled medical devices are not that easy either, as new haptic capabilities might take up a lot of time and safety and performance testing before clinical approval is granted.

1.7 Emerging technologies in haptic medical robotics

1.7.1 Artificial intelligence integration

The field of artificial intelligence will be able to fundamentally change the functionality of haptic feedback systems used in medical robotics. Machine learning methods to predict forces provide the possibility of estimating interaction forces using sensor modalities that are more readily incorporated into surgical tools than direct force sensors, including visual data from an endoscopic camera or electrical data from motor drivers. Neural networks that are trained on escalated force and visual datasets can be trained to compute the contact forces by visual data, which may allow force feedback in systems where direct force measurements are inconvenient (e.g., sterilization or size). The artificial intelligence-driven adaptive feedback systems are able to adjust the properties of haptic feedback in real-time depending on the learned models of the working environment, the interaction patterns of a particular surgeon with it, and the specifics of the current procedure. The reinforcement learning algorithms may be used to optimize the haptic rendering parameters to achieve the performance measures of the task, which include force accuracy and procedural safety to tailor the haptic experience to individual perceptual preferences of the surgeon and the circumstances of operation. Such intelligent adaptation is a great step forward compared to the fixed, programmed feedback strategies that are currently being used in haptic systems.

1.7.2 Smart haptic sensors

Tactile sensing technologies are now developing further, and sensor systems are now being made with much greater capabilities than those of normal force and pressure sensors. Smart haptic sensors are provided with on-board signal processing, multimodal sensing capability, and self-calibration to alleviate the data processing load on the central control system and enhance measurement robustness under variable operating conditions. Tactile sensor arrays made of organic electronics, conductive polymers, or liquid metal composites can be shaped to complex curved surfaces to provide an extensive coverage of the robotic instrument surfaces to a level that could not be obtained with rigid sensor substrates (Zhang et al., 2024).

1.7.3 Soft robotics

Soft robotics is a novel approach to the philosophy of robotic design, wherein the hard structural building blocks of traditional robots are replaced with compliant, deformable materials that display more biologically-like mechanical behaviors. The potential of flexible robotic manipulators based on elastomeric materials, shape-memory polymers, or hydraulically actuated soft structures has the benefit of compliance, which limits the damage to tissue in the case of unwanted interactions and allows controlled operation of sensitive forces of action. Flexible tactile sensing is a current area of research, integrating distributed tactile sensors into soft robotic bodies with the aim of creating robotic instruments capable of feeling and acting on contact forces throughout their bodies. Soft robotics being used in medical devices is introducing a new opportunity in the medical field of minimally invasive surgery, especially natural orifice surgeries and intraluminal surgeries, where traditional rigid tools are incapable of following the complex and curvilinear anatomy. Endoscopes, catheters, and surgical instruments with in-built haptic sensing capabilities will be able to offer less invasive and more agile access to previously hard-to-reach anatomical locations and deliver the force feedback of safe and effective intervention.

1.7.4 Augmented reality/virtual reality integration

Emerging opportunities in surgical guidance, surgical education, and surgical design are being offered by the convergence of augmented reality (AR) and VR technologies with those of haptic feedback systems. Immersive VR visuals plus haptic simulation training experiences are significantly more fidelic than training experiences attained through either visual feedback or a haptic feedback system. The next generation of surgical simulators will utilize high-resolution stereoscopic displays, real-time tissue simulation engines, and state-of-the-art haptic interfaces to generate near-photorealistic surgery settings in which the visual, spatial, and tactile experience of surgery can be trained more realistically than ever before. Haptic feedback in an AR surgical system can be synchronized on top of superimposed visual overlays that display subsurface anatomy and force distribution maps, or procedural guidance information itself, on a direct view of the operative field of the surgeon. This multimodal expansion of the surgical space complements situational awareness and aids in making better intraoperative decisions by giving perceptual information in more than one sensory channel at a time. Technical difficulties in the realization of accurate spatial and temporal correspondence between visual and haptic feedback devices are still active topics of research.

1.8 Future research directions

Further development of haptic feedback technology in medical robotics will necessitate long-standing research in a variety of interdependent fields. There are some directions that especially have a bright future and a highly significant impact on the future of the field. Tele-surgical systems with low latency are an important area of research focus because telemedicine and remote surgery are ever-increasing in clinical importance. It will take the development of network infrastructure and predictive haptic rendering algorithms, as well as control architectures capable of guaranteeing performance in the face of continuously changing network conditions, to achieve stable and transparent haptic interaction over very large and very long-range communication links. The introduction of fifth-generation telecommunication systems and their lower latency and higher bandwidth is an important enabling infrastructure advancement, but will require specific protocol and algorithm development to fully utilize these features in the context of surgical haptics.

Surgeon wearable haptic devices present an opportunity that allows changing the manner in which force and tactile experience are provided in surgery. The existing haptic systems are mostly chained-desk devices that limit the scope of movement of the surgeon, and they are not well integrated into clinical practices. The next-generation wearable haptic system, such as haptics-based surgical gloves and forearm tactile displays, has the potential to provide rich multimodal feedback and yet enable the surgeon to move freely and interact with the robotic interface in a natural manner. Other critical research issues are miniaturization of actuation components, sensing components, wireless communication protocol development with enough bandwidth and reliability, and ergonomic form factor design that fits surgery gloves and sterile field standards. The adaptive feedback control using AI is perhaps the most radical area of research in haptic medical robotics. The ability of systems to learn the specific preferences of each surgeon, adjust the nature of feedback to the unique aspects of a process, and react to abnormal force episodes in real time would have a significant positive impact on the results in terms of safety and performance. Such systems will need massive datasets of annotated surgical interactions, sophisticated machine learning systems that can be used to make inferences in real time, and robust validation procedures that can determine the clinical utility of adaptive haptic systems. State-of-the-art tactile sensor material is required to address the shortcomings of existing state-of-the-art rigid sensor technologies and to allow more extensive tactile perception based on biological principles in robotic systems. Studies of new materials such as piezoelectric polymers, ionic hydrogels, and stretchable electronic composites are yielding sensor platforms with incredible sensitivity, spatial resolution, and mechanical compliance. The combination of such materials into working tactile sensing skins on robotic instruments, combined with efficient signal processing models of high-density tactile data, will be needed to achieve the vision of real-world touch-aware surgical robots. The long-term horizon of haptic robotics studies is intelligent robotized surgical assistants that are capable of active participation in surgical decision-making. These systems would not be passive receivers of sensory data relayed to the surgeon; instead, they would interpret haptic data and identify clinically meaningful patterns and provide proactive guidance or remedial measures via artificial intelligence. These would necessitate the integration of haptic sensing with other intraoperative data streams, such as imaging, electrophysiological sensor monitoring, and genomic biomarker data collections, to develop comprehensive situational awareness models that would be useful in both human surgeon decision-making and autonomous protective responses.

1.9 Conclusion

The significance of haptic feedback in contemporary medical robotics can hardly be overestimated. With growing roles of robotic systems in surgery, rehabilitation, and medical training, the capacity to add meaningful, tactile, and force feedback to the human–robot interaction loop has become a defining challenge and opportunity to the field. This chapter has summarized the basic concepts, technological applications, clinical uses, and research opportunities of haptic feedback in medical robotics that have shown a technology domain of extraordinary breadth and depth. Advantech's technology has allowed the creation of haptic systems to reproduce the sense of touch in an increasing number of interaction conditions. Recent developments in MEMS sensing, smart materials, real-time computation, and machine learning have all increased the fidelity and reliability of the haptic feedback system as well as their usability to levels that were only possible 10 years ago. These advances have culminated in clinical benefits in terms of surgical accuracy, safety, training effectiveness, and rehabilitation outcomes. There are still considerable obstacles to overcome before haptic feedback in medical robotics can be fully utilized. The remaining factors that are hindering the application of haptic technologies in clinical practice are latency, fidelity of tissue modeling, stability of control systems, and their implementation cost. The solution to these issues will involve interdisciplinary teams that encompass robotics, materials science, computational biology, signal processing, and clinical medicine. The future of smart, AI-enabled, haptic in health-care is remarkably bright. With the continued maturation and convergence of AI, soft robotics, novel sensing materials, and AR systems, haptic feedback systems will transform into active, intelligent systems of collaborative human–robot surgical and therapeutic teams. The final beneficiaries of this technological advancement would be patients, who will be the beneficiaries of the power of robotic accuracy and capability without losing the touch and sense of the situation that characterize competent clinical practice.

AI disclosure

During the preparation of this work, the authors used ChatGPT to help structure and develop ideas into a theoretical framework, and Grammarly to correct grammatical errors and improve language clarity. After using this tool/service, the author(s) reviewed and edited the content as needed and take(s) full responsibility for the content of the publication.

References

Better Words Online, "Haptic – definition." [Online]. Retrieved March 5, 2026, from https://www.betterwordsonline.com/dictionary/haptic.

Castro, E., Lorente, D., & Olmos, D. (2025). PARP inhibitors and prostate cancer: The struggle to separate the grain from the Chaff. *European Urology, 87*(6), 641–642. https://www.sciencedirect.com/science/article/abs/pii/S0302283825000247.

Guo, X., McFall, F., Jiang, P., Liu, J., Lepora, N., & Zhang, D. (2024). A lightweight and affordable wearable haptic controller for robot-assisted microsurgery. *Sensors, 24*(9), 2676. https://doi.org/10.3390/s24092676.

Hou, C., Gao, H., Yang, X., Xue, G., Zuo, X., Li, Y., Li, D., Lu, B. o, Ren, H., Liu, H., & Sun, L. (2024). A piezoresistive-based 3-axial MEMS tactile sensor and integrated surgical forceps for gastrointestinal endoscopic minimally invasive surgery. *Microsystems & Nanoengineering, 10*(1), 2055–7434. https://doi.org/10.1038/s41378-024-00774-6.

Humanetics Group, Advancing robotic surgery with multi-axis force sensors. Retrieved March 5, 2026, from https://hitec.humaneticsgroup.com/insights/perspectives/advancing-robotic-surgery-multi-axis-force-sensors.

Iakovidis, D. K., & Vartholomeos, P. (2025). Medical and healthcare robotics: A roadmap for enhanced healthcare. *Measurement Science and Technology, 36*, 103001.

Kapoor, S., Arora, P., Kapoor, V., Jayachandran, M., & Tiwari, M. (2014). Haptics - Touchfeedback technology widening the horizon of medicine. *Journal of Clinical and Diagnostic Research., 8*(3), 294–299. https://doi.org/10.7860/JCDR/2014/7814.4191.

Laga Boul-Atarass, I., Rubio Manzanares Dorado, M., Padillo-Eguía, A., Racero-Moreno, J., Eguía-Salinas, I., Pereira-Arenas, S., Jiménez-Rodríguez, R. M., & Padillo-Ruiz, J. (2025). Role of haptic feedback technologies and novel engineering developments for surgical training and robot-assisted surgery. *Frontiers in Robotics and AI, 12*. https://doi.org/10.3389/frobt.2025.1567955.

Rae-Dupree, J. (2025). Feel the precision: Next-gen robotic surgery with haptic feedback. *IEEE Pulse, 16*(1), 12–15. https://doi.org/10.1109/mpuls.2025.3526484.

Siewert, M. B., et al. (2025). Recent advances in robotic surgery and intelligent surgical systems. *Journal of Pediatric Surgery*.

Takano, H., Ebihara, Y., Hirano, S., Shichinohe, T., Murakami, S., Morohashi, H., Oki, E., Hakamada, K., Taketomi, A., Ikeda, N., & Mori, M. (2025). Effectiveness of haptic feedback during local and remote robotic surgery: Single-blind cadaveric study. *Journal of Robotic Surgery, 19*(1), 1863–2491. https://doi.org/10.1007/s11701-025-02826-z.

Verma, A., Balian, J., Hadaya, J., Premji, A., Shimizu, T., Donahue, T., & Benharash, P. (2024). Machine learning–based prediction of postoperative pancreatic fistula following pancreaticoduodenectomy. *Annals of Surgery, 280*(2), 325–331. https://journals.lww.com/annalsofsurgery/abstract/2024/08000/machine_learning_based_prediction_of_postoperative.22.aspx.

Yoon, J. H., et al. (2024). Machine learning-based prediction of postoperative complications following surgery. *Annals of Surgery, 280*(2).

Zhang, Y., Shen, Y., Wang, Q., Song, C., Dai, N., & He, B. (2024). A novel hybrid swarm intelligence algorithm for solving complex optimization problems. *Robotics and Autonomous Systems, 177*, 104678. https://www.sciencedirect.com/science/article/pii/S0921889024000617.

CHAPTER 2

Fundamentals of human touch perception

Sanjana S[a], Raj Jaydeep Patel[b] and Arun Pratap Singh[c]
[a]Estonian Business School, Tallinn, Estonia
[b]Genpact, Hyderabad, Telangana, India
[c]Freshworks, California, CA, United States

2.1 Introduction to touch perception

2.1.1 What is touch perception? Definitions and scope

Perception of touch is perhaps the most basic yet perplexing of the body systems. Unlike sight or sound which can be experienced from a distance, touch requires the body to be in contact with another object. Touch is the perception of a contact force, how and where the force is applied, the temperature of the object, whether it is a solid, liquid, or gas, and the chemical makeup of the object. Touch perception is much more than simple contact perception. Touch can perceive the difference in texture between materials, such as silk and sandpaper, and can recognize certain objects without the aid of sight. Touch can perceive the vibration of certain tools and can also sense gentle pats, which can convey both emotional and social value. Touch also provides the body with the ability to control and coordinate fine movements and is critical in providing the feedback needed to maintain the body's posture and balance. Finally, touch alerts to the presence and location of harmful stimuli.

2.1.2 Touch as a multimodal sense: Beyond "Simple" contact

Reducing touch to simple tactile contact ignores its sophistication. Touch involves several different submodalities: mechanical pressure, vibration, skin stretch, temperature, and pain. Each of these submodalities is mediated by different receptors and distinct neural pathways. However, meaningful touch also involves integration with proprioception and kinesthesia. For example, to identify a coin by touch, a person receives multiple streams of information about the surface texture of the coin and also about finger position and finger movement. The active exploration of touch particularly illustrates its multimodal nature. Using a hand to touch an apple entails the brain considering several components, including: smoothness as detected by slowly-adapting mechanoreceptors, firmness as detected by pressure-sensitive receptors, and coolness as detected by thermoreceptors. In addition, the brain receives proprioceptive feedback as a result of the grip force exerted

Haptic Feedback Control in Medical Robotics
DOI: https://doi.org/10.1016/B978-0-443-27777-1.00004-4

by the person. The brain integrates all of these signals to produce a single perception that directs the appropriate action.

2.1.3 Historical perspectives: From Aristotle to modern neuroscience

Over 2000 years ago, Aristotle (350 BCE/1984) placed touch among the five primary senses and noted that it was the most vital for a being's survival. He noted the absence of sight or hearing in some animals, but said that none could survive without touch. However, Aristotle, oversimplifying, considered touch to be a single sense, and that view continued for many centuries. The foundations of contemporary science began forming in the 19th century. Weber's psychophysical studies (1834) showed that the sensitivity of different parts of the body to touch is not the same (Weber, 1834). He showed that people respond to different touch stimuli in terms of point discrimination at their fingertips, while at their back, several centimeters are needed to respond to touch. This was the first contrasting evidence of touch sensitivity (Weinstein, 1968). As histologists began to study skin, the specialization of nerve endings in the skin was elaborated. Between the 1850s and 1890s, Meissner (1853), Pacini (1840), Ruffini (1898), and Merkel (1875) described encapsulated nerve endings (now named after them) and established that touch is mediated not by a single receptor type, but by several.

The advent of the 20th century brought about the use of electrophysiology to understand how receptors and neurons code for the elements of touch. Mountcastle's work in the 1950s–60s was the first to study the columnar organization of the somatosensory cortex and to demonstrate hierarchical feature extraction from tactile input (Mountcastle, 1957). Today, touch processing in the human brain can be visualized using modern neuroimaging, while molecular biology has revealed the mechanotransduction ion channels involved in the conversion of mechanical energy (Coste et al, 2010). Touch is paramount biologically and behaviorally for people at all age levels and in all stages of development. For example, in the case of premature babies, tactile stimulation has been shown to improve developmental outcomes and result in a more rapid increase in weight. Touch is also a component in all manually skilled activities. For example, with respect to the pianist, the touch of the fingers is important. In the case of the surgeon, touch is also important, as well as in the case of the sculptor in forming clay. Touch is also fundamental to all social and emotional activities. For example, praise, support, and affection can be communicated with a touch, a hand, or an embrace. Touch can be more effective than words. People who are deprived of touch can suffer serious psychological and even physical harm. Appropriate touch, for example, can improve the immune system. In the case of touch deprivation, the levels of several important stress hormones increase, and blood pressure is increased. As a means of communication, touch has the ability to communicate several things simultaneously. In evolutionary terms, touch has also played an important role. For example, touch has the ability to convey vital or important information. Touch can also be used to perceive danger, for example, when there is a risk of injury. The hand can be important as a tool, for example, to grasp and manipulate food. For object recognition, the use of touch is important. For example, touch gives the ability to manipulative dexterity that is important for the hands. For all of the above reasons, touch has the ability to work great.

2.2 The somatosensory system: An overview

2.2.1 Elements of somatosensation: Cutaneous, proprioceptive, and interoceptive

The components of the somatosensory system perform three distinct functions. Cutaneous somatosensation originates from the receptors in the skin, which respond to surface contact, texture, vibration, and temperature, forming the basis for touch perception. The receptors in the muscle, tendons, and joints are responsible for proprioception, which signals the position, movement, and force of the limbs. Spindles respond to changes in length, Golgi tendon organs respond to changes in tension, and joint receptors respond to extreme positions. This is, in the absence of visual contact, body awareness. As for interoception, it is the perception of the internal organs, which are associated with visceral pain, hunger, and fullness. While interoceptive sensations have usually been studied in isolation, they also share neural pathways with cutaneous touch and are integral to our awareness of self and emotional experience.

2.2.2 Functional classifications: Exteroception, proprioception, and nociception/temperature

Another way to define somatosensation is through its functional organization. Exteroception is the perception of external stimuli through skin receptors and includes touch (discriminative and crude) and mechanical sense (mechanoreception). Proprioception pertains to the awareness of posture and movement of the body. While exploring an object through haptic sense, the distinction between exteroception and proprioception can become unclear, yet both are essential for the perception of the properties of an object through active touch. Nociception (pain) and thermosreception (temperature) are often studied separately due to their different receptors and protective functions. However, in natural behavior, these modalities are integrated with mechanical touch, as in the example of a hot stove, which is thought of as hard, smooth, and dangerous, and causes simultaneous sensations of extreme heat.

2.2.3 Comparison with other sensory modalities

The comparison of touch with sight and hearing brings out a number of defining differences. Touch, more than the other two, relies on direct contact with the environment. It is more egocentric than the other modalities, with reference to the position and contact of the body. Touch is also more limited than sight in modality and body coordination. Touch is more limited than sight in modality and body coordination. While sight can take in several objects at once in the distance, touch is more of a series of contacts and more of a body coordination. In touch, the contact is direct, so there is no room for misleading appearances. Most of the body feels touch more than sight, while the other senses are more limited. Unlike the fairly even distribution of receptor density found in the retina of the eye, the spatial resolution of touch receptors is highly specialized. For example, the fingers are very highly receptor-dense, while another part of the body is the back. These differences are due to the different functions of different parts of the body, fingers being specialized for manipulation and the back for protection. Touch can sense vibrations of several hundred hertz, providing a temporal resolution between that of vision and hearing. Touch perceives textures through vibrations as skin slides over different surfaces.

2.3 Anatomy and physiology of cutaneous receptors

2.3.1 The layers of skin and structures pertinent to touch

The majority of components of human skin are the dermis and epidermis, with each layer linked to unique populations of mechanoreceptors. The epidermis serves as a barrier and in the glabrous skin of the fingers and palms, exhibits a series of patterns that increase friction and increase the vibration while the epidermis is in contact with something. Touch receptors that are associated with the epidermis are positioned at the dermal-epidermal junction. The dermis contains collagen and elastin fibers that constitute the skin's mechanical properties and the variation of the receptors to respond. It contains blood vessels, nerve fibers, and the lower mechanoreceptors. Below this is the subcutaneous layer that contains fat, the deeper receptors, and the nerve pathways that make the connection of the peripheral inputs to the central nervous system. In glabrous skin and hairy skin, the skin structures differ greatly. Glabrous skin is specialized to discriminate fine tactile with a high density of receptors and with distinct skin ridges. Hairy skin has a low density of receptors, and there are, in addition, hair follicle receptors that are sensitive to the movement of hairs; thus, tactile sensitivity extends beyond the surface of the skin.

2.3.2 Mechanoreceptors

Four primary types of mechanoreceptors mediate discriminative touch. These four types differ in terms of morphology, location, receptive field size, and function. Moreover, they can be classified by adaptation rate (slowly adapting vs. rapidly adapting) and receptive field size (small vs. large) into four functional classes: slowly adapting type I (SAI), slowly

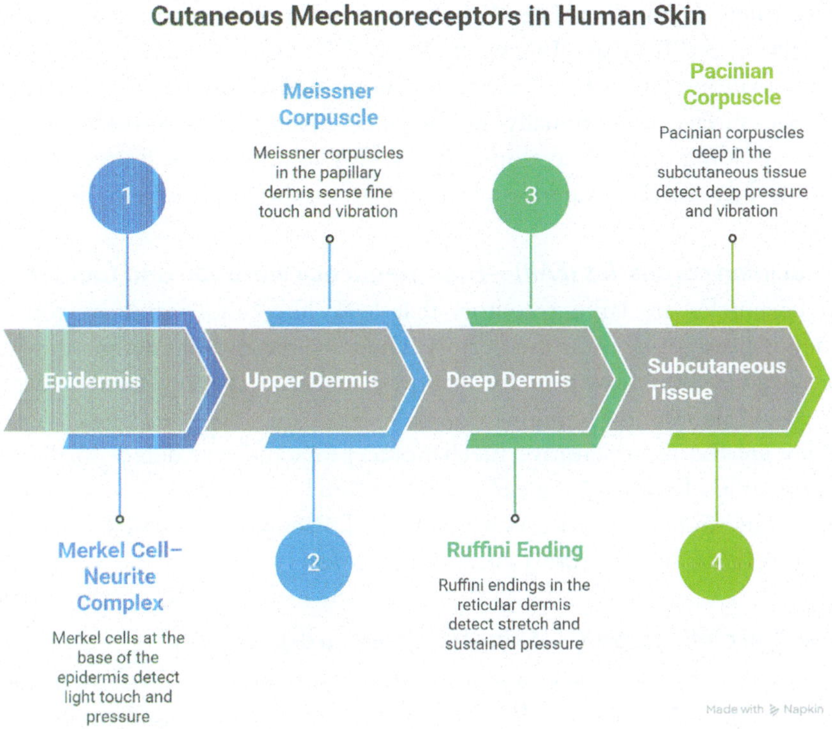

Figure 2.1 Cutaneous mechanoreceptors in human skin.

adapting type II (SAII), fast adapting type I (FAI/RAI), and fast adapting type II (FAII). Please refer to Fig. 2.1 (Data taken for image was from Claude AI).

2.3.2.1 Merkel cells/discs (SAI)—Sustained pressure and fine spatial detail

Clusters of touch domes (Merkel, 1875), the dermal-epidermal junction, and form Merkel cell-neurite complexes, slowly adapting touch receptors, and fire with sustained pressure. These receptors have small receptive fields (2–3 mm at the fingertips) and provide the highest spatial resolution (Johansson & Vallbo, 1979). Merkel cells are already mechanosensitive and contain mechanotransduction channels that convert the deformation of the skin into electrochemical signals within the cells and the nerve endings that are associated with them (Coste et al., 2010). SAI receptors (Johnson, 2001) encode and relay the fine spatial details of edges, corners, and textured patterns. Merkel cells are abundant at the fingertips, so spatial acuity is particularly high (Vallbo & Johansson, 1984).

2.3.2.2 Meissner corpuscles (FAI/RAI)—Low-frequency vibration and slip

Located in the dermal papillae, below the epidermis of skin without hair, Meissner's corpuscles are encapsulated receptors that are particularly concentrated in the fingers and palms. Their anatomy enables them to respond to changes in the skin. These are rapidly adapting and have small receptive fields (3–5 mm). They respond most to low-frequency vibrations (10–50 Hz). Such tuning is optimal for the detection of slip between skin and other surfaces. Functionally, the Meissner's corpuscles are involved in the control of the grip by aiding in the detection of initial slips, responding to fluttering movements of the skin to increase the awareness of the texture of the surfaces being interacted with.

2.3.2.3 Pacinian corpuscles (FAII)—High-frequency vibration and tool use

Pacinian corpuscles are large receptors that look like onions and are found in the subcutaneous tissue or deep in the dermis. Their layered capsule design amplifies rapid vibrations and sieves out slow pressure changes. Pacinian corpuscles are fast-adapting type II receptors and are sensitive to high-frequency vibrations (100–400 Hz) in large receptive fields. They are the most sensitive mechanoreceptors and can detect small amplitudes even for vibrations. Pacinian corpuscles mediate touch and encode fine textures through vibration, while tool-mediated touch is activated. Pacinian corpuscles and extends the perception of touch beyond direct contact to the skin.

2.3.2.4 Ruffini endings (SAII)—Skin stretch and grasp stability

Ruffini endings are type II receptors found in deeper layers of the dermis and around the joints that are slow adaptors. Because of their unique elongated structure, they are tuned for sensitivity for lateral stretch and deformation of the skin rather than vertical indentations. They are designed for large fields that will respond to prolonged signals of tangential stretching of the skin, which happens during movement of the hands and manipulation of objects. By encoding the configuration of skin stretch associated with grasp, Ruffini endings are crucial for hand shaping, stability of grip, and sense of finger position. Please refer to Table 2.1 (Source: Claude AI).

2.3.3 Distribution and density across body regions

The distribution and density of mechanoreceptors in the human body differ and correlate with touch sensitivity. The fingers, for example, have around 240 mechanoreceptors in a given area of 1 cm^2. The heightened concentration of Meissner and Merkel receptors in the area helps to facilitate the recognition of objects and fine detail. Other body regions have measurably lower receptor densities. The palm, for example, has about 100–140 mechanoreceptors in a 1 cm^2 area. The density of mechanoreceptors in the tongue and lips is comparable to that in the fingers. In contrast, the density of mechanoreceptors in the skin of the trunk and back is less than 10 cm^2. The receptor density in a body region helps to predict and explain the differences in the thresholds for two-point discrimination. The total receptor count in the peripheral system does not explain the level of touch acuity. The levels of cortical magnification, which are greater in the hands and face,

Table 2.1 Summary of mechanoreceptor types and their functional properties.

Receptor Type	Classification	Location	Receptive field size	Adaptation rate	Optimal stimulus	Primary function
Merkel cells/discs	SAI (Slowly adapting type I)	Dermal-epidermal junction	Small (2–3 mm on fingertips)	Slowly adapting	Sustained pressure, edges	Fine spatial detail, form, and texture perception, Braille reading
Meissner corpuscles	FAI/RAI (Fast adapting type I)	Dermal papillae of glabrous skin	Small (3–5 mm on fingertips)	Fast adapting	Low-frequency vibration (10–50 Hz)	Slip detection, grip control, dynamic touch
Pacinian corpuscles	FAII (Fast adapting type II)	Deep dermis and subcutaneous tissue	Large (entire finger or palm)	Fast adapting	High-frequency vibration (100–400 Hz)	Fine texture perception, tool-mediated touch
Ruffini endings	SAII (Slowly adapting type II)	Deep dermis, near joints	Large	Slowly adapting	Skin stretch, lateral forces	Hand shaping, grasp stability, proprioception

are responsible for amplifying the notable peripheral differences and explain the greater touch sensitivity in these areas.

2.3.4 Free nerve endings: Temperature and nociceptive contributions

Free nerve endings are unencapsulated mechanoreceptors located in the skin and are responsible for the mediation of temperature, pain, and some mechanical sensations. These nerve endings have no specialized structures, but have various ion channels which determine the modality of the nerve endings. There are cold receptors, thermoreceptors, and warm receptors. TRPM8 is the cold receptor, which responds to temperatures below 30 degrees, and TRPV responds to temperatures above 30 degrees. There is also a thermal neutral zone, which is the temperature around the skin. The free nerve endings also have nociceptors, which respond to mechanical, thermal, or chemical stimuli that are potentially damaging. There are high-threshold mechanonociceptors, which respond to strong pressures, and polymodal nociceptors that respond to heat, chemicals, and intense forces. Although the nociceptive signals are protective, they negatively impact the sense of touch. A class of unmyelinated low-threshold mechanoreceptors, C-tactile (CT) afferents, responds selectively to the kinds of gentle stroking that characterize social touch. These afferents are associated with the emotional aspects of touch rather than the discriminative aspects of touch and send their projections to the cortical areas concerned with emotions.

2.4 Afferent pathways and neural coding of touch

2.4.1 Primary afferent fibers: Aβ, Aδ, and C fibers

Primary afferent neurons, with cell bodies located within the dorsal root ganglia, relay touch information from peripheral receptors to the brain. They can be distinguished based on axonal diameter, type of myelination, and conduction speed; these factors also determine their functional role. Functionally, Aβ fibers are the largest, most myelinated, and fastest conducting, up to 30–70 cm/m. They transmit signals from receptors of touch (Meissner's, Merkel's, Pacinian, and Ruffini's) and support coding of touch with both time and space consideration. Touch receptors located on the fingertips can send and relay signals to the brain within 20–30 ms, which allows the brain to execute a rapid response as part of a reflex action. On the other hand, Aδ fibers have a thinner coat of myelin and are therefore slower conducting (5–30 cm/m); the receptors they innervate (via cold receptors) are responsible for sharp pain and temperature

Finally, C fibers have the slowest conduction of <2 cm/m. They are unmyelinated, have a small diameter, and innervate warm receptors, polymodal (or common) nociceptors, and CT afferents. The signals they transmit are responsible for the slow onset of burning pain as well as the gradual onset of touch reception, as a pleasant sensation. These fibers have a temporal hierarchy in the sense that they transmit Aβ fibers first, which means they will be the first to transmit touch signals, after which Aδ and C fibers transmit pain and touch signals, respectively, which adds a protective and desirable dimension to the touch sensation.

2.4.2 The two major ascending pathways

2.4.2.1 Dorsal column–medial lemniscal pathway

The dorsal column-medial lemniscal pathway (DCML) pathway communicates low-fidelity information about touch, vibration, and proprioception through Aβ afferents. Primary afferents enter the spinal cord and travel upwards ipsilaterally in the dorsal columns without synapsing. The input data is organized such that medial sacral fibers and lateral cervical fibers are arranged together. Inputs from the lower body are organized into the Gracile Tract, while those from the upper body are organized into the cuneate tract. The first synaptic event in the DCML pathway occurs in the gracile and cuneate nuclei of the medulla, where lateral inhibition occurs to increase the spatial resolution of the input data. The second-order neurons that cross the medial lemniscus of the contralateral thalamus and then go to the cortex. This pathway is capable and designed to complete rapid and accurate location of stimuli and integrates sensorimotor information in real time.

2.4.2.2 Anterolateral (Spinothalamic) pathway

Anterolateral pathway receptors (A?? and C fibers) receive stimuli related to pain, temperature, and gross touch. Primary afferents synapse in the dorsal horn, while second-order neurons cross at the same spinal level and travel contralaterally up to the thalamus. The lateral spinothalamic tract is organized to relay information to the specific thalamic sensory nuclei, while the medial spinothalamic tract ends in the thalamus and the limbic system, where the emotional and motivational components of pain are integrated. Unlike the DCML pathway, this system's slower conduction and reduced spatial resolution emphasize emotional and autonomic integration over detail.

2.4.3 Spinal cord and brainstem processing

Active processing of tactile signals occurs in the brainstem and spinal cord. For example, Kandel et al. (2013) describe gain control, temporal filtering, and inhibitory interactions that control the flow of the processing of stimuli in the dorsal horn circuits. Gate control theory explains the phenomenon, where large fiber tactile input can inhibit the transmission of pain (Melzack & Wall, 1965). This explains the relief of pain when one rubs the area in pain. Further studies have explored the specific population of interneurons and the modulatory effects of neuromodulators.

2.4.4 Thalamic relay: Ventral posterolateral and ventral posteromedial Nuclei

Somatosensory pathways to the cortex travel through the posterior part of the thalamus. The ventral posterolateral (VPL) and ventral posteromedial (VPM) are part of the posterior thalamus. The VPL gets input from the body via the medial lemniscus, and the VPM gets input from the face via the trigeminal pathways. Thalamic neurons are modality-specific, have well-defined receptive fields, and are organized in a somatotopic way. Thalamic neurons do not simply transmit information; they also play a role in the sensitization of the stimulus through the feedback that comes from the cortex to the thalamus and the arousal system. The mechanism of attention and expectation can amplify relevant stimuli and diminish the processing of the stimuli in the thalamus.

2.4.5 Neural coding principles

Neural coding involves many methods to convey features of a stimulus that are continuous. Rate coding is one of the methods and is used to convey the intensity of a stimulus. It is a good way to code for the pressure and temperature for a stimulus, but it does require temporal integration. Repetition of spikes demonstrates temporal coding of neurons. Dynamic stimuli feature acceleration and velocity. Neurons encode stimuli with high-frequency vibrations and phase-lock with stimulus oscillations. Information is represented across cultural neuron ensembles through population coding. Encoded position is determined by neurons that fire, and the level of recruitment can determine intensity. This leads to a representation that is both robust and of high dimensions. Firing rates and spike timing represent different intensities and frequencies, contributing to multiplexed coding. It is identified that the somatosensory system is able to capture large amounts of information even with limited neural resources through the use of parallelism.

2.5 Central processing: From thalamus to cortex

2.5.1 Primary somatosensory cortex (S1): Brodmann areas 3a, 3b, 1, and 2

The primary somatosensory cortex (S1) is located in the postcentral gyrus, situated directly posterior to the central sulcus, and is comprised of four parallel Brodmann areas, 3a, 3b, 1, and 2, arranged in a rostrocaudal formation.

- Area 3a (the most rostral and located within the central sulcus) is the first to receive proprioceptive information about muscles/joints and encodes the position and movement of the joints.
- Area 3b receives the most dense cutaneous input from the thalamus, and exhibits the finest somatotopy, which establishes it as the fundamental cortical area of touch.
- Area 1, which is located posterior to area 3b, possesses larger receptive fields and encodes more complex features such as the orientation, movement, and texture of stimuli.

- Area 2 combines cutaneous and proprioceptive inputs to facilitate haptic recognition of objects by merging the surface characteristics of objects with the position of the hand.

Lesions in the primary somatosensory cortex (S1) are seen to follow the above-stated hierarchy. In the case of area 3b, there is limited or no detection of stimulation, while in area 2, there is no recognition of the objects that are present, and no appreciation of the sensation.

2.5.2 Somatotopic organization: The sensory homunculus

S1 features a sensory homunculus (Penfield & Boldrey, 1937), an example of an ordered somatotopic map of the body's surface. The representation on the cortex depends on the density of the receptors, not on the size of the body part; so, the hands and face, for instance, are represented more than the trunk and legs. Somatotopy demonstrates plasticity. With training, injury, or loss of a limb, the cortical maps are reorganized, with the neighboring deprived regions being taken over by invading representations. This kind of plasticity is one of the contributors to phantom limb sensations (Ramachandran & Hirstein, 1998).

2.5.3 S1 processing: Columnar and laminar

Neurons in S1 areas are arranged in vertical columns. Each column contains neurons with a shared location of a receptive field as well as response properties. As such, adjacent columns respond to different adjacent areas of skin, and/or different submodalities.

Different cortical layers have different functions.

- Layer 4 obtains input from the thalamus and has a primary role in the encoding of uncomplicated, modality-specific features.
- Layers 2 and 3 have integrating functions across columns, and aid in the encoding of complex features.
- Layer 5 is responsible for projections to motor and subcortical areas, carrying out the functions of sensorimotor integration.
- Layer 6 projects to the thalamus, providing contextual and attentional modulation of sensory gain.

All of these layers function to create a uniform, complex, spatiotemporal structure.

2.5.4 The posterior parietal cortex and secondary somatosensory cortex

The secondary somatosensory cortex S2 (in the parietal operculum) receives inputs from all areas of the S1, and has bilateral receptive fields, allowing for integration across the two sides of the body. Compared to S1, the neurons in S2 are more complex, integrated with other sensory modalities (multisensory), and are more influenced by attention and task relevance, implying that S2 has a role in conscious perception and memory of touch.

In the posterior parietal cortex (areas 5 and 7), integration of sensory inputs (somatosensory, visual, and auditory) is used to direct motor actions. When these areas are damaged, the patient experiences tactile agnosia and extinction, which is a sign of poor sensory integration and not the absence of basic tactile sensations.

2.5.5 Integration with motor systems

Some mirror neurons in the premotor cortex fire both when an individual performs an action and when the individual observes the same action being performed by others. Also, some of these neurons fire in response to self-observed movement. During such movement, the motor cortex sends copies of commands to the relevant muscles. The purpose of these efference copies is to alter how the central nervous system interprets stimuli, which explains active tactile sensation and the ability to distinguish self-generated stimuli from those caused by another person (Wolpert et al., 1995).

2.6 Active versus passive touch: The role of haptic exploration

2.6.1 Passive touch: Stimulus-driven processing

Passive touch can be described as the process of making contact with unengaged skin. For example, getting rain on your arm, or someone tapping you on your shoulder. Or, it could be something like touching the skin on the surface of a table. The example of passive touch described here uses the environment's criteria to inform the brain of the skin's sensory input. The brain can only determine the stimuli's location, the extent of the pressure used, the frequency of the touch, or the pattern of the touching. The studies of psychophysics and passive touch combined have shown some of the most fundamental thresholds in human sensitivity. These amazing levels of sensitivity in human beings are a reflection of how highly developed the peripheral receptor system is in their bodies. Individuals can perceive the softest touch or smallest pressure, even detecting slight indentations in their skin. This remarkable sensitivity of the human body has been documented in studies reporting the phenomenon of haptic sensitivity. Skin contact with a stationary object limits the information that passive touch can provide. An object pressed against skin can provide information concerning the skin's surface, how it's able to provide insulation, and how hard it is. However, information pertaining to the surface area of the object, the shape, or the three-dimensional structure is absent. To understand complex features that go beyond the basic sensations of skin or bodily insulation, frequent and successive movements are needed, and even then, the information spans a wide area or is of a complex nature.

2.6.2 Active (haptic) touch: Exploratory procedures and sensorimotor integration

Active or haptic touch is a type of somatosensation that is altered by self-generated movement. When someone is active in exploration, such as running fingers across a surface or grabbing and hefting an object, the motor actions are organized in a way that the sensory apprehension is oriented toward a perceptual goal. This results in a loop caused by the sensors and the motors—the commands and the signals are dictated by the motors, while the feedback is dictated by the sensors. Lederman and Klatzky (1987) recognized the mechanism of some of the more specific, exploratory procedures, which are defined as stereotyped movements used to draw out certain properties of an object. Touching an object with lateral movement allows the person to perceive the texture of the object, while the use of pressure determines the compliance of the object. Hefting, or unsupported holding, allows a person to estimate the weight of the object, whereas shape contouring allows someone to perceive the shape of the object. Finally, enclosure allows a person to perceive the size and volume of an object (Klatzky & Lederman, 2003). When humans use exploratory procedures, they choose the movements that are most appropriate for a particular task. The use of lateral movements for texture, the use of pressure for hardness, and the use of contour tracing for shape all demonstrate that haptic perception is not a passive process.

Active touch is explained by tight sensorimotor coupling. The posterior parietal cortex issues predictions about the expected sensory feedback of a movement, which are continually compared to the actual feedback received. When the feedback does not match the prediction, it reveals an unexpected property of the object, triggering either a motor response or increased attention. This mechanism allows for efficient motor control and adaptive responses.

2.6.3 Kinesthetic (proprioceptive) contributions to touch perception

Proprioception provides information about the position and movement of limbs and is essential for haptic perception. While contacting surfaces, tactile feedback records surface contact, and haptic feedback enables the contact to be integrated into a spatial percept of the object's contour. Proprioceptive information is critical for integrating the individual haptic sensations, as experienced by individuals suffering from large-fiber sensory neuropathy. People with this condition can feel a touch, but they cannot identify objects haptically due to an absence of the position cues, which are necessary for integration. Neuroimaging has shown that the perceptual haptics of form activate the somatosensory and parietal spatial-processing areas. The resultant representations of objects are processed in relation to the external spatial coordinates, which is the reason for the overlap in tactile and visual recognition of objects in the brain.

2.6.4 Illustrations: Object recognition by hand, texture discrimination

Active touch provides the basis for perceptual principles, and an illustration of this is texture perception. Discriminating between surfaces requires relative motion between the skin and the surface. Without this movement, information is greatly limited. As fingers slide over a surface, the subcutaneous structures modulate the produced vibrations, and this is detected by the mechanoreceptors. Coarse textures (>200 μm) are spatially encoded by Merkel and Meissner receptors, while fine textures (which are <200 μm) are encoded temporally by Pacinian receptors due to the high-frequency vibrations. Discrimination between textures can be achieved with variations of less than 10 μm, enabled by nanometer-scale vibrations and the sensitivity of Pacinian corpuscles. The scanning speed of the fingers is crucial: very slow movement reduces sensitivity to vibrations, while excessive speed impairs spatial resolution. People typically adopt an optimal scanning speed of 5–10 cm/s. Object recognition by touch involves multiple senses, with the brain analyzing size, shape, weight, texture, and pliability. These processes engage different brain regions and typically require 2–5 s. This is slower than a person's visual system, but from studies, the touch system is doing sampling very efficiently. Studies looking at vision and touch recognition of objects show that there is a large degree of overlap in areas of the brain used to process the two modalities, further confirming that there is a system of brain areas that processes objects regardless of the input stream. The brain exhibits remarkable plasticity, particularly in regions adjacent to the visual system. In individuals who are blind, these areas often become enhanced, supporting compensatory sensory functions such as touch and auditory processing.

2.7 Perceptual phenomena and psychophysics of touch

2.7.1 Thresholds for basic psychophysical functions: Detection, discrimination, and scaling

Detection thresholds simply show the stimulus required to get a sensory reaction. Depending on the site, skin indentations of 2–10 μm can be felt, fingertip vibrational changes of less than 1 μm can be detected, and temperature fluctuations of 0.01°C can be felt under controlled conditions (Stevens & Choo, 1998; Verrillo, 1963). The discrimination process is guided by the Weber law: for a change to be noticed, a proportional change is required on the baseline (Weber, 1834). With respect to pressure, the Weber fraction is approximately 0.14. This means that to adequately distinguish two stimuli, one of the stimuli must be 14% stronger than the other (Gescheider et al., 1990). Magnitude estimation attempts to analyze how the perception of pressure changes as the intensity of the stimulus is altered. In most cases of the human body, perception changes almost linearly (i.e., power exponent ≈ 1.0), although the body region and the type of stimuli influence the perception in varying degrees.

2.7.2 Spatial acuity: Discrimination of two points and orientation of grating

Fingertips can discern gratings with pitches below 1 mm, demonstrating a more refined spatial perception than that shown by two-point discrimination tests (Johnson, 2001). This level of sensitivity of the fingertip is due to the abundant presence of a type of mechanoreceptor, the Merkel's disc, whose small receptive fields allow for strong cortical magnification. Furthermore, there is evidence of lateral inhibition at these receptors, which is thought to be responsible for the sharpening of spatial contrast (Hsiao & Bensmaia, 2008). Two-point discriminatory measures gap cutting, not actual spatial resolution, whereas grating orientation discrimination measures spatial resolution more appropriately. In this case, subjects determine whether vertically or horizontally oriented ridged gratings have been presented. Fingertips resolve gratings with spacing <1 mm, indicating higher spatial sensitivity over two-point discrimination. High spatial acuity arises from many tightly packed Merkel receptors, small receptive fields, strong cortical magnification, and enhanced lateral inhibition, which sharpens or improves spatial resolution.

2.7.3 Temporal resolution: Vibration perception

The definition of touch temporal acuity is sensitivity to timing. Humans can feel vibrations in the range of about 5–700 Hz, with the highest sensitivity in the range of about 200–300 Hz, which corresponds with the peak sensitivity of Pacinian receptors. If the vibrations are in the range of about 5 Hz, stimulation will feel discrete. The range of oscillations that will feel discrete is up to 40-50Hz. In this range, it will feel distinct Meissner mediated, and above this, it will feel like a continuous buzz, which is Pacinian mediated. Touch temporal acuity is not as fine as the acuity of vision or hearing, but adequate to serve the daily needs of touch at around 20–30 ms.

2.7.4 Tactile illusions

By considering a variety of different methods, researchers are interested in how touch is formulated rather than directly measured from the skin.

- Aristotle's illusion: Due to crossing fingers and touching one object, an individual may feel as though two different objects are being touched because of conflicting spatial mappings.
- Thermal grill illusion: Warm and cool interlaced bars lead to burning pain without damage to the tissue. Thermal pain and the pain pathway are working in parallel and can lead to an injury.
- Cutaneous rabbit illusion: Rapid tapping in separate locations is a phenomenon known in perceptual Interpolation.
- The sensation of a phantom limb is a phenomenon that can occur if touch is applied to the area of the body that is intact. The person experiences touch in the missing limb, which is a result of cortical remapping that occurs postamputation.

2.7.5 Multimodal interactions

Touch is the most integrated with the other senses. Visual information makes touch more sensitive to the area that is being touched visually, and it's more sensitive to touch in the area that is visually being touched. In the rubber hand illusion, there is multisensory visceral body ownership—synchronized visual and tactile stimulation means a rubber hand feels like it is part of one's body (Botvinick & Cohen, 1998; Tsakiris & Haggard, 2005). In the skin of parchment illusion, high frequencies are amplified, and the skin feels rough. Touch perception, together with sound, is heightened (Zampini & Spence, 2004).

2.8 Affective and social dimensions of touch

2.8.1 C-tactile afferents and gentle touch

A CT afferent is a type of mechanoreceptor located in hairy skin, which is unmyelinated and has low thresholds and responds best to gentle stroking at a rate of 1–10 cm/s (Loken et al., 2009). In contrast to Aβ afferents that transmit to the sensory cortex, CT afferents send projections to the secondary/posterior insular cortex, which is responsible for the processing of inner bodily sensations and emotions (Olausson et al., 2002). CT afferent activity coincides with the subjective feeling of pleasantness associated with light touch (Ackerley et al., 2014; McGlone et al., 2014). This indicates that CT afferents are part of a unique pathway, which is proposed to represent the social and emotional qualities of touch (Liljencrantz & Olausson, 2014). Patients with rare large-fiber neuropathy, a condition in which a person loses the ability to sense Aβ fibers, but with C-fiber sensations preserved, cannot sense the location of the touch, cannot feel objects through the sense of touch, and cannot feel objects through the sense of touch. In contrast, patients can still feel a gentle touch and describe the sensations as pleasant. On the other hand, when CT activity is inhibited, gentle touch will not be perceived as pleasant, but instead will be perceived as neutral, which will still leave the ability to detect the touch. This proves the CT afferents' function is to mediate affective touch.

2.8.2 Emotional and social functions of touch

Various species, including humans, use touch for important social purposes. In the primate order, the stress-relieving benefits of grooming are bidirectional, as cortisol is reduced and oxytocin is elevated for both the groomer and the groomed. Touch in humans is no different; it conveys messages, establishes relationships, and soothes. Hertenstein et al. (2009) found that people are able to identify certain emotions through touch alone. Touching someone's arm to convey sympathy, love, gratitude, and anger is a message that can be received accurately beyond random guessing. This suggests emotions can be conveyed through a system of touch. Touch affects trust and the propensity to cooperate. In economic games, the touched participants cooperated more (Crusco & Wetzel, 1984). Oxytocin, the social bonding neuropeptide, is the biological mechanism (Craig, 2002). Pleasant touch is one of the mechanisms that triggers the social bonding neuropeptide oxytocin; thus, it is shown to decrease the responsiveness level of the amygdala (where perceived social threats are prevalent) and increase the level of trust. CT (C-fiber) stimulation increases the release of oxytocin, thereby linking the pathways of affective touch to the social neurochemistry.

2.8.3 The role of touch in development and interpersonal communication

Touch is important for proper development. Harlow's research from 1958 showed us that infant primates preferred to be around soft cloth surrogate mothers rather than ones that were made of wire, but also provided food. This demonstrates that even in infants, the primary role of forming an attachment is through touch. Field (2001) showed that preterm infants who were in the hospital and received massages did better than infants who received normal care. They slept better, gained weight faster, and showed better development. Some of the benefits of touch are related to decreased stress, the body's response to touch, and hormones. Touch provides both emotional and physiological support during early development and remains vital in adulthood. In adult relationships, affectionate touch is positively correlated with relationship satisfaction. It also alleviates stress and pain, while contributing to improved mental and physical health. The ability to touch and express emotion through it seems to be universal, even if the norms around physically touching are determined by the culture. This shows we can communicate through touch on a social level.

2.9 Summary and open questions

2.9.1 Fundamental aspects of touch sensation

Touch perception relies on several principles. Specific mechanoreceptors are organized in parallel streams to capture distinct details: Merkel cells encode fine spatial information, Meissner corpuscles respond to movement, Pacinian corpuscles detect vibrations, and Ruffini endings register skin stretch.

Secondly, touch processing is organized into a hierarchy from peripheral receptors to the spinal cord, brainstem, thalamus, and cortex. Each stage of processing touch sensation applies different mechanisms to transform and refine the signal, such as lateral inhibition, temporal filtering, and a combination of different senses to result in a high-level cortical representation.

Third, while passive touch stimulation might involve only feelings of touch, active (haptic) stimulation is different. With active touch, as you move to sample different surfaces, a combination of touch and proprioceptive senses is used to achieve a specific purpose.

Fourth, the two functions of touch, which are discriminative (for object recognition, texture, and localization) and affective (for social communication and emotional regulation), operate partly in distinct touch pathways, but sensory discrimination and emotional regulation are both at work during everyday activities.

Lastly, more training in touch perception is more adaptable to injury and sensory loss. Cortical maps reorganize to dynamically adjust areas of sensory receptors that are used and how they are represented in the cortex in relation to the tasks performed.

2.9.2 Open questions and path forward

Despite the extensive work done in understanding the neural mechanisms of tactile attention, there are still numerous open questions. Most of the mechanisms are not well understood when compared to tactile attention mechanisms of visual attention; neural mechanisms of touch are not well understood. Multisensory integration still presents a challenge to standard models for understanding this phenomenon. Convergence that is visuo-tactile from an early sensory cortex suggests that multisensory processing is more extensive than previously thought. The relationship between the discriminative and affective touch pathways also needs more consideration. Especially the role of CT afferent and their interactions with cortical networks during social touch involves an intricate mix of sensory and emotional components. The field of touch perception is still relatively unexplored. Atypical somatosensory processing has associations with tactile sensitivity and texture aversion that can help identify a focus for understanding and addressing complications arising from atypical somatosensory processing. The field has also seen continued advancement of mechanotransduction studies. Major breakthroughs have come with the development of Piezo channels. However, receptor-specific transduction mechanisms and their modulation still present active research possibilities.

2.9.3 Applications: Haptics in technology, rehabilitation, and virtual reality

The technology and clinical possibilities that come from an understanding of touch perception are extensive. Haptic technology, which is capable of programmable tactile feedback, has improved virtual reality, teleoperation, and human-computer interfaces. Current technology is predominantly focused on the use of vibration motors. However, future technology is set to achieve the greater tactile feedback that includes the sensation of texture, force, and temperature. Tactile technologies aid in sensory substitution as well. For example, tactile displays using vibrating pins provide a Braille alternative for blind users. However, the information bandwidth is restricted. Somatic neural prosthetics that stimulate users' somatosensory cortex will potentially provide users' prosthetic limbs with tactile feedback, thereby enhancing the users' prosthetic limb control and embodiment. The potential of tactile plasticity is utilized in rehabilitation. For example, the recovery of function and the reorganization of the cortex after a stroke or nerve injury are promoted by the retraining process. Contrarily, tools for tactile substitution provide access to sensory information that is not available by transforming auditory or visual information to touch. The appreciation of the clinical implications of understanding affective touch is of great relevance. Pain, anxiety, and depression, which respond to touch, are improved by touch therapy. These effects are the result of CT afferents and the release of oxytocin as part of the neuro-modulatory mechanism.

Haptics are increasingly integrated into virtual reality, but natural touch is not satisfactorily duplicated at present. Potentially, future technologies will provide more realistic touch environments in which touch, pressure, and heat can be integrated with textures to create a more realistic experience. The rapid advancement of molecular biology, neural recording, optogenetics, and computational modeling has greatly strengthened the knowledge of tactile mechanisms. At the same time, the application of this knowledge has increased the availability and improved the recovery of the ability to use touch after the loss of sensory perception.

AI disclosure

During the preparation of this work, the author(s) used Claude.ai for table content formatting. After using this tool/service, the author(s) reviewed and edited the content as needed and take(s) full responsibility for the content of the publication.

References

Ackerley, R., Backlund Wasling, H., Liljencrantz, J., Olausson, H., Johnson, R. D., & Wessberg, J. (2014). Human C-tactile afferents are tuned to the temperature of a skin-stroking caress. *Journal of Neuroscience, 34*(8), 2879–2883. https://doi.org/10.1523/JNEUROSCI.2847-13.2014Sweden.

Botvinick, M., & Cohen, J. (1998). Rubber hands 'feel' touch that eyes see. *Nature, 391*(6669), 756. https://doi.org/10.1038/35784.

Coste, B., Mathur, J., Schmidt, M., Earley, T. J., Ranade, S., Petrus, M. J., Dubin, A. E., & Patapoutian, A. (2010). Piezo1 and Piezo2 are essential components of distinct mechanically activated cation channels. *Science, 330*(6000), 55–60. https://doi.org/10.1126/science.1193270.

Craig, A. D. (2002). How do you feel? Interoception: The sense of the physiological condition of the body. *Nature Reviews Neuroscience, 3*(8), 655–666. https://doi.org/10.1038/nrn894.

Crusco, A. H., & Wetzel, C. G. (1984). The Midas touch: The effects of interpersonal touch on restaurant tipping. *Personality and Social Psychology Bulletin, 10*(4), 512–517. https://doi.org/10.1177/0146167284104003.

Field, T. (2001). *Touch*. MIT Press.

Gescheider, G. A., Bolanowski, S. J., & Verrillo, R. T. (2004). Some characteristics of tactile channels. *Behavioural brain research, 148*(1-2), 35–40. https://doi.org/10.1016/s0166-4328(03)00177-3.

Hertenstein, M. J., Holmes, R., McCullough, M., & Keltner, D. (2009). The communication of emotion via touch. *Emotion, 9*(4), 566–573. https://doi.org/10.1037/a0016108.

Hsiao, S. S., & Bensmaia, S. J. (2008). Coding of object shape and texture. In E. P. Gardner, & J. H. Kaas (Eds.). In *The senses: A comprehensive reference: 6* (pp. 55–68). Academic Press.

Johansson, R. S., & Vallbo, A. B. (1979). Tactile sensibility in the human hand: relative and absolute densities of four types of mechanoreceptive units in glabrous skin. *The Journal of Physiology, 286*(1), 283–300. https://doi.org/10.1113/jphysiol.1979.sp012619.

Johnson, K. (2001). The roles and functions of cutaneous mechanoreceptors. *Current Opinion in Neurobiology, 11*(4), 455–461. https://doi.org/10.1016/s0959-4388(00)00234-8.

Kandel, E. R., Schwartz, J. H., Jessell, T. M., Siegelbaum, S. A., & Hudspeth, A. J. (2013). *Principles of neural science* (5th ed.). McGraw-Hill Education.

Klatzky, R. L., & Lederman, S. J. (2003). Touch. In A. F. Healy, & R. W. Proctor (Eds.). In *Handbook of psychology: Experimental psychology: 4* (pp. 147–176). John Wiley & Sons.

Lederman, S. J., & Klatzky, R. L. (1987). Hand movements: A window into haptic object recognition. *Cognitive Psychology, 19*(3), 342–368. https://doi.org/10.1016/0010-0285(87)90008-9.
Liljencrantz, J., & Olausson, H. (2014). Tactile C fibers and their contributions to pleasant sensations and to tactile allodynia. *Frontiers in Behavioral Neuroscience, 8*, 1662–5153. https://doi.org/10.3389/fnbeh.2014.00037.
Löken, L. S., Wessberg, J., Morrison, I., McGlone, F., & Olausson, H. (2009). Coding of pleasant touch by unmyelinated afferents in humans. *Nature Neuroscience, 12*(5), 547–548. https://doi.org/10.1038/nn.2312.
McGlone, F., Wessberg, J., & Olausson, H. (2014). Discriminative and affective touch: Sensing and feeling. *Neuron, 82*(4), 737–755. https://doi.org/10.1016/j.neuron.2014.05.001.
Meissner, G. (1853). *Beiträge zur Anatomie und Physiologie der Haut*. Rostock.
Melzack, R., & Wall, P. D. (1965). Pain mechanisms: A new theory. *Science, 150*(3699), 971–979. https://doi.org/10.1126/science.150.3699.971.
Merkel, F. r. (1875). Tastzellen und Tastkörperchen bei den Hausthieren und beim Menschen. *Archiv für Mikroskopische Anatomie, 11*(S1), 636–652. https://doi.org/10.1007/bf02933819.
Mountcastle, V. B. (1957). Modality and topographic properties of single neurons of cat's somatic sensory cortex. *Journal of Neurophysiology, 20*(4), 408–434. https://doi.org/10.1152/jn.1957.20.4.408.
Olausson, H., Lamarre, Y., Backlund, H., Morin, C., Wallin, B. G., Starck, G., Ekholm, S., Strigo, I., Worsley, K., Vallbo, Å. B., & Bushnell, M. C. (2002). Unmyelinated tactile afferents signal touch and project to insular cortex. *Nature Neuroscience, 5*(9), 900–904. https://doi.org/10.1038/nn896.
Pacini, F. (1840). *Nuovi organi scoperti nel corpo umano*. Tipografia Cino.
Penfield, W., & Boldrey, E. (1937). Somatic motor and sensory representation in the cerebral cortex of man as studied by electrical stimulation. *Brain, 60*(4), 389–443. https://doi.org/10.1093/brain/60.4.389.
Ramachandran, V. S., & Hirstein, W. (1998). The perception of phantom limbs. *Brain, 121*(9), 1603–1630. https://doi.org/10.1093/brain/121.9.1603.
Ruffini, A. (1898). On the minute anatomy of the neuromuscular spindles of the cat, and on their physiological significance. *The Journal of Physiology, 23*(3), 190–208. https://doi.org/10.1113/jphysiol.1898.sp000723.
Stevens, J. C., & Choo, K. K. (1998). Temperature sensitivity of the body surface over the life span. *Somatosensory and Motor Research, 15*(1), 13–28. https://doi.org/10.1080/08990229870925.
Tsakiris, M., & Haggard, P. (2005). The rubber hand illusion revisited: Visuotactile integration and self-attribution. *Journal of Experimental Psychology: Human Perception and Performance, 31*(1), 80–91. https://doi.org/10.1037/0096-1523.31.1.80.
Vallbo, A. B., & Johansson, R. S. (1984). Properties of cutaneous mechanoreceptors in the human hand related to touch sensation. *Human Neurobiology, 3*(1), 3–14.
Verrillo, R. T. (1963). Effect of contactor area on the vibrotactile threshold. *Journal of the Acoustical Society of America, 35*(12), 1962–1966. https://doi.org/10.1121/1.1918868.
Weber, E. H. (1834). *De pulsu, resorptione, auditu et tactu: Annotationes anatomicae et physiologicae*. Koehler.
Weinstein, S. (1968). In D. R. Kenshalo (Ed.), *Intensive and extensive aspects of tactile sensitivity as a function of body part, sex, and laterality* (pp. 195–222).
Wolpert, D. H. (1995). Horizontal generalization. *Machine Learning Proceedings 1995*, 566–574. https://doi.org/10.1016/B978-1-55860-377-6.50076-1.
Zampini, M., & Spence, C. (2004). The role of auditory cues in modulating the perceived crispness and staleness of potato chips. *Journal of Sensory Studies, 19*(5), 347–363. https://doi.org/10.1111/j.1745-459x.2004.080403.x.

CHAPTER 3

Haptic sensing technologies

Sanskar Agrawal and Anusha Rani
School of Business, Woxsen University, Hyderabad, Telangana, India

3.1 Introduction

Innovative learning systems have been proposed already at the beginning of the 21st century, deployed over virtual environments (Cantoni et al., 2004). Designing is completely visual and graphical for these environments. These systems have recently evolved into "Augmented Reality (AR)" and "Virtual Reality (VR)" environments that anticipate different modalities like haptic interactions and eye tracking, which is a very underrated modality (Donmez, 2023; Papanastasiou et al., 2019). Digital environments have a lot of tools to improve decision-making and user immersion processes, such as haptic devices (Fig. 3.1). Transforming educational needs is not at all limited to digital environments. For example, training children with special needs with voice has been a serious issue (Lewis-Dagnell et al., 2023).

Promoting education is a highly multidisciplinary endeavor, which consists of psychology, pedagogy, engineering, electronics, art, and medicine (Almusawi et al., 2021; Lv et al., 2021). Additionally, special interaction tools are very important for people with special needs because of poorly developed behavioral, sensory, or cognitive potential (Gallud et al., 2023; Trudel et al., 2023). For example, quiet rooms and other standalone sources may be ideal for maladaptive behaviors in children, while self-tracking apps or devices may improve focus of students (Marwati et al., 2023; Scheibel et al., 2023).

Haptic technologies are focused on duration, intensity, displacement, location, and arrangement of sensing, such as creating semantic images of sensory data or generating dermic sensations. Even though there is a huge research gap in haptic sensation as compared to visual connection, it has gained great significance over the years. It has been observed that combining haptic and visual interfaces develops motor and visual skills (Bara & Gentaz, 2011). Haptic sensations have been proposed by various studies as forerunners of data based on application, showing the significance of touch in searches in libraries or in gamified edtech services or products to promote sustainable behaviors (Zhang & Song, 2022).

In addition, it has been observed that users can interpret tactile systems in different channels (Wersényi, 2022). In typical STEM curricula, haptic interaction has a positive impact as observed in recent reviews (Sanfilippo et al., 2022; Zhu et al., 2022). Even when it comes to learning gymnastics, haptic augmentation helps users with verbal data

Haptic Feedback Control in Medical Robotics
DOI: https://doi.org/10.1016/B978-0-443-27777-1.00003-2

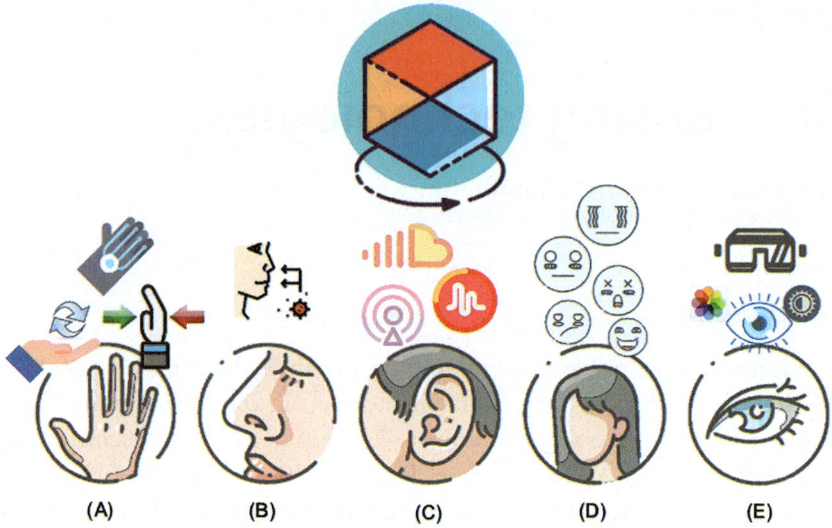

Figure 3.1 Haptic interactions with virtual environments.

(Frikha et al., 2019). At the end, communication can be improved with blind and deaf people by discussing haptics (Sorgini et al., 2018). Haptic technologies are actually a part of VR applications, making them a strong tool to improve immersion. On the other hand, the level of engagement of an individual with a haptic sensor can be given with grip, vibration, and pressure. Such evaluations of contact quality enhanced model clarification in some areas. For instance, the level of certainty to select objects can be assessed in a virtual environment by the user.

Haptics are very important to make virtual worlds feel more realistic, as proposed by large social media giants (Bandiaky et al., 2024). They are also a vital aspect of virtual environments and learning spaces, which are becoming standard tools in different areas of education and medicine (Chahal et al., 2024). This narrative review provides a deep insight into haptic applications and technologies, while focusing on training people with special needs. This review is aimed at promoting the development of a haptic system for this cause. There is a lack of innovative systems for people who deserve those systems the most.

Some leading tech companies have played a vital role in developing haptic sensing technologies for clinical applications (Table 3.1).

3.2 Overview of haptic technologies

Fig. 3.2 illustrates some closely related disciplines in haptic technology, either promoting the rise of haptic devices or synergy with overall properties to add immersive experience

Table 3.1 Top companies focused on haptic sensors for healthcare.

Company	Founded in	Based in	Technologies developed
Haption	2001	France	Haptic interface for diagnostics and surgical training.
Immersion Corporation	1993	USA	Haptic sensors for training and simulation in healthcare
Synaptics Inc.	1986	USA	Offers display, biometrics, and touch sensors for medical devices
ON Semiconductor	1999	USA	Developing controllers and sensors for power-saving haptic interfaces for healthcare
Microchip Technology Inc.	1989	USA	Designs vital components like IC chips and microcontrollers for medical haptic feedback
Ultraleap Holdings	2013	UK	Hand tracking and gesture-based haptic tech for VR programs in clinical settings
Johnson Electric	1959	Hong Kong	Programs haptic parts and motion subsystems in different devices
DOT	2014	South Korea	Immersive haptic programs to create devices accessible by visually impaired people
SMK Corp.	1925	Japan	Develops components used for haptic interfaces in healthcare

in virtual scenarios. This universe of digital services and interaction tools is entwined with advancements of computational and artificial intelligence (AI) (Górriz et al., 2020, 2023).

3.2.1 Artificial skin

Along with tactile insights, human skin is full of details about its surroundings. It senses vibration, pressure, temperature, and in-depth properties of surfaces of objects, like roughness, contact area, and texture. For example, a chemically-induced system of skin-heating was used to enhance immersion and give the impression to the user about rising heat signatures in a VR environment (Lu et al., 2021). In addition, it is possible to

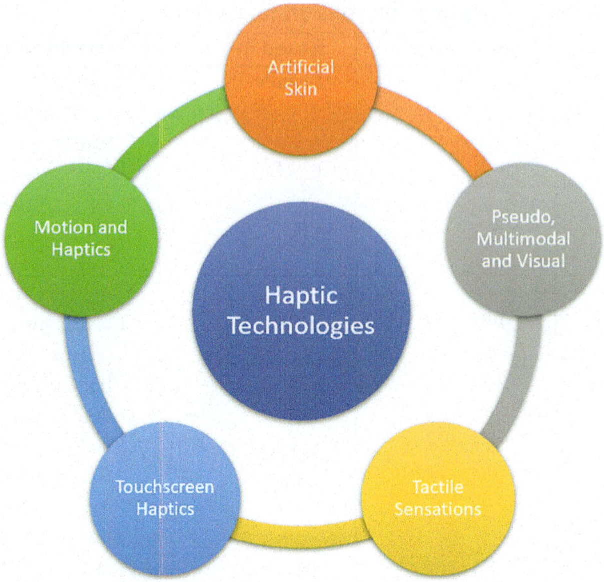

Figure 3.2 Components of haptic technology.

evaluate the confidence or activeness of a person by measuring human skin temperature while performing some tasks.

The complex sensory properties of real human skin are now replicated by flexible biological components with embedded micro and nano sensory systems, thanks to advancements in artificial skin technologies. The new artificial skins work by collecting temperature, humidity, and pressure variations in real time and sending them straight to medical robotics control systems. Piezoelectric sensors integrated into synthetic skin enable mechanical stress detection, which generates electrical impulses to regulate the surgical precision of robotic limbs.

Current artificial skin technology combines autonomous healing functions with wireless connectivity features, which enhance its suitability for extended hospital and medical purposes. Artificial skin technology serves as a vital element in telesurgery and remote diagnostics by enabling surgeons to receive physical touch information, which reduces mistakes and delays in surgical responses. Prosthetic devices using artificial skin technology enable users to detect touch sensations similar to natural body experiences, thus promoting their environmental physical interaction.

Studies in materials science that focus on e-skin made from graphene, liquid metal, and nanowire composites continue to develop innovative advances in artificial skin development. Medical robotics stands to benefit greatly from technologies that aspire to reproduce human touch both sensationally and emotionally, and in entire environmental contexts.

3.2.2 Pseudo, multimodal, and visual

With other modalities, integrating haptic feedback has been widely discovered in the interaction between humans and computers. Haptic impressions are induced by pseudo-haptics from other sensory signals, like images or sounds, or both. They have been observed to respond to people's interaction needs, which will be sheltered because of upcoming and existing pandemics (Kapralos et al., 2022).

Multimodal systems have a combination of haptics and sensors to generate cognitive synergies to improve the ideal experience. For example, combining visual deformation and haptic force feedback is applied to improve examination and identification of soft tissues (Li et al., 2015). In the same way, the mouse pointer's vibration can increase the vibro-tactile view of coarse surfaces (Ujitoko et al., 2019). Visual haptics include inducing sensations with only visual effects (Lecuyer, 2017). For example, mimicking visuals of an avatar adds haptic sensation in the task of weight lifting (Jauregui et al., 2014).

3.2.3 Tactile

Haptic devices can simulate different tactile perceptions that can be used for immersive experiences in VR gaming. For example, stimulating liquids can be delivered to the user's skin to form various sensations (Lu et al., 2021). The technology has established itself in medical robotics for simulating human tissue textures during remote surgical procedures. The medical instruments that consist of electrotactile or vibrotactile arrays transmit sensory feedback to surgeons, which helps them perceive the operation tissues and enhances their precision and self-assurance. To train medical professionals in minimally invasive surgical techniques, modern tactile feedback technology uses mini-actuator arrays to imitate tissue pressure, together with softness, and warmth sensations.

3.2.4 Touchscreen

Interacting with a touchscreen reveals different levels of maturity with different patterns of dragging and tapping icons or objects. It can be a certain haptic experience that can easily be measured for scientific reasons (Russo-Johnson et al., 2017). Smartphone vibration or tactile stimulation can convert interaction with special education to reduce embarrassing interactions (Dubuque et al., 2021).

The integration of haptic touchscreens serves medical training simulators by enabling users to practice procedures that include ultrasound navigation and drug administration through sensory feedback systems. The systems deliver immediate haptic feedback to users for both their incorrect moves and proper movements, which improves their ability to retain learned information. Touchscreen haptic systems are being transformed for visually impaired users to enable spatialized digital feedback, which drives better medical device inclusion for visually impaired users.

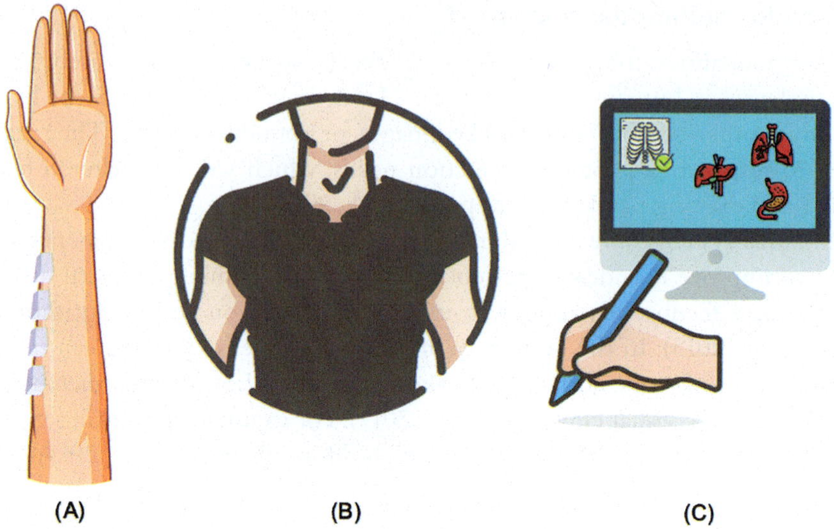

Figure 3.3 Medical advances based on actuators and sensors.

3.2.5 Motion and haptics

There is some influence of gestures similar to subjective or virtual haptic representation to manipulate objects (Pouw et al., 2020). The relation between gesture, haptics, and vision has been observed in the training to reinterpret ambiguous images (Kamermans et al., 2019a, 2019b). In robotic-assisted operations, where the surgeon's movements are converted into mechanical actions and haptic feedback provides information regarding tissue engagement or resistance, this interaction is especially helpful. The accuracy of gesture-based haptic systems has increased due to recent advancements in motion-capture technology, enabling more precise control and enhanced spatial orientation. To help patients regain muscular control and coordination following an injury or stroke, motion-enabled haptic devices are also being utilized in rehabilitation therapy. In this setting, patients can execute directed movements and receive force feedback.

3.3 Key application areas

This section explores some of the major areas where haptic technologies are widely used. A lot of these areas are associated with using virtual environments for training, learning, or rehabilitation. Here are some of those areas:

3.3.1 Medical

Fig. 3.3 illustrates several areas where haptic devices are being deployed for medical training in various fields. It proposes a new kind of method known as haptic medicine,

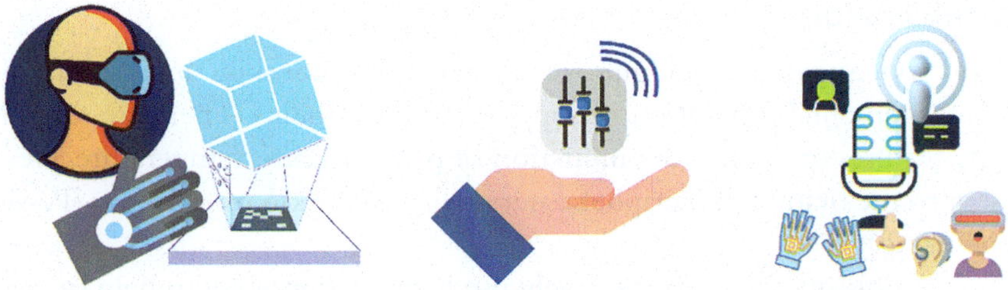

(A) (B) (C)

Figure 3.4 User interaction with haptic responses in the virtual environment.

where touch can be involved in curative care (Mason & Mason, 2009). Wearable solutions that have been developed improve user perception for individuals with some disabilities, like wearable devices having small units embedded with actuators inducing light, sound, and haptic or any stimuli to the users (Schiatti et al., 2020).

Haptic feedback technology in surgical simulations enables medical trainees to feel tissue structure resistance, enabling them to better execute actual medical procedures. Robotic procedures and the da Vinci robotic surgery system now integrate haptic feedback mechanisms to supply surgeons with essential tactile feedback that helps them recognize differences between soft tissues and dense obstructions. Doctors performing remote diagnostics can deliver better assessments of patients' physical health thanks to haptic feedback systems enabling remote physical examinations. Tactile sensor development enables robotic fingers to perform complete tactile examinations, which help medical professionals detect tumors early through robotic palpation.

3.3.1 Gamification

Fig. 3.4 illustrates a new interaction in the game from VR labs like Meta Touch (Meta, 2021). It has been ideal to improve sensory responses and interact with the user to open new frontiers for modern edtech solutions. Another example of haptic technology advancement is Teslasuit, offering different solutions like gloves and suits (Teslasuit, n.d.). These days, new ideas and proposals are given in the haptic world like Hapticast (Hapticast, 2023).

Haptic technology-enabled gamification elevates therapeutic and educational games above mere amusement. In phobia therapy, haptic-enhanced games are now used as a psychological therapeutic tool that offers controlled assessments of tactile stimuli. Students and underprivileged learners benefit from the tactile aspects in educational platforms, which improve their motor skills and memory performance. To improve realism and immersion across several platforms, a growing number of haptic businesses are concentrating on meeting cross-platform compatibility requirements to synchronize touch feedback during multiplayer and collaborative experiences.

3.3.2 Rehabilitation

With haptic devices, patients can improve sensory and motor skills in rehabilitation, either through environmental perception or support for movements. There are several devices with handy features to control the induction of pain, pleasure, temperature, pressure, etc. that are sent to the brain with peripheral nerves like SPA-skin or e-dermis (Pessina, 2019; Tech Briefs, 2020).

Healthcare institutions now extensively use haptic technology to help stroke patients in their recovery through repeated activities that generate haptic feedback to rebuild motor control pathways in their brains. Exoskeleton devices containing haptic actuators help restore joint movement while giving real-time haptic signals that guide proper movements. Haptic interfaces in cognitive rehabilitation systems present tactile puzzles as well as give response-based feedback to optimize mental coordination. Researchers are looking at collaborative studies that combine brain stimulation and haptic feedback since this method promises to accelerate healing by providing motor neurons with focused stimulation.

3.3.3 Personal assistant

Haptic technologies are also very important to develop personal assistants for people with special needs when performing daily tasks. There are some nonintrusive devices in the market to improve the lives of people with special needs, for the visually impaired. It vibrates to alert the user about obstacles or a sensory alert by feeding data through any other pathway (Wewalk, 2025). Haptic personal assistants now utilize AI to tailor their inputs through user behavior patterns, which leads to more authentic interactions between them and users. The directional vibrations emitted by haptic wristbands function as navigation indicators that work most effectively when users spend time outside. The conversion of speech into real-time vibrations helps hearing-impaired people maintain better communication with others. The recently developed haptic shoes use foot sole vibrations to guide users for spatial orientation through the shoes. Smart home systems adopt these technologies, which function to deliver dual-purpose manual feedback for spoken command processing and to distribute notifications.

The creation of haptic-enabled pen devices, together with both games and tablets, offers virtual paper writing options to users. Immediate resistance features, together with vibrating signals, deliver directional hand motion to students for building muscle memory and proper stroke pattern development. The haptic devices feature adaptive features that customize their feedback according to student skill level capabilities and deliver precise support to prevent both under-challenging situations and overwhelming experiences in the learning process.

Research reveals that early classrooms that adopt touch-based feedback systems will assist students in developing refined muscle control through building their interest levels and confidence, particularly after they overcome motor delays. Children with dysgraphia, along with other learning difficulties, benefit from haptic technologies by using multisensory inputs that improve both the ease and enjoyment of writing education.

Therapeutic practitioners and classroom educators use these technological resources in their clinical and pedagogical work to include them within individualized education plans, which help children with special educational requirements. These advanced tools have strong potential to transform handwriting teaching practices along with rehabilitation methods in digital learning contexts. These innovative resources prove capable of developing new approaches toward handwriting rehabilitation and instruction as they support current education practices in online classrooms.

3.4 Haptic technologies for special needs

For educational tools and resources, diverse needs are a rising concern because of the social consciousness of students of all ages. Technological solutions are widely needed for addressing those concerns. For instance, interaction between robots and humans opens up a new frontier for training sessions where gamification can have therapeutic benefits on kids with developmental syndromes (Fitter & Kuchenbecker, 2018). Usually, traditional approaches are out of question in a lot of events. For example, real interactions were prohibited during the pandemic. This way, haptic interfaces have been valuable in those cases and also in real-life situations where people definitely need haptic communication with special value. Sensory deprivation usually opens up unrelated and new communication channels. Kids with hearing problems may enjoy benefits from psychomotor growth with the use of haptic interfaces in computer-aided systems (Peñeñory et al., 2018). Fusion of haptic movement and data has a key impact on the sensation of embodiment in VR for kids (Dewe et al., 2022).

3.4.1 Visual impairment

Tactile data is important for blind people for interaction. For example, intelligence tests have been developed for blind children as per early haptic interfaces, which were soon proposed to experience VR scenarios by visually impaired individuals (Dekker, 1993; Max & Gonzalez, 1997). Blind people must learn Braille reading. Adaptive Braille screens with adaptable cell sizes can be used with haptic feedback and tactile sensitivities (Saikot & Sanim, 2022), and their learning models can be improved with various range of Braille displays (Kim et al., 2019). Facial expressions and other nonverbal cues can be transmitted into vibrotactile data, and it can be translated to the blind person with artificial vision (Buimer et al., 2019). Pin-array matrices represent maps that can be customized to improve self-localization (Brayda et al., 2018). New haptic solutions are also proposed for blind and deaf people to improve their modes of communication. Automated speech-recognition devices can be used to activate a glove with vibrotactile cells, enabling almost realistic translation of vocals into haptic signs (Giulia et al., 2019).

Current haptic devices, which include Global Positioning System (GPS) functionality, offer vibrotactile feedback as a navigation aid for blind users moving in unfamiliar terrains. Security measures through directional vibrations in these systems enable users to traverse pathways and receive warnings and direction-based information, which enhances mobility and independence levels. AI, alongside haptic belts and wristbands, enables users to understand space better because these devices eliminate unnecessary environmental noises. Industrial accessibility products now enable screen readers and haptic facilities to deliver an integrated digital platform for academic resources as well as graphic elements and artistic works.

3.4.2 Writing

It is the most important motor skill. It is possible to improve and acquire handwriting skills with haptic feedback for children (Palsbo et al., 2011). Because of the increasing use of screens, there have been concerns over developing these kinds of skills because of the negative impact on the overall presentation of mental imagery (Suggate & Martzog, 2020). Hence, the researchers have proposed new gaming applications so that children can be engaged in handwriting (Ozgur et al., 2020). The haptic pen, which operates from a stylus base, gives children interactive pen control feedback, thereby helping them improve their handwriting techniques. The systems detect speed and trajectory patterns until they provide feedback resistance whenever the system detects deviations from established patterns. Users writing through tablet applications now receive digital pen-to-paper simulation capabilities because of the implemented haptic feedback technology. Research shows haptic-enhanced writing tools facilitate dysgraphic children's motor coordination development alongside improvement of their character shaping abilities. The combination of adaptive learning tools which incorporate gamification in writing exercises and haptic difficulty adaptation mechanics, proved successful in keeping student abilities strong and creating student engagement. The haptic writing tools enhance penmanship development by building necessary cognitive-motor abilities which result in enhanced written message delivery.

3.4.3 Motor deficits

Motor impairments adversely affect educational outcomes in different ways. For example, kids facing coordination issues may find it harder to perform simple tasks like using pure haptic sensation to determine the size of rods (Wade et al., 2016). In cases of motor deficits, haptic perception is affected, especially for kids with language disorders. So, robotic-based haptic training is very important for them (Tseng & Hsu, 2023; Tseng et al., 2019). Table tennis has been very useful for motor skill training (Tseng et al., 2023).

Repetitive tasks can be performed easily with computer-aided training to provide handwriting training. Robotic systems are widely proposed for regular improvement of routine skills in kids with motor problems (Shire et al., 2016). The training process can be tailored according to the eccentricity of children. It is possible to apply different haptic supports like partial haptic guidance, full haptic guidance, and disturbance haptic guidance for various tasks (Park et al., 2019). In addition, 3D haptic tracing can be assigned to train them to improve their drawing capabilities and deal with their motor challenges (Snapp-Childs et al., 2016). Children who have "hemiplegic cerebral palsy" can perform rehabilitation work by playing VR games with haptic feedback to engage better and recover (Goyal et al., 2022).

Exoskeleton systems function as haptic support devices to help children who suffer substantial disabilities in their motor function develop movement capability. The haptic robots send physical guidance commands to direct movements as they physically assist users within specific training sessions that require minimal human effort. Haptic interfaces integrated with mirror therapy allow children to exercise their functional hand while getting sensory feedback on their affected side. The popularity of educational games employing haptic feedback generation has risen recently due to their real-time feedback which helps students develop better practice consistency. Research indicates rhythmic haptic cues benefit children with developmental coordination disorder since these cues help them plan their movements for schoolwork and physical activities.

3.4.4 Cognitive impairment

There has been a rise in interest in haptic response for educational purposes to train kids with special needs, with the emergence of digitalization. A lot of studies have proposed using these technologies to treat children with unusual mental growth (Yoshida, 1973), such as early recognition of passive and active touching to build perception of actual objects (Derevensky, 1979). It has been more important to develop a sense of touch when it comes to overcoming the problems of a priori knowledge. In a survey, kids were guided to recognize objects that are unknown and known through haptic feedback (Alexander et al., 2002).

Using VR techniques, which are powered by haptic devices, has been widely used to cure different mental health disorders like cerebral palsy, attention-deficit/hyperactivity disorder (ADHD), and "autism spectrum disorder (ASD)" (Wang & Reid, 2011). They are applicable as tools for diagnostic assistance for some health conditions, like ADHD. For treating the condition, research is still at a nascent stage. For instance, games are used in a study on the response of children with ASD toward rewards on tactile platforms (Valori et al., 2023). When it comes to going commercial, there are plenty of apps in the market targeting children with ASD with little value (Rehman et al., 2021).

Children with cognitive impairments produce better learning outcomes when their multisensory environments merge haptic feedback with auditory, together with visual signals. The combination of touch-based sensory discovery enhances memory recall and understanding performance in children using this type of environment. Educational haptic tasks direct students with Down syndrome through their educational activities to develop their eye-hand coordination and increase task persistence. The combination of haptic actuators along with text elements enables storybook interaction features that improve children's comprehension of stories. Researchers analyze wearable haptic bands that use programmed rhythms via patterns to support adult autistic patients during emotional crises.

3.5 Discussion and conclusion

There are two directions of processing haptic information–from and to the user, while interacting with automated and artificial systems like robots, computers, or other devices, as haptics and related technologies are a part of plenty of interactive designs. In the vast literature, a lot of systems consider only one of these directions. For gaming applications, new haptic suits are developed based on giving an impression of effects or events to impress the user, such as adding stress or feeling an artificial heartbeat. But getting feedback on user status is not considered.

When it comes to motor responses or the physiology of the user, data can be used from inertial sensors to determine the user engagement and prevent anomalous situations like cognitive issues or heart conditions. It may not be feasible to implement a two-way flow of data in a lot of haptic system designs. For instance, there is no real way to get haptic feedback from such users in case of visual haptics. Building a complex mental representation is challenging for supporting blind and deaf people, which can be explained in language from the complete haptic experience. Without a perception of objects one over the other, there is no true meaning of below or above. This way, haptic sensing can provide stimuli related to obscure terms.

Conveying meanings to General Artificial Intelligence (GAI) programs by integrating haptic and other sensations is another challenge in the near future. A lot of cognitive categories related to pleasure and pain are grounded in haptic sensations that GAI programs should advance to approximate intelligence. In gaming systems, improved immersion can be used, and it has garnered a lot of attention due to the high economic value of small improvements and advances. Rapid evolution of gaming communication is supposed to reduce the cost of separate units and provide a significant rise in features. Deploying gamification in training programs is another challenge for individuals who are at risk of inclusion. For example, Hight Quality (HQ) gaming suits could generate virtual training scenarios for blind-deaf individuals by preparing artificial skin.

This study has covered major areas of applications of haptic interfaces, while focusing on education and training. Haptics have already shown great value in some training areas, especially surgical and medical, where they can improve virtual experience and improve the outcome of training. This study suggests the usefulness of haptic systems in educating individuals who are at risk of exclusion, like blind-deaf people. There is still a lot of scope for innovation and development. There is also a lack of research on using these technologies for kids with developmental syndromes.

AI disclosure

During the preparation of this work, the author(s) used Claude AI and Grammarly to structure and develop our ideas into a theoretical framework for this manuscript, and Grammarly to correct grammatical errors and improve language clarity. After using this tool/service, the author(s) reviewed and edited the content as needed and take(s) full responsibility for the content of the publication.

References

Alexander, J. M., Johnson, K. E., & Schreiber, J. B. (2002). Knowledge is not everything: Analysis of children's performance on a haptic comparison task. *Journal of Experimental Child Psychology, 82*(4), 341–366. https://doi.org/10.1016/S0022-0965(02)00100-5.

Almusawi, H. A., Durugbo, C. M., & Bugawa, A. M. (2021). Wearable technology in education: A systematic review. *IEEE Transactions on Learning Technologies, 14*(4), 540–554. https://doi.org/10.1109/TLT.2021.3107459.

Bandiaky, O. N., Lopez, S., Hamon, L., Clouet, R., Soueidan, A., & Guehennec, L. L. (2024). Impact of haptic simulators in preclinical dental education: A systematic review. *Journal of Dental Education, 88*(3), 366–379. https://doi.org/10.1002/jdd.13426.

Bara, F., & Gentaz, E. (2011). Haptics in teaching handwriting: The role of perceptual and visuo-motor skills. *Human Movement Science, 30*(4), 745–759. https://doi.org/10.1016/j.humov.2010.05.015.

Brayda, L., Leo, F., Baccelliere, C., Ferrari, E., & Vigini, C. (2018). Updated tactile feedback with a pin array matrix helps blind people to reduce self-location errors. *Micromachines, 9*(7), 351. https://doi.org/10.3390/mi9070351.

Buimer, H., Schellens, R., Kostelijk, T., Nemri, A., Zhao, Y., Van der Geest, T., & Van Wezel, R. (2019). Opportunities and pitfalls in applying emotion recognition software for persons with a visual impairment: Simulated real life conversations. *JMIR mHealth and uHealth,* 7(11), e13722. https://doi.org/10.2196/13722.

Cantoni, V., Cellario, M., & Porta, M. (2004). Perspectives and challenges in e-learning: Towards natural interaction paradigms. *Journal of Visual Languages & Computing., 15*(5), 333–345. https://doi.org/10.1016/j.jvlc.2003.10.002.

Chahal, B., Aydin, A., & Ahmed, K. (2024). Virtual reality vs. physical models in surgical skills training. An update of the evidence. *Current Opinion in Urology, 34*(1), 32–36. https://doi.org/10.1097/mou.0000000000001145.

Dekker, R. (1993). Visually impaired children and haptic intelligence test scores: Intelligence test for visually impaired children (ITVIC). *Developmental Medicine & Child Neurology, 35*(6), 478–489. https://doi.org/10.1111/j.1469-8749.1993.tb11678.x.

Derevensky, J. L. (1979). Relative contributions of active and passive touch to a child's knowledge of physical objects. *Perceptual and Motor Skills, 48*(3_suppl), 1331–1346. https://doi.org/10.2466/pms.1979.48.3c.1331.

Dewe, H., Gottwald, J. M., Bird, L. A., Brenton, H., Gillies, M., & Cowie, D. (2022). My virtual self: The role of movement in children's sense of embodiment. *IEEE Transactions on Visualization and Computer Graphics, 28*(12), 4061–4072. https://doi.org/10.1109/TVCG.2021.3073906.

Donmez, M. (2023). A systematic literature review for the use of eye-tracking in special education. *Education and Information Technologies, 28*(6), 6515–6540. https://doi.org/10.1007/s10639-022-11456-z.

Dubuque, E. M., Collins, L., & Dubuque, M. L. (2021). Improving performance covertly and remotely with tactile stimulation. *Behavior Analysis in Practice, 14*(1), 203–207. https://doi.org/10.1007/s40617-020-00493-0.

Pessina, L.-A.,2019. Artificial skin could help rehabilitation and enhance virtual reality. EPFL.

Fitter, N. T., & Kuchenbecker, K. J. (2018). Teaching a robot bimanual hand-clapping games via wrist-worn IMUs. *Frontiers in Robotics and AI, 5*, 2296–9144. https://doi.org/10.3389/frobt.2018.00085.

Frikha, M., Chaâri, N., Elghoul, Y., Mohamed-Ali, H. H., & Zinkovsky, A. V. (2019). Effects of combined versus singular verbal or haptic feedback on acquisition, retention, difficulty, and competence perceptions in motor learning. *Perceptual and Motor Skills, 126*(4), 713–732. https://doi.org/10.1177/0031512519842759.

Gallud, J. A., Carreño, M., Tesoriero, R., Sandoval, A., Lozano, M. D., Durán, I., Penichet, V. M. R., & Cosio, R. (2023). Technology-enhanced and game based learning for children with special needs: a systematic mapping study. *Universal Access in the Information Society, 22*(1), 227–240. https://doi.org/10.1007/s10209-021-00824-0.

Giulia, C., Chiara, D. V., & Esmailbeigi, H. (2019). GLOS: GLOve for speech recognition. In *2019 41st annual international conference of the IEEE engineering in medicine and biology society (EMBC)* (pp. 3319–3322). IEEE. https://doi.org/10.1109/EMBC.2019.8857927.

Górriz, J. M., Álvarez-Illán, I., Álvarez-Marquina, A., Arco, J. E., Atzmueller, M., Ballarini, F., Barakova, E., Bologna, G., Bonomini, P., Castellanos-Dominguez, G., Castillo-Barnes, D., Cho, S. B., Contreras, R., Cuadra, J. M., Domínguez, E., Domínguez-Mateos, F., Duro, R. J., Elizondo, D., Fernández-Caballero, A., Fernandez-Jover, E., ... Ferrández-Vicente, J. M. (2023). Computational approaches to explainable artificial intelligence: Advances in theory, applications and trends. *Information Fusion, 100*, 101945. https://doi.org/10.1016/j.inffus.2023.101945.

Górriz, J. M., Ramírez, J., Ortíz, A., Martínez-Murcia, F. J., Segovia, F., Suckling, J., Leming, M., Zhang, Y. D., Álvarez-Sánchez, J. R., Bologna, G., Bonomini, P., Casado, F. E., Charte, D., Charte, F., Contreras, R., Cuesta-Infante, A., Duro, R. J., Fernández-Caballero, A., Fernández-Jover, E., Gómez-Vilda, P., ... Ferrández, J. M. (2020). Artificial intelligence within the interplay between natural and artificial computation: Advances in data science, trends and applications. *Neurocomputing, 410*, 237–270. https://doi.org/10.1016/j.neucom.2020.05.078.

Goyal, C., Vardhan, V., & Naqvi, W. (2022). Virtual reality-based intervention for enhancing upper extremity function in children with hemiplegic cerebral palsy: A literature review. *Cureus, 14*(1). https://doi.org/10.7759/cureus.21693.

Jauregui, D. A. G., Argelaguet, F., Olivier, A. H., Marchal, M., Multon, F., & Lecuyer, A. (2014). Toward 'Pseudo-haptic avatars': Modifying the visual animation of self-avatar can simulate the perception of weight lifting. *IEEE Transactions on Visualization and Computer Graphics, 20*(4), 654–661. https://doi.org/10.1109/TVCG.2014.45.

Kamermans, K. L., Pouw, W., Fassi, L., Aslanidou, A., Paas, F., & Hostetter, A. B. (2019a). The role of gesture as simulated action in reinterpretation of mental imagery. *Acta Psychologica, 197*, 131–142. https://doi.org/10.1016/j.actpsy.2019.05.004.

Kamermans, K. L., Pouw, W., Mast, F. W., & Paas, F. (2019b). Reinterpretation in visual imagery is possible without visual cues: A validation of previous research. *Psychological Research, 83*(6), 1237–1250. https://doi.org/10.1007/s00426-017-0956-5.

Kapralos, B., Quevedo, A., Silva, C. D, Peisachovich, E., Collins, K. C, Kanev, K., & Dubrowski, A. (2022). Revisiting pseudo-haptics for psychomotor skills development in online teaching. *Cureus, 14*(3). https://doi.org/10.7759/cureus.23664.

Kim, S., Ryu, Y., Cho, J., & Ryu, E.-S. (2019). Towards tangible vision for the visually impaired through 2D multiarray braille display. *Sensors, 19*(23), 5319. https://doi.org/10.3390/s19235319.

Lecuyer, A. (2017). Playing with Senses in VR: Alternate perceptions combining vision and touch. *IEEE Computer Graphics and Applications, 37*(1), 20–26. https://doi.org/10.1109/MCG.2017.14.

Lewis-Dagnell, S., Parsons, S., & Kovshoff, H. (2023). Creative methods developed to facilitate the voices of children and young people with complex needs about their education: A systematic review and conceptual analysis of voice. *Educational Research Review, 39*, 100529. https://doi.org/10.1016/j.edurev.2023.100529.

Li, M., Konstantinova, J., Secco, E. L., Jiang, A., Liu, H., Nanayakkara, T., Seneviratne, L. D., Dasgupta, P., Althoefer, K., & Wurdemann, H. A. (2015). Using visual cues to enhance haptic feedback for palpation on virtual model of soft tissue. *Medical & Biological Engineering & Computing, 53*(11), 1177–1186. https://doi.org/10.1007/s11517-015-1309-4.

Lu, J., Liu, Z., Brooks, J., & Lopes, P. (2021). Chemical haptics: Rendering haptic sensations via topical stimulants. In *UIST 2021 - Proceedings of the 34th annual ACM symposium on user interface software and technology* (pp. 239–257). Association for Computing Machinery, Inc. https://doi.org/10.1145/3472749.3474747.

Lv, P., Wang, X., Xu, J., & Wang, J. (2021). Intelligent personalised exercise recommendation: A weighted knowledge graph-based approach. *Computer Applications in Engineering Education, 29*(5), 1403–1419. https://doi.org/10.1002/cae.22395.

Marwati, A., Dewi, O. C., Wiguna, T., & Aisyah, A. (2023). Visual-sensory-based quiet room: A study of visual comfort, lighting, and safe space in reducing maladaptive behaviour and emotion for autistic users. *Journal of Accessibility and Design for All, 13*(1), 69–93. https://doi.org/10.17411/jacces.v13i1.318.

Mason, C., & Mason, E. (2009). Haptic medicine. Strategy for the future of health. *Studies in health technology and informatics*, 368–385. https://doi.org/10.3233/978-1-60750-050-6-368. IOS Press.

Max, M. L., & Gonzalez, J. R. (1997). Blind persons navigate in virtual reality (VR); Hearing and feeling communicates 'Reality'. *Studies in Health Technology and Informatics, 39*, 54–59. https://doi.org/10.3233/978-1-60750-883-0-54.

Meta. (2021). *Inside reality labs research: Bringing touch to the virtual world*. Meta.

Ozgur, A. G., Özgür, A., Asselborn, T., Johal, W., Yadollahi, E., Bruno, B., Skweres, M., & Dillenbourg, P. (2020). Iterative design and evaluation of a tangible robot-assisted handwriting activity for special education. *Frontiers in Robotics and AI, 7*. https://doi.org/10.3389/frobt.2020.00029.

Palsbo, S. E., Marr, D., Streng, T., Bay, B. K., & Norblad, A. W. (2011). Towards a modified consumer haptic device for robotic-assisted fine-motor repetitive motion training. *Disability and Rehabilitation: Assistive Technology, 6*(6), 546–551. https://doi.org/10.3109/17483107.2010.532287.

Papanastasiou, G., Drigas, A., Skianis, C., Lytras, M., & Papanastasiou, E. (2019). Virtual and augmented reality effects on K-12, higher and tertiary education students' twenty-first century skills. *Virtual Reality, 23*(4), 425–436. https://doi.org/10.1007/s10055-018-0363-2.

Park, W., Korres, G., Moonesinghe, T., & Eid, M. (2019). Investigating haptic guidance methods for teaching children handwriting skills. *IEEE Transactions on Haptics, 12*(4), 461–469. https://doi.org/10.1109/TOH.2019.2922284.

Peñeñory, V. M., Manresa-Yee, C., Riquelme, I., Collazos, C. A., & Fardoun, H. M. (2018). Scoping review of systems to train psychomotor skills in hearing impaired children. *Sensors (Switzerland), 18*(8). https://doi.org/10.3390/s18082546.

Haptic Intel. (2023). Hapticast. *Apple Podcasts*. https://podcasts.apple.com/us/podcast/haptic/id1679488220.

Pouw, W., Wassenburg, S. I., Hostetter, A. B., de Koning, B. B., & Paas, F. (2020). Does gesture strengthen sensorimotor knowledge of objects? The case of the size-weight illusion. *Psychological Research, 84*(4), 966–980. https://doi.org/10.1007/s00426-018-1128-y.

Rehman, I. U, Sobnath, D., Nasralla, M. M., Winnett, M., Anwar, A., Asif, W., & Sherazi, H. H. R. (2021). Features of mobile apps for people with autism in a post COVID-19 scenario: Current status and recommendations for apps using AI. *Diagnostics, 11*(10), 1923. https://doi.org/10.3390/diagnostics11101923.

Russo-Johnson, C., Troseth, G., Duncan, C., & Mesghina, A. (2017). All tapped out: Touchscreen interactivity and young children's word learning. *Frontiers in Psychology, 8*, 1664–1678. https://doi.org/10.3389/fpsyg.2017.00578.

Saikot, M. M. H., & Sanim, K. R. I. (2022). Refreshable braille display with adjustable cell size for learners with different tactile sensitivity. *IEEE Transactions on Haptics, 15*(3), 582–591. https://doi.org/10.1109/TOH.2022.3184265.

Sanfilippo, F., Blazauskas, T., Salvietti, G., Ramos, I., Vert, S., Radianti, J., Majchrzak, T., & Oliveira, D. (2022). A Perspective review on integrating VR/AR with haptics into STEM education for multi-sensory learning. *Robotics, 11*(2), 41. https://doi.org/10.3390/robotics11020041.

Scheibel, G., Zimmerman, K. N., & Wills, H. P. (2023). Increasing on-task behavior using technology-based self-monitoring: A meta-analysis of I-Connect. *Journal of Special Education Technology, 38*(2), 146–160. https://doi.org/10.1177/01626434221085554.

Schiatti, L., Cappagli, G., Martolini, C., Maviglia, A., Signorini, S., Gori, M., & Crepaldi, M. (2020). A novel wearable and wireless device to investigate perception in interactive scenarios. In *Proceedings of the annual international conference of the IEEE engineering in medicine and biology society, EMBS* (pp. 3252–3255). Institute of Electrical and Electronics Engineers Inc. https://doi.org/10.1109/EMBC44109.2020.9176167.

Shire, K. A., Hill, L. J. B., Snapp-Childs, W., Bingham, G. P., Kountouriotis, G. K., Barber, S., & Mon-Williams, M. (2016). Robot guided 'pen skill' training in children with motor difficulties. *PLoS ONE, 11*(3). https://doi.org/10.1371/journal.pone.0151354.

Snapp-Childs, W., Shire, K., Hill, L., Mon-Williams, M., & Bingham, G. P. (2016). Training compliance control yields improved drawing in 5-11 year old children with motor difficulties. *Human Movement Science, 48*, 171–183. https://doi.org/10.1016/j.humov.2016.05.006.

Sorgini, F., Caliò, R., Carrozza, M. C., & Oddo, C. M. (2018). Haptic-assistive technologies for audition and vision sensory disabilities. *Disability and Rehabilitation: Assistive Technology, 13*(4), 394–421. https://doi.org/10.1080/17483107.2017.1385100.

Suggate, S. P., & Martzog, P. (2020). Screen-time influences children's mental imagery performance. *Developmental Science, 23*(6). https://doi.org/10.1111/desc.12978.

Tech, Briefs. (2020). Electronic "skin" brings sense of touch and pain to prosthetic hands. *Tech Briefs Magazine*: 44. SAE Media Group https://www.techbriefs.com/component/content/article/37624-electronic-skin-brings-sense-of-touch-and-pain-to-prosthetic-hands.

Teslasuit. (2020). TESLASUIT Dev Kit. *Teslasuit.*

Trudel, S. M., Winter, E. L., Fitzmaurice, B., Norman, G., & Bray, C. R. (2023). Integration of physical health and sensory processing assessment for children with autism spectrum disorder in schools. *Psychology in the Schools, 60*(2), 378–400. https://doi.org/10.1002/pits.22704.

Tseng, Y. T., Holst-Wolf, J. M., Tsai, C. L., Chen, F. C., & Konczak, J. (2019). Haptic perception is altered in children with developmental coordination disorder. *Neuropsychologia, 127*, 29–34. https://doi.org/10.1016/j.neuropsychologia.2019.02.004.

Tseng, Y. T., & Hsu, H. j. (2023). Not only motor skill performance but also haptic function is impaired in children with developmental language disorder. *Research in Developmental Disabilities, 134*. https://doi.org/10.1016/j.ridd.2022.104412.

Tseng, Y. u-T., Tsai, C.-L., Wu, T. H., Chen, Y. i-W., & Lin, Y. i-H. (2023). Table Tennis, as a method of sensorimotor training, induces haptic and motor gains in children with a probable developmental coordination disorder. *Motor Control, 27*(3), 480–497. https://doi.org/10.1123/mc.2021-0123.

Ujitoko, Y., Ban, Y., & Hirota, K. (2019). Modulating fine roughness perception of vibrotactile textured surface using pseudo-haptic effect. *IEEE Transactions on Visualization and Computer Graphics, 25*(5), 1981–1990. https://doi.org/10.1109/TVCG.2019.2898820.

Valori, I., Carnevali, L., Farroni, T., & Goto, Y. (2023). Agency and reward across development and in autism: A free-choice paradigm. *PLOS ONE, 18*(4), e0284407. https://doi.org/10.1371/journal.pone.0284407.

Wade, M. G., Tsai, C. L., Stoffregen, T., Chang, C. H., & Chen, F. C. (2016). Perception of object length via manual wielding in children with and without developmental coordination disorder. *Journal of Motor Behavior, 48*(1), 13–19. https://doi.org/10.1080/00222895.2015.1025902.

Wang, M., & Reid, D. (2011). Virtual reality in pediatric neurorehabilitation: Attention deficit hyperactivity disorder, autism and cerebral palsy. *Neuroepidemiology, 36*(1), 2–18. https://doi.org/10.1159/000320847.

Wersényi, G. (2022). Perception accuracy of a multi-channel tactile feedback system for assistive technology. *Sensors, 22*(22), 8962. https://doi.org/10.3390/s22228962.

Wewalk, WeWALK Smart cane. (2025),

Yoshida, R. K. (1973). A Guttman scalogram analysis of haptic perception for trainable mentally retarded children. *American Journal of Mental Deficiency, 77*(4), 439–444.

Zhang, Y., & Song, Y. (2022). The effects of sensory cues on immersive experiences for fostering technology-assisted sustainable behavior: A systematic review. *Behavioral Sciences, 12*(10), 361. https://doi.org/10.3390/bs12100361.

Zhu, M., Biswas, S., Dinulescu, S. I., Kastor, N., Hawkes, E. W., & Visell, Y. (2022). Soft, wearable robotics and haptics: Technologies, trends, and emerging applications. *Proceedings of the IEEE, 110*(2), 246–272. https://doi.org/10.1109/JPROC.2021.3140049.

CHAPTER 4

Haptic feedback mechanisms

Nasrulla K. Khan, Jitha Janardhanan and Simna Shanavas
Department of Computer Applications, Presidency College (Autonomous), Bangalore, Karnataka, India

4.1 Introduction

Human haptics, machine haptics, and computer haptics are three different subfields in haptic technology. Interaction forces are applied to the skin of an individual when he or she touches an object, stimulating the sensory systems to provide information to the brain and triggering haptic perception (Olugbade et al., 2023; Lee et al., 2024). The brain then creates signals that cause muscle activation and lead to the movement of the hand or arm. This process is referred to as the human haptic system. Machine haptics is the use of machines to simulate human touch independently or perform the same through telerobotics and haptic interfaces. Computer haptics deals with the generation of forces, torques, and tactile feedback, while computer graphics deals with haptic rendering as well as visual rendering (Giri et al., 2021; Lee et al., 2024). Interaction with the virtual environment is provided through joysticks, robotic arms, and actuation systems (Girbés-Juan et al., 2022; Lee et al., 2024).

4.2 How sense of touch works

The skin is a complex organ that contains a high number of nerve endings and touch receptors known as the somatosensory system. The somatosensory system is in charge of bringing all the sensations we feel to the brain (Martín-Alguacil et al., 2022; Silvera-Tawil et al., 2015). When something is too hot or cold, when it hurts, or when it tickles, the system sends the message to your brain. It does this through the use of tactile receptors. The touch receptors stimulate the closest neuron, which sends the message to the next closest neuron, and this goes on until the message is received by the brain. The brain tells the body how to react to the sensation, and the whole process occurs in a split second. Haptics stimulate a person's somatosensory system to give context and communicate specific information, as shown in Figs. 4.1 and 4.2. Haptic technology may be used to teach a person to instruct an individual to create something or just to entertain the user. They allow lifeless objects to make them feel like holding real objects, which further enables these lifeless objects to communicate in response. Haptic feedback is the process of using tactile sensations and vibratory feedback to communicate feelings or emotions to the user, adding more to the virtual experience. This is done by software reacting to user

Haptic Feedback Control in Medical Robotics
DOI: https://doi.org/10.1016/B978-0-443-27777-1.00009-3

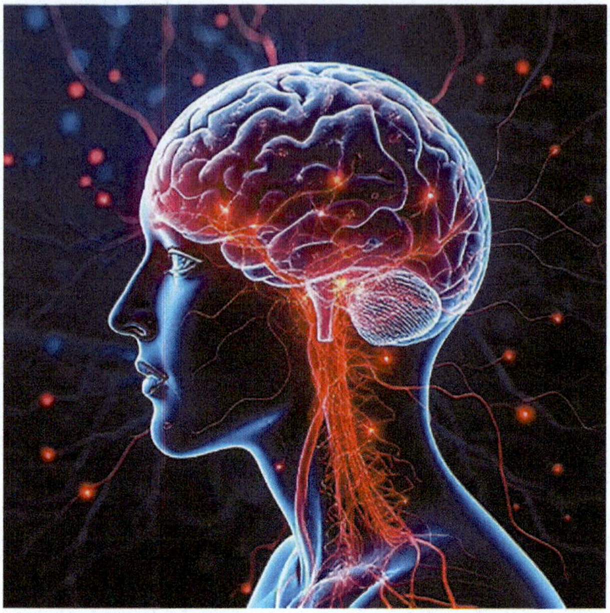

Figure 4.1 Somatosensory system-The brain communicates through neurons.

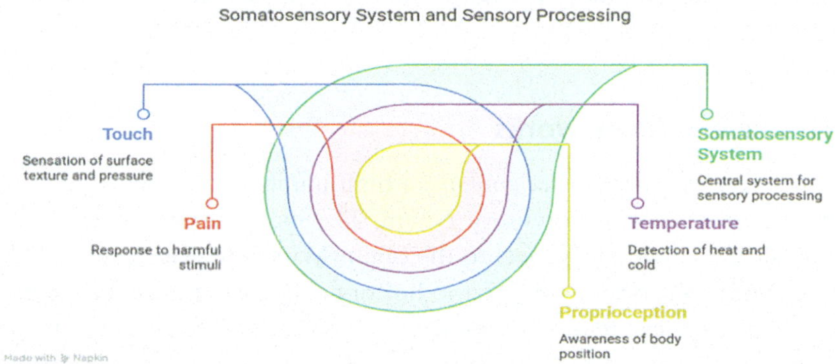

Figure 4.2 Somatosensory system and sensory processing.

input as felt through controller vibration through some actions in a video game, or the artificial button-pressing sensation given by a smartphone as the user moves its screen. Haptic feedback uses physical sensations such as unique vibratory patterns to simulate tactile sensations. Haptic devices that give haptic feedback generally render a variety of sensations according to different visual and audio input, such as a smartphone rendering the sensation of a tap as the user taps on its screen.

Figure 4.3 Perception of information via the sense of touch.

Haptics may be described as the science of transmission and perception of information via the sense of touch. Technology that uses the sense of touch to facilitate information exchange is known as haptic technology, while the information exchanged is known as haptic feedback. The main goal of haptics is to enable tactile communication between living and inanimate objects. Haptic technology is now implemented in devices like smartphones, laptops, vests, chairs, robots, simulators, and other devices, as shown in Figs. 4.3 and 4.4.

4.2.1 Haptic feedback functioning

Haptics or haptic technology is defined as the technology of applying touch sensation while interacting with a physical or virtual environment (Giri et al., 2021; Lee et al., 2024). Haptic devices utilize components such as motors, sensors, and speakers to generate haptic feedback. Devices are designed to provide haptic feedback when a specific action is executed, as shown in Fig. 4.5.

The mechanical sensation experienced by the user can be produced by various technologies such as skin indenting devices, exoskeleton systems, or vibrotactile methods. Devices that create skin indentations are present in numerous haptic technologies, such as haptic gloves and various wearables, as shown in Figs. 4.6 and 4.7. These systems constrict the skin to mimic the feeling of touching or manipulating an object (Biswas & Visell, 2021; Emami et al., 2024).

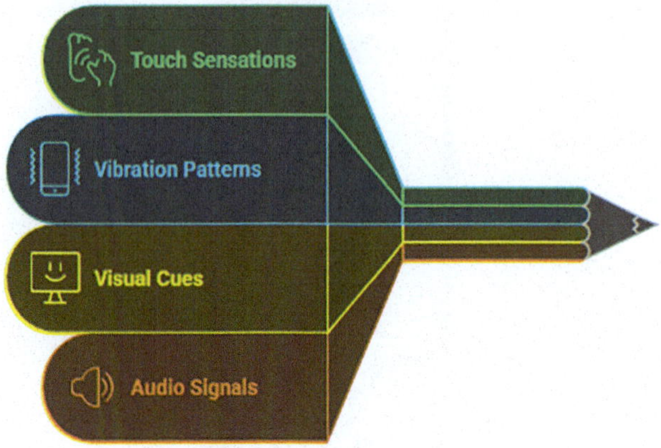

Figure 4.4 Convergence of sensory feedback.

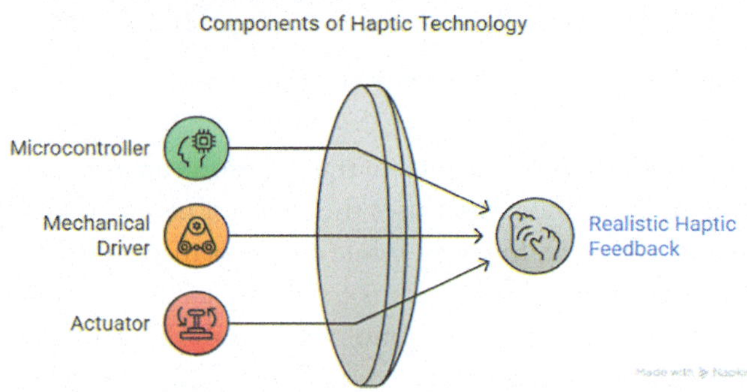

Figure 4.5 Haptic feedback functioning.

4.3 Technical specifications

Haptic technology creates tactile feedback by applying force, vibration, and motion. These effects stimulate the user's sense of touch when interacting with objects that have haptic technology. There are several methods of generating the mechanical forces, vibrations, and motions. The most widely used method is by applying an eccentric rotating mass actuator. As the eccentric rotating mass vibration motor's(ERM) spins at high speeds, the force generated by its mass is unbalanced, causing the motor to move. This movement of the motor causes the creation of haptic feedback. Linear resonant actuators (LRAs) consisting of a spring-mounted magnet surrounded by a coil and encapsulated with an external layer have been used as an alternative to eccentric rotating mass vibration motor's (ERM) . When the electromagnetic coil is energized, it causes oscillation of the

Figure 4.6 Virbotactile.

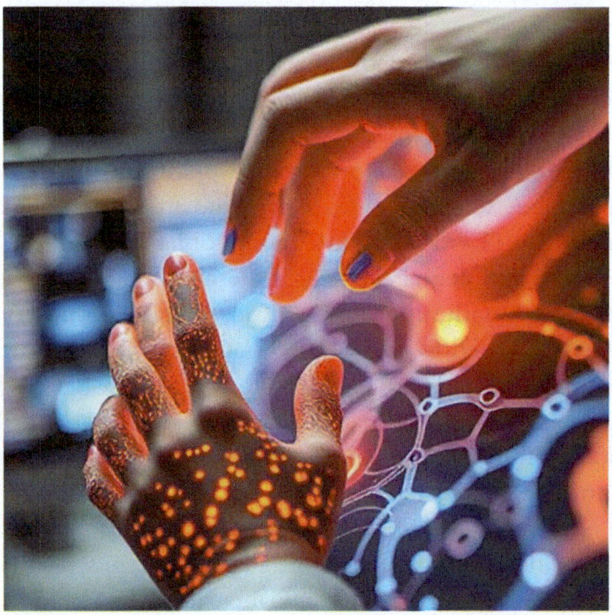

Figure 4.7 Haptic glove.

magnetic mass within it, creating the feeling of feedback. The Nintendo Switch, for instance, utilizes an LRA to generate a better player's immersive experience.

Haptic interfaces have applications across several industries, from the entertainment industry to specialized medical equipment, wearable gloves, and surgical procedures (Giri et al., 2021; Lee et al., 2024). Other applications that are augmented by haptic technology are telemanipulators, exoskeletal systems, advanced prosthetic limbs, physical rehabilitation devices, intelligent assistance devices, and near-field robotic systems (Giri et al., 2021; Lee et al., 2024). Additionally, new applications of haptic devices are computerized forensics, e.g., 3D facial reconstruction, radiological cross-sections, and analytical methods.

- *Control systems microcontrollers*: Haptic feedback systems include microcontrollers that translate user input, like touch and gestures, to find the appropriate feedback to provide. This can involve the application of algorithms that are programmed to translate user actions into specific haptic responses.
- *Software integration*: Haptic feedback can be incorporated into software programs and gaming systems, allowing developers to create personalized feedback experiences that react to user input.

4.4 Critical components and factors to operate and enhance the user experience

Actuation mechanisms: At the core of the working of haptic feedback are the actuators, including vibration motors, LRAs, electromagnetic actuators, and ultrasonic or paired actuators.

Haptic feedback algorithms: These algorithms decide how actuators respond to various inputs, which may include waveforms, frequency, and amplitude regulation patterns.

User interface integration: Harmonious integration of haptic feedback in the overall user experience is essential. This integration includes design suggestions that enable visual and audio signal improvement without user saturation. Context Similarity also suggests that input must be specifically tied to the ongoing task at hand. In addition, sensor inputs are required to facilitate dynamic response. Haptic interfaces often include sensor inputs to monitor user activity. These can be touch sensors sensing touch through capacitive or resistive technology, force sensors sensing pressure level, and motion sensors sensing orientation and movement, especially in virtual and augmented reality (AR) settings.

User input and personalization: Users need to be able to personalize haptic feedback, as personal preferences can be very varied. Sensitivity adjustments, feedback categories, and compatibility and standardization. The performance functionality and evolution of haptic interfaces rely on the type of feedback, degree of freedom of movement, and end effector manipulation, haptic stimulation, and actuator technologies (Giri et al., 2021; Lee et al., 2024). Differences between haptic systems (shown in Table 4.1)

Table 4.1 Differences between Haptic feedback, vibrations, and tactile.

Difference between haptic feedback and vibrations	Difference between haptic haptics vs. tactile
The primary purpose of haptic feedback is to convey information, distinguishing it from vibration, which typically presents a single waveform and maintains a consistent intensity over its entire duration. Haptics, in contrast, employ various sophisticated waveforms to transmit information to the user.	Tactile refers to anything associated with the sense of touch, making it impossible to experience haptic feedback without first having some kind of tactile feedback. Tactile feedback is a mild type of haptics. However, it offers minimal force because it is restricted to the skins surface.

4.5 Challenges in haptic feedback

The development of haptic technology has been plagued by a number of challenges, such as the complexity of simulating the subtle and sophisticated human sense of touch, overcoming technological limitations, and filling the gaps in our understanding of the human sensory system. Advances in materials science, sensor technology, computational algorithms, and design methods continue to drive this development (Biswas & Visell, 2021; Emami et al., 2024). One major disadvantage of haptic interface technology development is its limited workspace and poor dexterity. It is critical to take variables like responsiveness of the device, tactile feedback, and interaction dynamics into consideration during the design process to achieve maximum functionality. Common hurdles in the design, personalization, and manufacturing of haptic devices are further analyzed in different applications. The current paper will discuss some problems faced in medical and virtual reality (VR) applications regarding haptic feedback (Giri et al., 2021; Lee et al., 2024).

In medical use, patients' safety, reproduction of realistic haptic feedback, affordability, probe control, and feedback training are some of the challenges that must be overcome before haptic devices can be used in medical procedures. Virtual simulators have an important role in medical training, and constant research has demonstrated evidence of many potential challenges. Challenges in medical applications include system safety in terms of stability, quality of force feedback, and transparency, regulatory approvals, economic considerations, and challenges related to the environment in which the haptic device will be used, such as MRI compatibility. Haptic devices used in surgical environments provided insufficient realism, leaving future physicians inadequately prepared for complex procedures. The haptic devices did not have adequate force feedback, and the benefits of these devices are poorly documented. Haptic devices are far from being used in medical fields. One of the major disadvantages of haptic devices in VR systems is the inability to deliver high-fidelity haptic feedback. Human–computer interaction (HCI) plays an important role in assessing the performance of haptic devices in VR systems. The three

Table 4.2 Types of haptic feedback.

Vibration feedback	Force feedback	Tactile feedback
The typical type of haptic feedback in which small motors or actuators generate vibrations to replicate touch experiences. This is frequently utilized in mobile gadgets for alerts and notifications.	This form of feedback offers resistance or force to the user, mimicking the sensation of actual interactions. It is frequently found in gaming controllers, flight simulators, and robotic systems.	Provides subtle sensations, frequently utilizing cutting-edge technologies to produce diverse textures or feelings. This may involve simulating various surfaces or the sensation of pressing buttons.

principal factors influencing HCI are the user, the interface device, and the digitally constructed virtual world. Over time, the mainstream issues of haptic devices have been the precise simulation and sensing of interactive objects in the computer-generated world (Giri et al., 2021; Lee et al., 2024). Types of haptic feedback are shown in Table 4.2.

4.6 Types of haptic device technologies

Haptics come in numerous types that are classified based on usage feedback and modality.

4.6.1 Based on usage

Graspable: Physical hardware, such as joysticks as shown in Fig. 4.8, is a well-known example of haptic technology providing kinesthetic feedback. Such devices, haptic feedback movements, and emotions allow individuals to become more immersed in games and make controlling robots more efficient in remote or virtual settings (Giri et al., 2021). The main applications of this technology include bomb defusal and space exploration. Using space exploration, astronauts or ground-based personnel use haptics-controlled robots to maintain equipment, for instance, spacecraft parts or satellites still inside the ship or from the Earth (Giri et al., 2021; Silvera-Tawil et al., 2015).

Touchable: Touchable haptic technology is prevalent in consumer products, for instance, take smartphones that respond to taps, rotations, and other user inputs. Companies could use programmable textures to allow customers to feel fabric materials such as cotton or silk before purchasing them all from the comfort of their homes (Giri et al., 2021; Silvera-Tawil et al., 2015).

Wearable: Wearable haptic technology (shown in Fig. 4.9) replicates the sense of touch by delivering tactile feedback like pressure, vibration, and temperature. One of the fast-evolving uses of wearable haptics is VR gloves that duplicate real sensations and send and receive information from users who control their virtual avatars or remote robots (Giri et al., 2021; Silvera-Tawil et al., 2015).

Figure 4.8 Joystick.

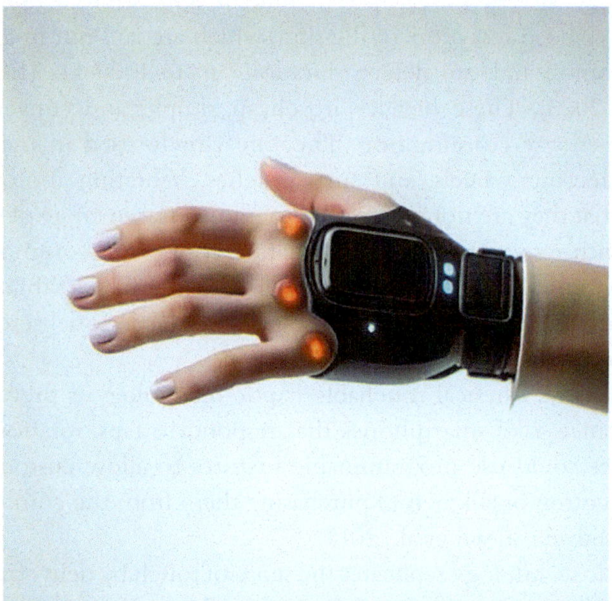

Figure 4.9 Wearable glove.

4.6.2 Based on feedback

Force feedback: This type of haptics began in the late 1960s and hence is one of the oldest and most studied types of this technology. It stimulates the skin muscles and ligaments, unlike other types of haptics, which typically stimulate the outer layers of skin receptors. This type of haptics exists in two forms to simulate human body parts: biomimetic and nonbiomimetic. Biomimetic devices are shaped like human limbs and simulate the motion of their human counterparts. An example includes exoskeletons (Giri et al., 2021; Silvera-Tawil et al., 2015). One of the disadvantages encountered while designing biomimetic devices is the complexity involved in their design. The equipment has to exactly mimic human limb motion and function according to different body sizes without limiting freedom of motion. Contrary to these nonbiomimetic devices, which do not simulate the human body structure do not have this particular disadvantage. Apart from their structure, force feedback devices may also be divided on the basis of the direction of the force applied. This division includes active and resistive devices (Giri et al., 2021; Silvera-Tawil et al., 2015). Active devices limit user motion and use motors to enable interaction. They are capable of simulating a large variety of types of interaction and are usually long-run, though they are often difficult to control. On the other hand, resistive devices limit user motion by applying a braking force.

Vibrotactile feedback: This most popular form of haptic technology, which uses vibrostimulators that press against the surface of human skin. Vibrotactile feedback directly stimulates the specialized receptors in the skin, which are laid out in a pattern similar to the layers of an onion and can detect vibrations up to 1000 Hz (Emami et al., 2024; Olugbade et al., 2023). These devices are cheap, simple, and convenient in terms of user control and energy consumption. They are largely used in mobile phone game controllers, car steering wheels, and smartwatches. Vibrating motors do have some disadvantages in that they are not well-suited to mimic a large range of sensations and can face challenges with respect to efficient miniaturization. One common example of this form of feedback in use is the smartphone, where the user feels a vibration that simulates the action of pressing a physical button when touching the touchscreen (Emami et al., 2024; Olugbade et al., 2023).

Electrotactile feedback: Physical touchable haptic technology is prevalent in consumer products, for instance, take smartphones that respond to taps, rotations, and other user inputs. Companies could use programmable textures to allow customers to feel fabric materials such as cotton or silk before purchasing them from the comfort of their homes (Giri et al., 2021; Silvera-Tawil et al., 2015).

Wearable haptic technology replicates the sense of touch by delivering tactile feedback like pressure, vibration, and temperature. One of the fast-evolving uses of wearable haptics is VR gloves that duplicate real sensations and send and receive information from users who control their virtual avatars or remote robots (Giri et al., 2021; Silvera-Tawil et al., 2015).

Figure 4.10 Ultrasonic tactile feedback.

Ultrasonic tactile feedback: This haptic technology employs the use of ultrasound emitters, which produce high-frequency sound waves to produce subtle feedback sensations. The devices, as shown in Fig. 4.10, employ a principle called acoustic time reversal, which causes the position of the emitter to shift from the target location on the human body (Emami et al., 2024; Olugbade et al., 2023). Haptic feedback areas are beneficial when the need to provide ultrasound signals is required for those parts of the body that have larger surface areas. Such areas include more than one emitter that creates intangible but palpable interfaces of ultrasound waves in the air. These interfaces produce turbulence that human skin can perceive. One of the major benefits of this haptic technology is that it is devoid of user-worn accessories. The setup, however, tends to have more costs than any other form of haptic feedback systems (Emami et al., 2024; Olugbade et al., 2023).

Thermal feedback: Thermal feedback haptics (Fig. 4.11) utilize actuator arrays that come into direct contact with the human body (Emami et al., 2024; Olugbade et al., 2023). The systems use thermoelectric diodes that are based on the Peltier effect. It does not have to use a large number of small units, nor does the stimulus have to be positioned with great clarity to achieve the desired simulation effect. Control of temperature may be problematic, however, because thermal energy in the form of heat or cold cannot just vanish from any surface and must be transferred based on the principles of energy conservation. Furthermore, the transfer must be achieved quickly to achieve

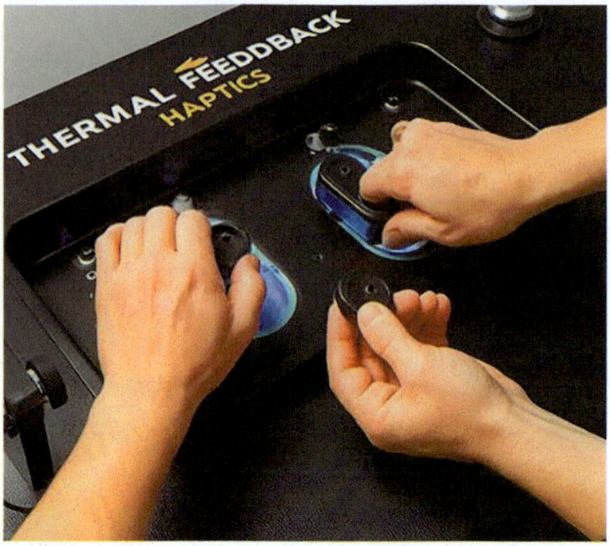

Figure 4.11 Thermal feedback.

a correct simulation. These devices are thus highly energy-consumptive and complex (Emami et al., 2024; Olugbade et al., 2023).

4.6.3 Based on modality

Vibration: Vibration is a common method in most of the haptic technologies. The aforementioned technologies, such as eccentric rotating mass and LRAs, belong to this type. This modality is applied in a wide range of devices such as wearables, mobile phones, controllers, and many others (Giri et al., 2021; Silvera-Tawil et al., 2015). It is worth mentioning that being a vibrating device does not necessarily make it a haptic device. The main difference is the goal and complexity of the vibration patterns. Typical vibrating devices produce a single waveform with fixed and constant intensity throughout the interaction. In contrast, haptic technology transfers information via complex waveforms. In other words, a sensation that transfers general information instead of a specific aim can be considered as just a basic vibration (Emami et al., 2024; Olugbade et al., 2023). For instance, when a smartphone vibrates during a phone call, it is an example of a simple vibration. On the other hand, a vibration of fixed strength within a particular area of the device while playing a game can send specific information, such as the impact of a crash in a racing game (Giri et al., 2021; Silvera-Tawil et al., 2015).

Kinesthetic: Kinesthetic feedback (Fig. 4.12) is another form of feedback that also makes a significant contribution to the area of haptics (Lee et al., 2024). Kinesthetic feedback is also referred to as proprioception and is defined as touch awareness or perception made by muscle tensions with the help of sensory receptors. Unlike tactile feedback, the sensory

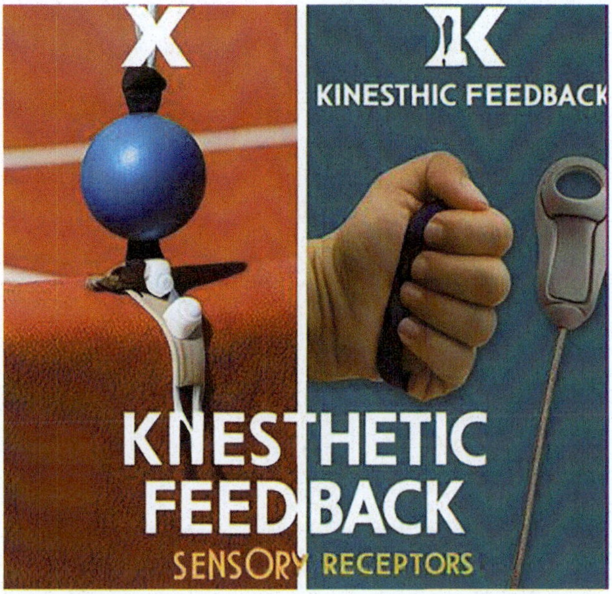

Figure 4.12 Kinesthetic feedback.

receptors are employed in muscles and joints instead of on the body surface of the user (Lee et al., 2024).

Button: Smart screens do not have the familiar tactile feedback of mechanical buttons. Simulated buttons employ sound and haptic feedback to mimic the feel of a mechanical pressure pad under the user's fingertip (Gage & Marzullo, 2022).

4.7 Applications of haptic feedback

Haptic technology has evolved as a crucial part of several popular devices like smartphones, smartwatches, and game controllers. This integration enables people to interact with the virtual world more physically and realistically (Carrasco, 2023). The use of haptic technology has promoted interdisciplinary research across fields like neurobiology, physics, engineering, and computer science. Interdisciplinary studies have provided breakthrough findings and accelerated growth in the field. Haptic devices are used in numerous applications across industries like medical training, rehabilitation, and teleoperated robotic surgeries.

4.7.1 Wearable devices

Breakthroughs in wearable haptic technology directly interfacing with human epidermis have given rise to a new generation of true tactile perceptions (Carrasco, 2023). Among

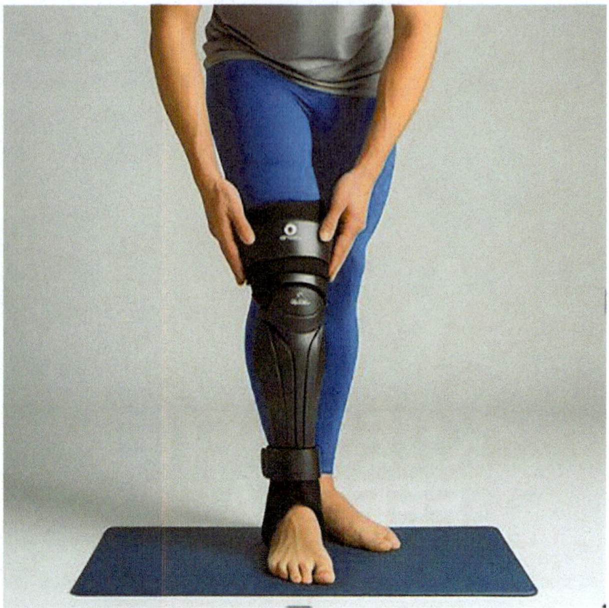

Figure 4.13 Wearable devices.

the significant highlights of such breakthroughs is the implementation of haptic gloves simulating tactile feedback within VR and thereby establishing an unprecedented level of interaction. Recent developments in wearable haptic tech directly interfacing with human dermal surfaces have introduced a new generation of true tactile feedback. One of the major developments of such advancements is the creation of haptic gloves to mimic the feeling of touch in virtual worlds, thus leading to the level of interaction. Additionally, haptic technology has also been incorporated into advanced prosthetic limbs, allowing users to experience the sensory feedback to improve the performance and acceptance of such devices. Developments and applications of electrical stimulation have transformed the haptic world, opening the door for true touch feedback and improving user development and accessibility. Moreover, such technology is used in smartwatches and fitness trackers for notification and alerting, thus providing users with soft feedback. In terms of wearable technologies, the haptic modality is described in terms of (1) tactile feedback to treat the tactile perception from the skin, e.g., vibrations, and (2) kinesthetic feedback to treat the kinesthetic perception related to our muscular effort (Lee et al., 2024; Raisamo et al., 2022). Such a context aids clinicians and therapists in assessing a patient's performance during a therapy session by using data gathered from two major components: skin and muscle (Fig. 4.13).

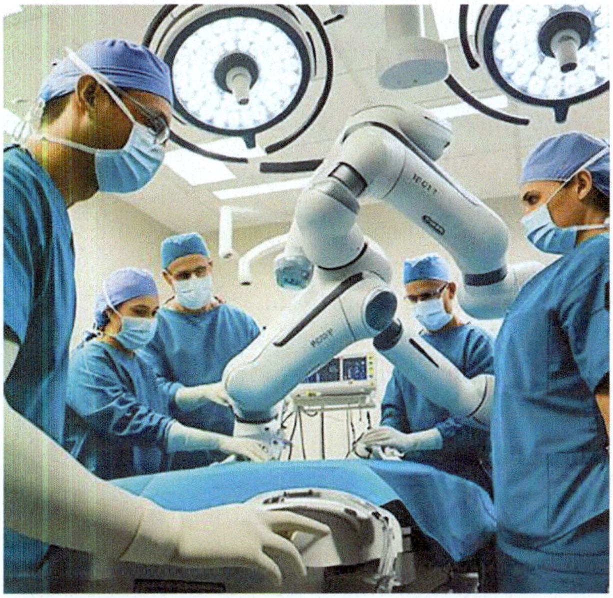

Figure 4.14 Invasive surgery laparoscopic instruments.

4.7.2 Healthcare

Haptics is at the center of much of contemporary healthcare. For example, in minimally invasive surgery, laparoscopic instruments, as shown in Fig. 4.14, are designed specifically to give force and tactile feedback. This cutting-edge technology allows doctors to assess tissues and identify anomalies with high accuracy from a distance. Haptic technology also improves the control surgeons have in robotic-assisted medical procedures (Emami et al., 2024). Surgical robots can conduct operations in inaccessible areas not accessible by human hands with tiny instruments, or even allow surgeons to do procedures remotely. Adding haptic feedback to robotic teleoperations enhances precision and decreases the duration of surgical procedures, thus significantly reducing the risk of tissue damage. Moreover, haptics is also important in the education and training of healthcare professionals. For example, medical students can train on virtual patients, learning hands-on skills in incision making and suturing without putting real patients at risk. Dental simulators also demonstrate the use of this technology, allowing dental students to drill and cut in a virtual environment where haptic feedback simulates real sensations and results (Silvera-Tawil et al., 2015).

4.7.3 Medicine and dentistry

An orthosis is a medical appliance that is designed to assist in the alignment, prevention, or correction of deformities or enhancement of the function of the moving body parts. The devices can accomplish several functions, including support and stabilization, which may aid in injured or compromised extremities by providing joint stabilization and reducing pain during movement.

Deformity correction: Orthoses may be employed to correct or manage bone or joint deformities like scoliosis or foot deformities.

-Improvement of function: Orthoses can significantly improve the functional capability of the disabled or injured to enable greater mobility and independence.

-Postsurgical rehabilitation: Following surgical procedures, orthoses can be employed to immobilize the surgical area and allow healing.

Varieties of orthoses: Braces accommodating joints like the knee, ankle, or wrist, which offer needed support and stability. Splints are most often applied to immobilize portions of the body, such as a finger or a wrist, thereby facilitating healing. Foot orthotics consist of customized insoles or inserts designed to rectify foot alignment while simultaneously enhancing comfort.

Cervical collars: Worn to support the neck and limit motion after injury or surgery. Orthoses may be mass-produced or custom-made based on the patients' particular requirements and the condition being addressed. In most instances, these units are prescribed by healthcare personnel, including orthopedic surgeons, physical therapists, or podiatrists.

4.7.4 Smartphones and tablets

Haptic feedback refers to a form of sensory response that allows individuals to perceive vibrations caused by their mobile phones. In smartphones, haptic feedback has functions beyond notification of the activation of buttons. The haptic technology used in smartphones can change the intensity of vibrations, which is helpful to game developers while generating interactive user experiences. Haptic feedback assists such game developers in incorporating various sensations from light patterns of vibration that indicate a light touch to violent shocks that mimic the sensation of impact. As technology in haptics continues to improve, we can expect even more realistic and interactive haptic experiences in the future. The smartphone produces vibration by using a small motor that is activated when a button is tapped. The motor produces a light vibration pattern that passes through the phone casing and to the user's hand. This specific form of haptic feedback is referred to as tactile feedback. Haptic view allows people to perceive their actions while using buttons, icons, or links on their mobile phones.

- Encourages a more comfortable and natural user experience.
- Offers better control of one's mobile phone.
- Provides an engaging experience by stimulating the tactile sensory modalities.
- Brings mobile gaming to the next level.

Also, haptic feedback is a unique feature. Force feedback is used in video games to increase the simulation quality of the gaming experience. For example, in the control of an arcade game, the controller that one is using can vibrate when one bumps into an object on the screen. The specific vibration pattern offers a true haptic experience that adds to one's overall entertainment from the game.

4.7.5 Video games/gaming controllers

Haptic technology provides a sensory revolution in gaming and entertainment. In the past, gamers used to rely only on their sight and sound to experience the game environment. However, because of the adoption of haptic feedback technology, gamers are now able to interact with the game through somatosensory channels. This multisensory interaction technology improves player immersion but also the game's realism and attractiveness. Haptic feedback technology is utilized through various devices such as haptic gloves, vests, feedback chairs, and specially designed controllers. The devices are used to give haptic feedback to the users through techniques such as vibration, pressure, and motion. For instance, when the player picks up a stone while gaming in a VR game, the haptic glove can replicate the weight and texture of the stone. Additionally, if the character is punched within the game, the haptic undershirt can impart an impact feel in the player's respective area, thus generating a sense of urgency and reality.

Tesla suit is a high-tech full-body haptic feedback suit that provides precise tactile sensations that simulate a variety of experiences from temperature shifts to physical contact. The wearer of the suit can sense various textures in the virtual environment, like the soft touch of a feather and the powerful blast of air (Fig. 4.15).

VR Glove is an advanced haptic device that seeks to offer precise hand tracking coupled with tactile feedback experiences. While gamers are wearing the glove, it enables them to feel the size, shape, and texture of virtual objects, thus improving their sense of agency and enjoyment. The DXRacer gaming chair utilizes haptic feedback technology to make sure that gamers are always in a comfortable position while sitting. The chair can synchronize with the game dynamics to provide vibrational feedback, thereby enabling gamers to distinctly feel explosions in the game or experience the vibrations felt on the racetrack (Fig. 4.16).

4.7.6 Virtual reality and augmented reality

Haptic rendering and visual rendering are important parts of creating a virtual haptic system. Haptic rendering deals with the calculation of the haptic force, while visual rendering allows us to display the interaction with a virtual object visually (Lee et al., 2024). A VR or AR system consists of head-mounted display devices like goggles, accelerometers, and speakers, which rely on these devices to create a computer-generated environment from which a user can interact with the environment, as shown in Fig. 4.17. Besides audio and video sensory systems, touch-based systems are not well developed in VR and AR technologies, which have been shown to enhance experiences in a variety of applications like entertainment and healthcare (Lee et al., 2024). Wireless

Figure 4.15 Tesla suit.

Figure 4.16 DXRacer gaming chair.

Figure 4.17 Virtual reality and augmented reality.

modes of operation are now feasible due to compact, lightweight epidermal VR systems. These systems are lightweight, flexible architectures that can be tied to the skin and programmed to perform haptic operations. The operator's interaction with a virtual environment can be enabled by texture and pressure difference-based rendering (Lee et al., 2024).

4.7.7 Entertainment

The use of haptic technology in cinema chairs and immersive game installations in theme parks and shopping malls is to create sensations, such as explosions, meteorological conditions, and diverse environmental and social experiences. The tactile feedback devices, such as gamepads, joysticks, jet seats, and steering wheels, give tactile feedback to the user via electro-tactile or force feedback systems since video games aim to create the realism of virtual worlds (Silvera-Tawil et al., 2015). Besides, the use of haptic technology extends to home consumers beyond VR headsets and gaming controllers. For instance, haptic vests on sale on the internet transmit low-frequency vibrations to body parts when worn. These vests are easily harmonious with compatible home entertainment systems to increase the sensory experience when interacting with media, for instance, video games and film. It also increases immersion as it mimics touch and interaction with virtual objects, thereby rendering experiences more real (Silvera-Tawil et al., 2015).

4.7.8 Robotics and teleoperation

Haptic feedback is essential in robotic applications because it delivers operators with a sense of touch and control while handling objects from a distance. Teleoperators use haptic technology to obtain critical feedback from robotic tools operating at a distance. In other operators, get instantaneous feedback about the forces applied to the robot (Emami et al., 2024). This feature allows them to perform tasks with high precision and accuracy. Robots, for example, are commonly used in the handling of dangerous materials and the deactivation of bombs (Emami et al., 2024). Intuitive interactions and accessibility, good haptic feedback have the potential to improve the user experience by making interactions more natural and compelling (Zhu et al., 2022). It can convey significant information like alerts or confirmations independent of visual or audio signals. Haptic feedback can also significantly enhance the accessibility of visually impaired or hearing-impaired users by giving them an alternative means of information reception. Instruction and learning in different fields. This technology offers a more comprehensive set of learning experiences and the opportunity to learn and practice difficult skills in a simulated, safe environment. Students studying with haptic devices when learning on their own performed better in memory and comprehension than students learning through images and text alone, according to a large haptic technology VR study.

In aviation, haptic devices provide pilots with sensations that are very similar to the sensations felt when actually navigating using improved flight simulators. These simulators are capable of simulating an airplane's reaction to flying in various meteorological conditions, including turbulence and hurricanes, thus enabling pilots to train and get used to handling a variety of difficult scenarios on solid ground. Haptic technology is gaining prominence in the context of industrial training. For example, the petroleum and gas sector uses VR in combination with haptic technology to mimic drilling and servicing operations. Through these, engineers and technicians are able to practice with high-pressure machinery without restricting actual physical exposure to risks and hence be capable of managing risk factors effectively. Such simulation-based training reduces the chances of accidents as well as the costs involved in the training process. Haptic equipment is also applied in military training. The system covers both simple firearms operations and complex strategic operations. Haptic undershirts, gloves, and other devices provide feedback when a soldier is hit, which not only brings soldiers a greater sense of realism in the training but also improves the soldier's ability to survive in real battle. The use of haptic technology in learning spaces has transformed the conventional way of learning. In medical education, medical students can learn surgical skills from virtual surgical simulators. Simulators ensure the truest sense of surgical procedures, particularly the haptic feedback of the instrument against the biological tissue. Likewise, in physics and chemistry, students can perform virtual experiments with the assistance of haptic devices. The experiments have been proven to be safe and allow the students to understand

Figure 4.18 Flight simulators.

abstract scientific concepts in a very easy way. Aerospace and aeronautics in aviation haptic technology significantly enhances the ability of flight crews to rapidly detect operational problems. For example, steering systems possess haptic features that notify pilots of key flight conditions. In addition, even in situations that are not a matter of urgent threat, haptic feedback is used to expand the pilot's situational awareness in general and to notify them of the status of the aircraft. This technology gives useful information about flight control, and it is simple and effective to communicate the flight rules in a safe manner. Haptic technology also assists with adherence to flight envelope protection procedures. Haptic actuators are embedded in cockpit parts and controls, allowing physical contact with pilots while delivering required information promptly and effectively (Fig. 4.18). In addition, haptics goes beyond real flights; it is also used to provide real sensations for flight simulations. This functionality enables train pilots to feel what they would otherwise feel in real situations, such as intense weather conditions, such as rain and storms, or engine failure, all simulated using haptics-equipped training tools. Haptics are changing dramatically not only within the metaverse scenario but also in space exploration. Ground teams as well as astronauts make use of the technology for several applications, such as the exploration of outer space. One major example of this is the European Space Agency's METERON project, which is focused on developing robotic interface communication systems and the hardware and software required for the remote control of robots in extra-terrestrial space (Emami et al., 2024). Space agencies can also make use of haptic

Figure 4.19 Metaverse.

technology in assisting the development of infrastructure on other planets or moons by employing robots that are remotely controlled by humans on Earth. Although this might appear to be a distant possibility, currently it should be pointed out that space agencies have already incorporated advanced haptic technology in ongoing space missions.

The metaverse: The term metaverse (Fig. 4.19) is increasingly becoming popular in technology news today (Emami et al., 2024). The concept is quickly gaining traction with businesses across sectors, preparing for its potential to transform our day-to-day lives. Haptic technology is likely to be one of the most popularly adopted technologies in this area that will facilitate the creation of the metaverse (Emami et al., 2024). The ultimate goal of the metaverse is to create a virtual environment that closely simulates reality and makes it nearly indistinguishable from reality. Achieving this calls for an immersive use of all human senses beyond the visual and auditory senses. The effectiveness of haptic technology is paramount to the realization of this shared vision in the technology world. Haptic technology is particularly best for those who desire to engage with virtual environments in a realistic and unconstrained manner. The technology substitutes the traditional user inputs, like tapping on a touchscreen or buttons on a handheld remote, with smooth gestures like zooming, pinching, pushing, touching, dragging, and other object-oriented manipulations. The equipment, most likely in the form of gloves that facilitate these interactions, will utilize more than haptic feedback. It will also utilize electromyography to convert the electrical signals generated by human thoughts into inputs that can be interpreted by computers. Haptic technology in the metaverse will function by giving users feedback or a response every time they interact. Consider the case when a user pushes an object in the metaverse; they can feel its weight and resistance. If a rock is thrown across a virtual pond, the user would feel it slipping from their hands. This is the amazing capability of haptics.

4.7.9 Automotive

Haptic technology has significant scope for improving the driver-vehicle interaction in addition to enhancing the vehicle's overall performance. These haptic components can be incorporated in various parts of the vehicle's user interface, including the steering wheel, pedals, seat belts, dashboard, and seating (Silvera-Tawil et al., 2015). These force or touch feedback interfaces are tangible interfaces. For instance, a vibrating seat can warn the driver about the possible presence of pedestrians walking onto the road ahead of the vehicle (Silvera-Tawil et al., 2015).

4.8 Security concerns

Applications for haptic technology, which gives users tactile feedback, are numerous and include everything from VR and gaming to medical simulations and remote communication. Our imaginations have been captured by the prospect of interacting, feeling, and touching virtual items or other people in a digital setting. It does, however, also raise issues that require attention.

4.8.1 Issues with haptic technology

Invasion of privacy: As these systems become more immersive, they require access to sensitive data about users' bodies and movements. The fear is that this data could be misused or exploited, leading to privacy breaches or unauthorized surveillance.

Security vulnerabilities: The risk of security flaws rises with the complexity of haptic systems. Users may experience discomfort or bodily danger if hackers manage to infiltrate these systems. Imagine a scenario where an attacker takes control of a full-body haptic suit and subjects the user to intense or harmful sensations. Such scenarios are, understandably, a cause for concern.

Addiction and overstimulation: Although full-body haptics can offer very immersive experiences, there is concern that this degree of immersion may cause overstimulation and addiction. Individuals may get so engrossed with these experiences that they overlook their relationships and obligations in the real world, which could negatively impact their lives.

4.8.2 Security practices in haptic technology

Protection of data privacy: Full-body haptic devices gather private information on the movements and bodies of their users. Strong data privacy safeguards must be put in place to guarantee that this information is not abused or accessed without authorization. Adherence to data protection laws, such as the General Data Protection Regulation(GDPR) is essential.

Verification and permission: Strong authorization and authentication procedures should be in place to stop illegal access to and control of haptic systems. It must be possible for users to manage who can access and alter their haptic experiences.

Safe interaction: Wireless communication is a common component of haptic systems. To avoid being intercepted or altered by malevolent actors, it is crucial to make sure that these conversations are safe and encrypted.

Frequent audits of security: To find and fix such flaws in their systems, haptic technology providers should regularly do security audits and vulnerability assessments. By being proactive, security breaches can be avoided.

Instruction for users: Users should receive instructions on how to utilize full-body haptics properly, as well as the risks involved. This entails being aware of the risk of addiction and overstimulation, as well as being able to identify and report any security issues.

4.9 Future of haptics

Efforts continue to advance haptic technologies that can duplicate intricate textures and sensations while also incorporating haptics with other sensory feedback to create a more immersive experience. The creation of wearable haptic devices capable of delivering feedback in different scenarios, like fitness training or navigation, is a field of ongoing research. Improper sensory feedback is recognized as one of the reasons for prosthesis rejection, which affects the efficiency of the system, as noises and disturbances are not removed properly (Lee et al., 2024). A frequent drawback of using haptic devices is the restriction in workspace and spatial limitations, which is especially examined during surgical procedures. The importance of the workspace and the concept of various contact points in a haptic interface call for further research and advancements in the future (Lee et al., 2024). The discrepancies occurring due to improper feedback, high contact speeds, and stiff environment setups in cable-driven teleoperation systems require more enhancements. The disadvantage of various types of haptic devices emphasizes the necessity for further research and development to deliver high-quality haptic feedback to users. Although many individuals are not yet familiar with haptic technology, it has infiltrated daily life for most of society and is poised to swiftly grow with the advancement of the Metaverse (Lee et al., 2024). With uses in healthcare, aviation, entertainment, and beyond, there are no limits to what haptic technology will enable us to achieve in the future.

4.10 Conclusion

Haptic technology is undoubtedly a valuable inclusion when it comes to giving additional user friendliness and direction to any device. Before this, though, when incorporating haptic feedback response into a technological program, one must seriously consider what elements would be most appropriate for their particular system. This would be some added time and energy, but the inclusion of haptic feedback will provide significant convenience to the device's human interaction functionality. The availability of haptic devices has made it possible to have widespread interaction between hardware and users. Different applications, especially in medical education, have been of significant interest by employing haptic technology. The medical sector has been advantaged by the few available haptic devices. However, there is doubt in the medical sector regarding the application of the devices in surgery and education. The haptic feedback system is an innovative technology that fills the gap between physical and digital sensations. The system improves user interaction in gaming, VR robotics, and medical devices by offering tactile feedback to virtual actions. The development of the system has progressed from basic vibrations to advanced and complex sensations that simulate textures, forces, and motions, thus providing a more immersive and intuitive user interface. With continued advancements in technology, mainly in wearable technology and advanced actuators, one can anticipate interactions to become even more seamless and realistic. Challenges remain in improving the accuracy, energy efficiency, and cost effectiveness of haptic systems. Despite these ongoing research and development in this area provide great potential for the development of more interactive, adaptive, and accessible user interfaces in different industries.

This research argues that haptic devices are gradually replacing traditional devices. The usability of haptic devices is emphasized by this chapter. In summary, there needs to be more investigation of the technical advantages of haptic devices, and people need to get to know haptic devices to enhance the use of the technology.

AI disclosure

During the preparation of this work, the author(s) used Grammarly to correct grammatical errors and improve language clarity. After using this tool/service, the author(s) reviewed and edited the content as needed and take(s) full responsibility for the content of the publication.

References

Biswas, S., & Visell, Y. (2021). Haptic perception, mechanics, and material technologies for virtual reality. *Advanced Functional Materials, 31*(39). https://doi.org/10.1002/adfm.202008186.

Carrasco, A. (2023). The feeling of control: The psychology behind immersive controls in video games and their real-world effects [Senior project, Bard College]. *Bard Digital Commons*.

Emami, M., Bayat, A., Tafazolli, R., & Quddus, A. (2024). A survey on haptics: Communication, sensing and feedback. *IEEE Communications Surveys & Tutorials*. 1–1 https://doi.org/10.1109/comst.2024.3444051 .

Gage, G., & Marzullo, T. (2022). *How your brain works: Neuroscience experiments for everyone*. The MIT Press https://doi.org/10.7551/mitpress/12429.001.0001.

Girbés-Juan, V., Schettino, V., Gracia, L., Solanes, J. E., Demiris, Y., & Tornero, J. (2022). Combining haptics and inertial motion capture to enhance remote control of a dual-arm robot. *Journal on Multimodal User Interfaces, 16*(2), 219–238. https://doi.org/10.1007/s12193-021-00386-8.

Giri, G. S., Maddahi, Y., & Zareinia, K. (2021). An application-based review of haptics technology. *Robotics, 10*(1), 1–18. https://doi.org/10.3390/robotics10010029.

Lee, S., Lee, H., Jung, G., Kwak, M. S., Kim, Y. R., Ko, H., Artificial flexible sensory electronics mimicking human somatosensory system. Korean Journal of Chemical Engineering. 42, 1977–1991.

Martín-Alguacil, N., Gaspar, I. d., Schober, J. M., Pfaff, D. W., & Vega, J. A. (2022). Somatosensation. In *Neuroscience in the 21st century: From basic to clinical* (pp. 1143–1182). Spain: Springer International Publishing. https://doi.org/10.1007/978-3-030-88832-9_27.

Olugbade, T., He, L., Maiolino, P., Heylen, D., & Bianchi-Berthouze, N. (2023). Touch technology in affective human–, robot–, and virtual–human interactions: A survey. *Proceedings of the IEEE, 111*(10), 1333–1354. https://doi.org/10.1109/jproc.2023.3272780.

Raisamo, R., Salminen, K., Rantala, J., Farooq, A., & Ziat, M. (2022). Interpersonal haptic communication: Review and directions for the future. *International Journal of Human-Computer Studies, 166*, 102881. https://doi.org/10.1016/j.ijhcs.2022.102881.

Silvera-Tawil, D., Rye, D., & Velonaki, M. (2015). Artificial skin and tactile sensing for socially interactive robots: A review. *Robotics and Autonomous Systems, 63*(3), 230–243. https://doi.org/10.1016/j.robot.2014.09.008.

Zhu, M., Biswas, S., Dinulescu, S. I., Kastor, N., Hawkes, E. W., & Visell, Y. (2022). Soft, wearable robotics and haptics: Technologies, trends, and emerging applications. In *Proceedings of the IEEE: 110* (pp. 246–272). Institute of Electrical and Electronics Engineers Inc. https://doi.org/10.1109/JPROC.2021.3140049.

CHAPTER 5

Advanced actuator technologies in medical haptics

Boddu Sanjana, Kosinepalli Kumar and Chirikonda Chandana
School of Business Analytics, Woxsen University, Hyderabad, Telangana, India

5.1 Introduction

With the inclusion of robotics in medicine, the scope of surgery, treatment, and patient recovery has been re-established. The da Vinci Surgical System, MIRO, and Raven are some examples of the systems that have proved that robotic help can be used to supplement the functions of surgeons because of the ability to eliminate tremor, scaling movements, and access to places of anatomy. However, the common shortcoming of existing surgical robotic systems is still the lack of trusted and high-quality haptic feedback—the capacity to operate the feeling of touch and tissue resistance on the surgical location and transfer it to the operating clinician. In medical robotics, haptic feedback includes two distinct categories: kinesthetic feedback, reporting force and torques to the robotic end-effectors, and tactile feedback, reporting local skin-level sensations (e.g., texture, temperature, vibration, etc.). Both modalities rely on actuators to move digital control signals to perceivable and meaningful physical mechanical output to the operator. The quality, realism, and safety of haptic feedback are thus inextricably linked to the performance properties of the underlying actuator technology. Traditionally, the medical robotic systems used the same actuator paradigms as the industrial robotics: direct current (DC) motors, servo drives, and pneumatic cylinders. Although powerful and well-established, these technologies are rather disadvantageous in the medical setting. Their stiffness is dangerous in cases of accidental collisions with delicate tissue. They are large and heavy and can not be miniaturized to be used in minimally invasive applications. Their noise, heat production, and electromagnetic signals make sterilization and compatibility with imaging difficult. This has seen the modern era of weapon diversification into novel actuator materials that have been driven by the advent of soft robotics, smart materials science, and micro-fabrication. Elastomeric soft pneumatic actuators are capable of safe deformation to human anatomy. Shape memory alloys (SMAs) make it possible to actuate without making noise, as well as compactly, due to reversible phase changes. Electroactive polymers (EAPs) are biological muscle analogs. Microelectromechanical systems (MEMS)-based actuators have unparalleled miniaturization for capsule endoscopy or implantable devices. All these technologies introduce

Haptic Feedback Control in Medical Robotics
DOI: https://doi.org/10.1016/B978-0-443-27777-1.00010-X

their unique trade-offs in force production, the speed of response, controllability, and biocompatibility, which have to be carefully weighed against particular clinical demands.

In this chapter, the author systematically addresses the traditional and modern technologies of actuators in the medical haptics field. Section 5.2 defines the theoretical premises of the haptic actuation and performance requirements. Section 5.3 discusses traditional methods. Section 5.4 is an overview of emerging technologies. Sections 5.5 and 5.6 it discusses integration with control systems and clinical applications. Sections 5.7 and 5.8 focus on design issues and future outlooks, and the chapter is summarized with an overview of key lessons.

5.2 Fundamentals of haptic actuation

5.2.1 Principles of force and tactile feedback

The system of haptic perception in human beings is two-fold. The kinesthetic system is based on the use of mechanoreceptors in the muscles, tendons, and joints to detect the position of the limbs, movement, as well as the force exerted on them. The cutaneous (tactile) system is based on four types of mechanoreceptors that are spread throughout the skin, such as Meissner corpuscles, Pacinian corpuscles, Merkow discs, and Ruffini endings, each responsive to a specific range of frequencies of the skin, as well as spatial resolution (Deflorio et al., 2022). Best haptic feedback interfaces should thus address the two subsystems to generate believable, clinically significant experiences. Robotic surgery Kinesthetic feedback requires actuators that are capable of simulating the stiffness of tissue, cutting resistance, and suture tension that robotic instruments feel. This requires actuators that can sustain a force over a physiologically significant range (usually, 0.1–50 N in surgical manipulation) at high bandwidth and low latency. In contrast, tactile feedback needs mechanical stimuli that are spatially resolved and rapidly modulated at the fingertips or palms, i.e., needs compact and high-frequency actuators like piezoelectric or vibrotactile motors and not large torque-generating drives. The control architecture of the haptic feedback is usually in the form of a closed loop consisting of three parts: the sensors to measure the state of the robotic environment (force sensors, position encoders, tactile arrays), the haptic controller that calculates the actuator output desired by a haptic rendering algorithm, and the actuator that actually provides the feedback. The integrity of the whole loop will be determined by the weakest link; a high-resolution sensor cannot be of use when the actuator is unable to faithfully reproduce the computed output and vice versa. The stability of the feedback loop is a vital issue: when the dynamics of actuators cause a phase delay or non-linear character, the system may be unstable, which manifests as unpleasant vibration or chattering to the operator.

5.2.2 Performance requirements in medical applications

The actuators targeted at medical haptic systems have to satisfy a set of performance requirements, which are, in many ways, conflicting and demanding, unlike those of an industrial or consumer system. Accuracy and force fidelity are the main things: surgical haptics is mandated to have a force resolution of millinewtons to provide a faithful transmission of the distinction between normal and diseased tissue (Chen et al., 2023). Temporal responsiveness. It should be sufficiently good (usually in terms of latency and bandwidth) to be below the human perception threshold of about 1 ms in the case of kinesthetic response and 300 Hz in the case of vibrotactile response. Mechanical compliance and safety are also imperative requirements. In contrast to industrial robots, where the robot is operated in a structured setting isolated from human beings, medical haptic systems have direct contact with the patient and clinicians. Actuators, therefore, need to be inherently compliant or controlled by compliant control architectures that limit forces in response to unanticipated contact. This property, commonly called safe physical human-robot interaction, does not allow stiff, high-impedance actuators to be used in most wearable and rehabilitation applications unless protective control layers are provided in large quantities. The chemically hostile and sterile nature of operating rooms puts material and surface demands on a firm. Depending on their sterilization procedure, actuator housings have to be resistant to repeated autoclaving (steam sterilization at 134°C and 2 bar), immersion in disinfectant solutions, or gamma irradiation. The surgery field must not be exposed to electronic parts. Also, actuators should be MRI-compatible in the case of MRI guidance procedures, i.e., they should be free of ferromagnetic substances that would disrupt the imaging field or cause projectile effects. Wearable and implantable applications are of special concern to compactness and energy efficiency. A rehabilitation glove aiding in the movement of fingers should be lightweight to allow a patient to use it over long durations, whereas a catheter-based micro-actuator needs to operate on a few microwatts of power delivered by a fine, biocompatible cable. Table 5.1 summarizes the key performance criteria in selected medical application areas of haptics.

5.3 Conventional actuator technologies in medical haptics

5.3.1 Electric motors (DC, servo, brushless)

The most prevalent form of actuator in modern surgical robots is electric motors. An example of such a system is the da Vinci Surgical System, which uses brushless DC motors in cable-driven patterns to operate its multi-degree-of-freedom wrist joints. The working principle of DC motors, namely the conversion of electrical energy into mechanical rotating energy by the interaction of the magnetic fields, is fully comprehended, highly controllable, and decades of servo control theory back the argument (Le et al. 2016). Brushless DC (BLDC) motors eliminate the commutation brushes, which add wear, electromagnetic interference, and maintenance needs, and are more favorable in high-duty-cycle surgical applications. High-resolution encoder-driven servo motors are used to provide the position and velocity feedback required to provide impedance or admittance

Table 5.1 Performance requirements and design criteria for actuators in medical haptic systems.

Parameter	Response time	Force output	Accuracy	Size	Safety level	Medical use case
Robotic Surgery	<1 ms	0.1–50 N	±0.01 N	Ultra-compact	Critical	da Vinci, MIRO
Rehabilitation	<5 ms	5–200 N	±0.1 N	Portable	High	Exoskeletons
Telemedicine	<2 ms	0.5–30 N	±0.05 N	Compact	High	Remote surgery
VR Training	<3 ms	1–20 N	±0.1 N	Moderate	Moderate	Surgical simulators
Wearable Haptics	<10 ms	0.1–5 N	±0.5 N	Miniature	Moderate	Tactile displays

control in haptic interfaces. Due to their good torque-to-weight ratio, controllability, and standard compatibility with commonly used industrial drive electronics, they are a natural choice to use in haptic display systems. Nonetheless, DC and servo motors have some limitations in the medical environment. Their mechanical impedance, or the ratio of pushing ability to the speed of the motion, is usually large; that is to say, they do not move easily. This is an undesirable safety attribute of human–robot interaction; a rigid motor can cause a surgical robot to apply massive force to delicate tissue in case of an error in control. Backdrivability: The operator may be restricted in their ability to manually move the actuator, and the drive ratio between the motor and the output must be relatively low, which is incompatible with the requirement of force amplification. Motor windings capable of producing electromagnetic interference can affect sensitive monitoring devices of the patient. Long periods of sustained high-torque operation necessitate heat management techniques that make system design difficult. These constraints notwithstanding, the maturity, reliability, and performance of electric motors guarantee their further supremacy in macro-scale surgical robots, especially with series elastic elements or variable impedance systems that have dampened the issue of rigidity (Table 5.2).

5.3.2 Pneumatic and hydraulic actuators

Compressed air drives pneumatic actuators, which have a very different performance profile from electric motors. They are naturally safer when they come into unintended contact due to their inherent compliance as a result of the compressibility of air and are better suited to rehabilitation purposes due to the desirable properties of the force profile of gentle and yielding forces (Surgical Robotics Technology, n.d.). Simple cylinder designs can give high force-to-weight ratios, and the technology is not sensitive to electromagnetic effects, which allows it to be used with MRI-guided interventions. Hydraulic actuators are based on the substitution of compressed air with incompressible fluid (usually hydraulic oil or biocompatible fluid) to achieve considerably higher densities of force as well as stiffer force control. By transmitting power using flexible tubing by hydraulic means, the actuator (as well as associated pump and reservoir) can be separated from the surgical end-effector over long distances, resulting in a compact tool design. Systems such as the Johns Hopkins University Steady Hand Robot and other platforms compatible with MRI have taken advantage of this by using hydraulic actuation. The main drawbacks of medical haptic pneumatic/hydraulic systems are that it requires support (compression, pressure regulator, fluid reservoirs, tubing) to be integrated with the device, which makes them large and limits mobility. Pneumatic systems are nonlinear and time-varying because of the compressibility of air, and therefore, haptic control at high bandwidth is challenging. Hydraulic systems bring threats of fluid leakage, which should be tightly sealed, and biocompatible fluids should be used. Haptic bandwidth is

Table 5.2 Comparison of advanced actuator technologies for medical haptic applications.

Technology	Working principle	Advantages	Limitations	Medical applications	TRL (1–9)
Soft robotics (Elastomers)	Pneumatic/hydraulic inflation of elastomer chambers	Compliant, safe human interaction; lightweight	Low force output; difficult control; bulky	Rehabilitation gloves, wearables	5–7
Artificial muscles (EAP)	Electroactive polymer deformation under electric field	Silent, lightweight, bio-inspired motion	Low force; susceptible to high voltage	Soft prosthetics, minimally invasive surgery	4–6
Shape memory alloy (SMA)	Thermal-induced martensitic phase transformation	High power density; silent; biocompatible	Slow response; hysteresis; heat management	Catheter steering, stents, and surgical tools	5–7
Magnetorhe-ological(MR) fluid	Magnetic field alters fluid viscosity	Rapid, variable damping; high bandwidth	Requires sealed housing; heavy	Force feedback joints, surgical brakes	6–8
Piezoelectric	Inverse piezoelectric effect converts voltage to strain	High precision; fast response; compact	Low stroke; brittle; high drive voltage	Ultrasonic surgical tools, tactile displays	7–9
MEMS actuators	Microelectromechanical electro-static/thermal actuation	Ultra-miniature; low power; integrable	Very low force output; fragile	Micro-needle arrays, capsule endoscopy	4–6
DC/servo motors	Electromagnetic torque via current-carrying conductors	Mature technology; high force/torque	Bulky; rigid; potential interference	Surgical robots, prosthetics	8–9
Pneumatic actuators	Compressed air pressure	High force output; compliant	Compressor needed; slow for haptics	Rehabilitation, surgical tools	7–8

restricted by the latency caused by the compliance of air or the viscosity of hydraulic fluid, as compared to electric motor systems.

5.3.3 Piezoelectric actuators

The piezoelectric actuators take advantage of the inverse piezoelectric effect—the mechanical deformation of some crystalline materials (such as lead zirconate titanate, PZT (Lead Zirconate Titanate), or polymer PVDF (Polyvinylidene Fluoride)) when exposed to an applied electric field. This allows an electromechanical coupling to be actuated with a very precise, repeatable, and fast actuation with a resolution of subnanometers (Fiveable Content Team, 2025). Piezoelectric actuators are able to achieve bandwidths in the kilohertz band, which is orders of magnitude more than most haptic applications need, and so are well-suited to vibrotactile feedback applications that need to recreate textural and vibratory stimulation in the 300 Hz band. Piezoelectric elements find a wide range of medical uses, including ultrasonic surgical tools, in which standing acoustic waves generate cutting and coagulation energy, and in tactile displays, in which small arrays of piezo elements directly stimulate skin mechanoreceptors. The technology is small, non-noisy, and can be made in thin-film formats to be incorporated into wearable haptic interfaces. Piezoelectric bimorphs can be packaged in either bending (large displacement, moderate force) or stacked (small displacement, very high force) modes. The main drawbacks of piezoelectric actuators are that they have a small stroke, normally a few micrometers to hundreds of micrometers with no mechanical amplification, and are brittle ceramic, which leads them to fracture under bending forces or under thermal shock. Drive electronics have to deal with large voltages (100–1000 V operating stack actuators) and have to deal with the hysteretic nonlinear characteristic between the voltage applied and the displacement. These make them limited to a few niche applications in the larger medical haptics environment.

5.4 Advanced and emerging actuator technologies

5.4.1 Soft robotics actuators

Elastomer-based systems: The use of soft pneumatic actuators (SPAs) made of silicone elastomers can be considered one of the most important paradigm shifts in rehabilitation and wearable medicine haptic actuation. In contrast to rigid actuators, which provide rigid kinematic constraints to the wearer, SPAs are deformable and conformable in nature, which intuitively adjust to the contours of the anatomy of the wearer. These designs allow bending, extending, twisting, or linear contraction with the motion of the fiber, depending on the geometry of the chamber and the fiber orientation, such as bellow-type, fiber-reinforced, and pleated designs. An example of clinical translation of elastomeric soft actuators in practice is the PneuNets developed by the Harvard Biodesign Laboratory and the finger rehabilitation gloves developed by Faboratory. These devices can be used to help or resist finger flexion with forces adequate to do therapeutic exercise with post-stroke or spinal cord injury patients by pressurizing embedded air chambers with modest pump pressures (usually 10–200 kPa). The adherence of silicone allows the overpressurization loads to be distributed on the finger surface instead of clustering at a rigid point of linkage, which significantly decreases the likelihood of accidentally injuring a person. Issues facing elastomeric SPAs are that they produce less force than motor-driven alternatives, they cannot produce precise position control, the materials are not linear, and portable pneumatic supply systems are needed. Current research is looking into the creation of a self-contained onboard pneumatic system that is integrated into the soft structure and the creation of models that are based on finite element analysis and machine learning to allow the closed-loop control to be performed correctly (Rodrigue & Kim, 2024).

Artificial muscles: Artificial muscles (also known as actuators) that replicate the contractile properties of biological skeletal muscle are a heterogeneous group that includes pneumatic artificial muscles (PAMs), twisted and coiled polymer (TCP) actuators, and electroactive polymer (EAP) artificial muscles. The McKibben pneumatic artificial muscles were invented in the 1950s and virtually forgotten until robotics researchers in the 1990s rediscovered and realized that the bladder is made of rubber and wrapped in a braided mesh; when inflated, the bladder contracts on the axial axis and expands on the radial axis, producing tensile forces similar to muscle contraction. EAP artificial muscles make use of the deformation of ionic or electronic polymers in any electric field. Ionic EAPs, including ionic polymer-metal composites (IPMCs), can be made to realize considerable deflection bending with low voltages (1–5 V) by electrochemical ion migration and produce weak forces, and necessitate the presence of hydration to operate. Dielectric elastomers and other electronic EAPs can obtain large area strains (10–150 MV/m) with a high electric field, which generates increased forces but high drive voltages. The two are being explored in the guise of prosthetic actuation, micro-surgical tool manipulation, and implantable therapeutic devices, where non-invasive, lightweight operation is vital.

5.4.2 Shape memory alloy actuators

The SMA, which is commonly nickel-titanium (Nitinol) alloys, displays a phenomenal thermo-induced change in phase between a low-temperature martensite phase and a high-temperature austenite phase. In the deformed state in martensites, the material can be heated above a transformation temperature (which is typically 40–90°C, which can be tuned by changing the alloy composition), and the material can then regain its original shape with significant force. This shape memory effect allows actuation without gears, coils, or magnets in compact, silent, and inherently biocompatible forms. Nitinol is biocompatible and has been used for decades of use as a cardiovascular stent material, orthodontic wires, and as a surgical guide wire. SMA wires have also been incorporated in a haptic actuation type system with catheter steering mechanisms, where the electrical Joule heating is used to bend a distal tip by bending SMA wires positioned on opposite sides of the catheter by differential contraction (Le et al., 2016). The noisy action of SMA actuators is specifically beneficial in the case of MRI-guided treatment, where a motorized actuator can cause acoustic noise, which disrupts imaging. The main setbacks of SMA actuators include high response time (also constrained by heating and cooling rates, typically 0.5–5 Hz in wire-based devices), hysteretic force-strain characteristic, and that they require active cooling to achieve favorable cycling rates. To at least partially overcome these limitations, research has been done on Joule-heated thin-film SMA actuators, lanthanide-based antagonistic SMA designs with forced convection cooling, and model-based hysteresis compensation to extend the available bandwidths and enhance positional performance.

5.4.3 Magnetorheological and electroactive polymer actuators

The magnetorheological (MR) fluids are composites of ferromagnetic particles in the form of micron-sized particles suspended in a carrier fluid. When a magnetic field is present, these particles will orient into a chain-like formation, which significantly raises the apparent viscosity and yield stress of the fluid; as the fluid is virtually at the viscosity of water in the off state, it forms a semisolid rheology that can oppose significant shear forces. Such a damping behavior that can be controlled by a field allows MR fluid-based actuators to act as variable-impedance devices in haptic systems and deliver tunable resistive forces in response to a magnetic field produced by electromagnet coils. MR actuators have been suggested in medical haptics for the joint resistance modules in robotic rehabilitation orthoses, and as controllable dampers in surgical tool handles that provide realistic tissue resistance feedback. They are clinically appealing due to their high bandwidth (>100 Hz), great force capability, and fail-safe operation, where the fluid will revert to a low-viscosity state when power is switched off. The issues are the sealed housings that will ensure no fluid is left to leak, the weight of the steel parts used to conduct the magnetic circuit, and the settling of the particles with time. EAPs in the wider context than artificial muscle applications are dielectric elastomer actuators (DEAs) and conjugated polymer actuators. DEAs are made of an elastomeric membrane that is cushioned between compliant electrodes; a high voltage compresses the elastomer, thickening it and swelling it in area to produce useful work. They are desirable due to their large energy density, silent operation, and the capability to generate large strains, and so are used as soft haptic displays with surgical gloves or rehabilitation wearable clothing (Cepolina & Razzoli, 2024).

5.4.4 Micro- and nano-scale actuation systems

MEMS-based haptics: Micro-electromechanical systems (MEMS) fabrication techniques, adapted from semiconductor manufacturing, enable the batch production of actuators with feature sizes ranging from micrometers to millimeters. Electrostatic, thermal bimorph, electromagnetic, and piezoelectric MEMS actuators have been demonstrated for haptic pin arrays, micro-needle stimulators, and tactile display panels. The primary advantage of MEMS-based haptic actuators lies in their spatial density: arrays of hundreds to thousands of independently controlled tactile pins can be fabricated on a chip area of a few square centimeters, enabling high-resolution spatial tactile displays that approach or exceed the spatial acuity of human fingertip mechanoreceptors (~1 mm separation threshold) (Liu et al., 2023). Applications in capsule endoscopy represent a compelling frontier for MEMS actuation. Passive capsule endoscopes currently traverse the gastrointestinal tract by peristalsis without active locomotion or tissue interaction capability. MEMS-actuated capsules under development incorporate electromagnetic, piezoelectric, or SMA-driven locomotion and biopsy mechanisms that would enable targeted tissue sampling under robotic teleoperation, with haptic feedback conveying tissue resistance to the operating gastroenterologist.

Wearable medical devices: The proliferation of wearable health monitoring devices has created a parallel demand for miniaturized haptic actuators capable of delivering clinically meaningful tactile feedback through smartwatches, biosensor patches, and therapeutic garments. Eccentric rotating mass (ERM) and linear resonant actuator (LRA) vibrotactile elements, already ubiquitous in consumer electronics, are being adapted for medical wearables that alert patients to physiological events (arrhythmia alerts, glucose level warnings) through coded tactile patterns. More sophisticated wearable haptic systems integrate arrays of pneumatic micro-chambers, shape memory polymer elements, or thermoelectric modules to deliver richer, multimodal tactile sensations relevant to biofeedback training and somatosensory substitution for prosthetic limb users.

5.5 Integration of actuators with haptic control systems

5.5.1 Control strategies for actuator-based haptics

The choice and adoption of a control system of haptic actuators is as paramount to the performance of the system as the technology of actuators. The haptic control is considered to have three main paradigms, namely, impedance control, admittance control, and adaptive control, depending on the actuator properties and clinical needs. The haptic interface is determined by impedance control, which makes the actuator a virtual mechanical impedance, a spring-mass-damper system, and instructs the actuator to provide forces that are proportional to the movement error in a desired position trajectory. This control is ideal in the case of the backdrivable actuators with low mechanical impedance (e.g., cable-driven DC motor systems or pneumatic actuators), a situation in which the operator is free to move the interface and forces are created in reaction. Impedance control provides intrinsic stability to analysis based on passivity and is common in surgical robotic systems, which imparts tissue hardening during tool-tissue interactions. Admittance control follows the opposite principle: the operator uses measured forces to drive a virtual dynamic model, which calculates the desired position or velocity of the interface, which is followed by the actuator. It is suitable with actuators (high-ratio geared motors) whose high impedance makes it difficult to drive back, but they respond accurately to position commands. In large rehabilitation robots such as the MIT-Manus and Lokomat, admission control is a common practice whereby strong motors act on articulated links in the exoskeleton, depending upon the measured forces of interaction with the patient. Adaptive control is used to handle the issue of time-varying dynamics in haptic applications—such as the varying stiffness of tissue during different stages of its excision, or the increasing fatigue of the muscles of a patient during their rehabilitation. The adaptive controllers adjust their parameters online according to observed system behavior and provide stable and realistic haptic rendering even under uncertainty or variation in the environment model. Active directions of research in the medical haptic application of haptics are model reference adaptive control, online parameter identification, and reinforcement learning.

5.5.2 Sensor–actuator coordination

Haptic feedback accuracy and realism require the quality of sensing modalities, which inform the actuator controller of the robotic environment. The interaction force direction between the surgical tools and tissue is measured using force and torque sensors, which are usually piezoelectric or strain-gauge bridge-based. The robot wrist six-axis force-torque sensors give detailed information on the wrench, which can be sent to the haptic interface once sufficient scaling and filtering are applied. Distributed contact information, such as pressure distribution, slip detection, and area of contact, is delivered to executing tactile actuator arrays on the operator side by tactile sensor arrays on the robotic finger pads. In bilateral teleoperation haptic systems, closed-loop stability is a prevailing issue because a time delay in the communication line between a master and the slave device generates phase lag, which disrupts the feedback loop. Established mechanisms of stability maintenance under time-varying delay with the preservation of haptic fidelity are wave variable transformation, time-domain passivity control, and model-mediated teleoperation. These methods control the flow of energy in the communication channel to impose passivity—a sufficient condition for stability—without any explicit delay compensation. Delay-tolerant haptic control architectures are gaining prominence as telemedicine and remote surgical applications are expanded to a broad range of distances (such as satellite-enabled remote surgery) and as both system and signal engineering communicate with each other at longer distances. Kalman filtering, Bayesian estimation, or neural network-based sensor fusion combines information from two or more sensing modalities (force sensors, vision systems, tissue models, and electromyography) and provides a more comprehensive and robust haptic rendering capability than is possible with any single sensor. To illustrate, force sensor measurements, combined with real-time tissue deformation measurements based on stereo endoscope images, can enhance the simulation of tissue stiffness in robotic surgery, to overcome the short bandwidth and noise of purely mechanical force sensors.

5.6 Applications of advanced actuator technologies in medical robotics

5.6.1 Robot-assisted surgery

The most challenging area of use of medical haptic actuators is robot-assisted minimally invasive surgery (MIS). The environment of working is highly demanding: the tools should go through readmission ports of 5–12 mm diameter, work within insufflated body cavities, withstand multiple rounds of sterilization, and have a tactile feedback to allow performing delicate manipulations such as the tension of suture, palpation of the lymph nodes, and evaluation of tumor margins. The commercially preeminent da Vinci surgical system now does not incorporate any haptic feedback to the operating surgeon into its offering. Force sensing and haptic display have been incorporated into research platforms such as the MIRO surgical robot (DLR), the Raven II, and the flex robotic system. Wrist mechanism cable-driven operating under compact BLDC motors passes tissue forces to master controllers through force-reflecting teleoperation. Natural orifice transluminal endoscopic surgery (NOTES) has also been demonstrated with SMA-based tool wrists, in which the tip has to go through torturous anatomical paths, with the tip providing force feedback via a flexible shaft. Recent studies have investigated piezoelectric arrays of tactile feedback integrated into the fingertip interface of surgeons that reproduce the local texture and stiffness changes of tissue experienced by robotic instruments. These constitute systems that combine arrays of deformable tactile pin arrays that are actuated by piezoelectric bimorphs to realize the virtual surface of the tissue dynamically in real time on the basis of finite element simulations, which are updated on the basis of measurements of forces at the instrument tip.

5.6.2 Rehabilitation and assistive devices

The field of haptic actuation in rehabilitation engineering has a very broad spectrum of applications, including both powered exoskeletons as an aid to patients with spinal cord injuries and also the soft glove used in hand rehabilitation after a stroke. One of the earliest rehabilitation haptic devices is the MIT-Manus planar robot, which has low-friction transmissions, allowing the device to give assist-as-needed haptic feedback during reaching exercises in stroke patients. Lokomat gait rehabilitation robot refers to a gait rehabilitation robot based on large actuated exoskeleton joints with brushless motors with series elastic elements that measure and control joint torques to retrain the gait.

A more recent and promising method of finger and hand therapy is the soft robotic rehabilitation gloves, which are powered with pneumatic elastomeric actuators. Not only are they safe against overworking the arthritic or postoperative joints, but because they are lightweight, they can be worn over a long period of time when patients perform the functions of daily living; this is relevant to the neuroplasticity-driven repetition that is the basis of effective motor rehabilitation. Each finger can be individually pressurized to follow a therapist-prescribed motion path and exert small resistive forces that demand to be surmounted by recovering muscles, with the pneumatic bending actuators. The enhanced haptic actuation capacity would be a big advantage to prosthetic limb interfaces since it offers users sensory information to reduce mental load and enhance dexterity in handling objects. Vibrotactile actuators are vibrotactile stimulators on the prosthetic sockets or skin-contacting bands, providing the coded tactile feedback that indicates the grip force, slips, or contact position of the prosthetic hand. More sophisticated systems apply EAP-based actuators or pneumatic micro-chambers to provide a more realistic and richer tactile experience by providing directed stimulation on the residual limb surface.

5.6.3 Telemedicine and remote operation

Telemedicine and teleoperated surgery allow bringing the specialized surgical knowledge to geographically distant or underserved areas. In 2001, due to the transatlantic Lindbergh operation, a laparoscopic cholecystectomy operation in Strasbourg, France, was done on a patient in New York, via a transatlantic robotic connection. A demonstration of the technical feasibility of remote surgery was achieved, pointing to the challenges of communication latency and fidelity of haptic feedback. Clinically practicable remote surgery has since been revived by high-bandwidth and low-latency communication network technologies, such as 5G cellular and software-defined networking. Greater attention is required of haptic actuators in telemedicine applications since reasonable communication delays may vary over time, which can range between hundreds of milliseconds across intercontinental paths, to milliseconds across local area networks. Master controllers that are force reflecting: force feedback. Master controllers that provide bilateral force feedback use DC motor drives with high-quality torque control. New methods based on predictive haptic models that are being held locally in the master and being updated through periodic synchronization with the slave can decrease the perceived impact of delay by causing locally produced force predictions to await feedback of the remote environment.

5.6.4 Training simulators and virtual reality medicine

Surgical training simulators combine haptic interfaces with virtual reality displays to provide the opportunity to safely and repeatedly practice surgical procedures without harm to patients. Multi-DOF haptic systems such as the LAP Mentor (Simbionix,

Cleveland, OH, USA), the V-Trainer (Mimic Technologies, Seattle, WA, USA), and the Fundamental Surgery VR platform are designed to provide multi-DOF tissue resistance, instrument collision, and procedure forces in training laparoscopy, endoscopy, and robotic surgery through multi-DOF haptic systems based on DC motor drives or hybrid cable-motor configurations. Accurate computational tissue models and high-performance actuator systems that have the ability to render the force signatures of various tissue types determine the fidelity of haptic rendering to surgical simulators. Soft tissue finite element models with precomputed haptic rendering algorithms can be used to compute the forces in real-time and provide actuated haptic handles. Textural detail is added to the motor-driven force feedback using piezoelectric vibrotactile elements over them—the unique feel of cutting through fascia or muscle or vibration of an ultrasonic dissector—that enhances the training experience and increases skill acquisition.

5.7 Design challenges and engineering considerations

Engineering of actuator systems, medical haptics, and navigating the landscape of competing constraints and interdisciplinary challenges in medical haptics engineering requires a holistic approach to systems engineering. The challenge of safety compliance and human interaction is the leading one. Physically interacting medical devices, such as industrial collaborative robots reconfigured for medical use, are regulated under regulatory frameworks, such as ISO 10218, IEC 60601, or FDA 510(k), which require strict safety certification. Actuators should exhibit limited force production no matter the failure mode; it can be the loss of power, the fault of the controller, or the loss of communications. DEA designs Inherent force limiting and backdrivability Series elastic actuator (SEA) designs, that place a compliant spring element between the motor and the output, achieve both safety compliance and ease of compliance. Compliance requirements are typically met by soft actuator materials, which are, in fact, required to be verified against both long-term mechanical fatigue and aging in clinical settings. One constraint in the design of medical haptics is miniaturization and power efficiency. The surgical instruments that have a minimum invasive size of 5 mm shafts are limited in size to the actuators of the same size, but they need to produce clinically significant forces (1–10 N at the tool tip). The high-performance materials (rare-earth permanent magnets, which can be used to make miniature motors, or Nitinol, which can be used to make SMA wires), the ability to create high-quality miniature fabrication technologies, and good power electronics are required to achieve sufficiently large force output of miniature actuators. The wearable rehabilitation devices that are powered by batteries have a severe energy budget that supports the low-power actuation strategies, like shape-locking mechanisms, that support force without an ongoing supply of energy. The choice of materials and design of structures is limited by biocompatibility and sterilization requirements. Contact assemblies to patients or to nonsterile surgical areas should be made of biocompatible materials (USP Class VI plastics, titanium, Nitinol, medical grade silicones) and tested to be suitable for the desired sterilization modality. Steam autoclaving subjects elements to 134°C saturated vapor and can destroy electronic assemblies, polymer insulators, and adhesive bonds. The ethylene oxide (EtO) sterilization passes through intricate geometries but has prolonged aerations. Designs of actuators that incorporate biocompatible hermetically sealed electronics and biocompatible materials, which are rated to support repeated sterilization cycles, are fundamental in clinical applications. The latency and stability of haptic feedback loops are the essential control engineering issues. Actuator dynamics, sensor noise, and control loop sampling rate each limit the Z-width—the range of impedances that a haptic device may be capable of stably rendering. There are higher loop rates (>1 kHz) and low-latency actuator drives that are making the possible Z-width much higher, so that both very stiff (high impedance) virtual environments and very free (low impedance) environments can be rendered without becoming unstable. The wireless communication links, which are becoming more attractive to untethered wearable haptic systems, provide variable latency and packet loss, which need to be handled by means of sound control architectures.

5.8 Future trends and research directions

Medical haptic actuation is an area undergoing dramatic change due to the convergence that takes place in the areas of artificial intelligence, advanced manufacturing, and soft matter science. AI-based adaptive actuators will be an interesting direction in which machine learning models incorporated within actuator control systems will allow real-time adjustment to the tissue characteristics of the patient, the skill level of the operator, and the conditions of the procedure. The agents of deep reinforcement learning that are trained on massive datasets of surgical interaction have the ability to optimize the parameters of impedance in robotic surgery to ensure that the force transparency is maximized, and the dynamics remain stable. These algorithms can be realized with ultralow-power neuromorphic computing architectures that can provide intelligence in controlling actuators of battery-powered wearable devices that are not connected to the cloud. The soft wearable haptic interfaces will reduce the line between therapeutic devices and wearable everyday devices more and more. Multilayer smart textile garments incorporating EAP actuator fibers, piezoelectric sensing threads, and flexible printed electronics are being developed that may be able to deliver continuous sensory feedback to amputees, spinal cord injury patients, or postsurgical rehabilitation patients in everyday life as opposed to specific therapy sessions. Washing, durability, and long-wear biocompatibility of such textile-integrated systems is a problem that promotes the continuous development of materials research.

Another practical way of tackling the shortcomings of individual technologies is to use hybrid actuator architectures that incorporate several principles of actuation into a single device. Such hybrid SMA-pneumatic catheters, as an example, may involve SMA wires to precisely steer the tip with large force delivery and pneumatic inflation of micro-sized chambers at the tip to provide compliant, low-force contact with tissue. Motor-SMA hybrid finger prostheses are fast motor-controlled to perform gross grasping and SMA-controlled to perform fine grip control. Co-optimization of hybrid actuator designs. Systematic methodologies to co-optimize multiobjective optimization of actuator design are an emerging field of work. Actuator optimization digital twins provide a game-changing ability in designing and clinical implementation of medical haptic systems. By deploying a digital twin—a high—quality computational model with real-time updates on sensor data provided by a physical actuator—one can real-time monitor actuator state, perform predictive maintenance (monitoring early signs of wear or degradation before clinical failure), and optimize model control parameters. In soft actuators with nonlinear material behavior of interest, neural network-based digital twins, which, on experimental deformation data, can offer accurate inverse models to allow precise closed-loop control without physics models that are analytically solvable. Lastly, the combination of haptic feedback and augmented and extended reality (AR/XR) systems can be seen as an interesting trend in surgical navigation and training. With the current trend of surgical robots to actively engage the AI scene recognition and anatomy recognition, the haptic actuator may be used to coordinate with the AR overlay to deliver spatially matched force feedback, such as guiding a surgeon on the best resection planes or issuing a warning because the instrument is approaching a vital neural or vascular region. Such a coming together of haptic actuation, computational intelligence, and mixed reality may re-establish the interface between human surgical experience and robotic accuracy.

5.9 Conclusion

This chapter has provided an overview of all actuator technologies that are the basis of haptic feedback systems in medical robotics, not only of traditional techniques in the field of electric motor and pneumatic engineering, but also of new technologies in the field of smart materials, soft robotics, and micro-scale fabrication. As it has been examined, there is no single actuator technology that can meet the multifarious and even contradictory demands of surgical robotics, rehabilitation engineering, telemedicine, and training simulation. The choice of a suitable actuator—or a set of actuators—should instead be informed by a critical consideration of the unique clinical environment, force and bandwidth needs, safety considerations, and environmental conditions of deployment. Electric motors are still the workhorse of the macro-scale robotic haptics, due to their maturity and controllability, but their rigidity, size, and electromagnetic interference are limiting factors that are driving the creation of alternatives. Pneumatic and hydraulic systems are compliant and MRI compatible at the expense of the complexity of the infrastructure. Piezoelectric actuators are highly precise and high-bandwidth vibrotactile actuators. New technologies—SMA, soft pneumatic actuators, EAPs, MR fluids, and MEMS—each has particular performance gaps, but they have new engineering challenges of nonlinearity, speed, power, and manufacturability. The combination of sophisticated actuators and complex control methods—impedance control, admittance control, adaptive control, sensor fusion—is also essential to attaining clinically meaningful haptics. The haptic feedback quality achievable depends on the actuator dynamics, sensor properties, communication latency, and control loop stability, and all these factors have to be co-designed to achieve the best outcomes. The problem of closed-loop stability, especially with teleoperated systems that have communication delay, is a cornerstone problem that has led to innovation in control theory and network-transparent haptic architecture. In the future, it is hoped that the future of medical haptic actuation will be more intelligent, adaptive, soft, and multi-modal, such that it approaches biological sensorimotor experience more closely. Adaptive control via AI, combining actuators, optimizing haptic feedback through a digital twin, and fused with augmented reality are some of the research directions that have the potential to overhaul the quality and breadth of the haptic feedback in medicine. With the continued growth of these technologies and regulatory channels in place, next-generation medical robotic systems will also offer more and more surgeons, therapists, and patients with the intuitive, safe, and immersive haptic experiences that can make a difference in all aspects of clinical practice. This vision requires long-term interdisciplinary work between the materials scientists, control engineers, human factors researchers, clinicians, and regulatory specialists. The actuator technologies discussed in this chapter give the physical basis to this effort; the chapters that follow will discuss the sensing, control, and human-factors aspects, which complete the medical haptic systems architecture.

AI disclosure

During the preparation of this work, the authors used ChatGPT to help formulate the theoretical content of the manuscript and Grammarly to correct grammatical errors and enhance readability. After using this tool/service, the author(s) reviewed and edited the content as needed and take(s) full responsibility for the content of the publication.

References

Cepolina, F., & Razzoli, R. (2024). Review of robotic surgery platforms and end effectors. *Journal of Robotic Surgery, 18*(1), 1863–2491. https://doi.org/10.1007/s11701-023-01781-x.

Chen, J., Teo, E. H. T., & Yao, K. (2023). Electromechanical actuators for haptic feedback with fingertip contact. *Actuators, 12*(3). https://doi.org/10.3390/act12030104.

Deflorio, D., Di Luca, M., & Wing, A. M. (2022). Skin and mechanoreceptor contribution to tactile input for perception: A review of simulation models. *Frontiers in Human Neuroscience, 16.* 16625161 https://doi.org/10.3389/fnhum.2022.862344 .

Fiveable Content Team, 2025. Actuator technologies and selection criteria: Medical robotics, Retrieved from https://fiveable.me/medical-robotics-and-computer-assisted-surgery/unit-4/actuator-technologies-selection-criteria/study-guide/G9rBjaJk00rjef8K. Accessed on 01 June 2026.

Le, H. M., Do, T. N., & Phee, S. J. (2016). A survey on actuators-driven surgical robots. *Sensors and Actuators, A: Physical, 247*, 323–354. https://doi.org/10.1016/j.sna.2016.06.010.

Liu, Q., Ghodrat, S., Huisman, G., & Jansen, K. M. B. (2023). Shape memory alloy actuators for haptic wearables: A review. *Materials and Design, 233*. https://doi.org/10.1016/j.matdes.2023.112264.

Rodrigue, H., & Kim, J. (2024). Soft actuators in surgical robotics: A state-of-the-art review. *Intelligent Service Robotics, 17*(1), 3–17. https://doi.org/10.1007/s11370-023-00506-1.

Surgical Robotics Technology, Actuators for surgical robots." Retrieved from https://www.surgicalroboticstechnology.com/suppliers/modules-components/actuators/. Accessed on 01 June 2026.

CHAPTER 6

Haptic interface design for medical applications

Polireddy Vasavi Reddy, Lakshita Gandhi and Paramaticona Reddy
School of Business, Woxsen University, Hyderabad, Telangana, India

6.1 Introduction

6.1.1 Background and motivation

Surgical robotics history is, in a great measure, a narrative of the gradual removal of the human hand from the operating room. Starting with the first telemanipulator in the 1980s, the next generation of surgical robots was more precise, less tremulous, with access to anatomical areas of the human body that could not be accessed by a human hand. However, every step toward the patient moved the surgeon farther away and made the operating area an interface of cameras, monitors, and joysticks to interpret the intent into action and deprive the surgeons of the tactile conversation on which they rely to make real-time judgments. This shortage is not one of just comfort or preference. It is always observed that without the haptic feedback, there is a high probability of exerting excessive force on the delicate tissue, there is a loss of the possibility of distinguishing the types of tissue, and the cognitive workload as the surgeons overuse the visual channel to offset the lack of the sensory channel. In the case of trainees, the effects are even more dramatic: they are taught to work without touch sense, as well as to play the piano with big gloves on. A technique is an easy thing to learn, but the feel of a properly done maneuver is never quite there, and mistakes that more full-fledged senses would have checked are left to play out until one sees the results.

Haptic interfaces are designed to fill this gap and measure force and tactile data where surgical contact occurs and accurately convey it to the hands of a surgeon via actuated end-effectors, wearable devices, or dedicated controllers. In earlier implementations, most implementations were more mechanical and passive. Modern systems combine sensor arrays, digital signal processing, and more machine learning algorithms that may predict forces during tissue interaction, correct transmission delays, and adjust the strength of feedback to the surgical situation and surgeon preference. This chapter, and the overall research project it represents, is inspired by the fact that it is believed that, rather than being a luxury, the restoration of the sense of touch to robotic surgery will be a clinical requirement.

Haptic Feedback Control in Medical Robotics
DOI: https://doi.org/10.1016/B978-0-443-27777-1.00008-1

6.1.2 The use of haptic interfaces in health care is important

The usefulness of haptic input in hospitals is not limited to the operating theater. In the example of rehabilitation robotics, it uses tactile and kinesthetic feedback extensively, both to retrain motor pathways in stroke patients, to train the behaviors of users of prosthetic limbs toward more natural behaviors, and to offer measurable sensory stimulation during neurorehabilitation programs. Simulators with haptic feedback performance in medical training always show better training transfer, with training participants displaying more calibration of forces, as well as procedural accuracy, when trained on tactile-enabled systems. The haptic signal fidelity degradation is an established cause of negative events in minimally invasive surgery, where the instruments are introduced into the body via small holes and communicate with the tissue by way of long, rigid instruments. Mechanical friction and compliance inside the instrument shaft cause an instrument-tissue force distortion, and by the time any tactile information can be felt by the surgeon's hand, it is significantly dampened. Haptic-controlled robotics that replicate this signal, instead of merely relaying whichever signal remains after mechanical transmission, can yield information which on some laparoscopic conditions, is actually more precise than direct hand manipulation.

User experience matters too. Surgeons with high cognitive fatigue, which can be explained by the need to fill the gaps of sensory feedback, have a higher risk of attentional lapses and errors in decision-making when operating in a long procedure. Having a properly designed haptic interface, which helps to alleviate this compensatory cognitive load, helps not only to increase procedural safety but also to improve the well-being of a surgeon, which, in its turn, is a patient safety problem.

6.1.3 Scope and objectives of the chapter

The chapter is a thorough coverage of haptic interface design in medical applications, which is arranged in a way that graduates to the basics of the design, the technology, design methodology, applications, governance, and future outlook. It is composed in a manner that will appeal to graduate researchers, medical device engineers, clinical informaticists, and healthcare administrators interested in the new technologies in surgery. The main goals are to survey the current state of the art in the field of designing medical haptic interfaces, to offer the practitioners a systematic approach to the evaluation and choice of the design methods, and to explain the research directions where the most significant improvements are expected to be in the next decade.

6.2 Fundamentals of haptic technology

6.2.1 Principles of haptic perception

Human tactile perception is a complex and multichannel sensory system, of which most individuals do not consciously consider until a stage when it starts to malfunction. Four major categories of mechanoreceptors are found in the skin: Meissner corpuscles, which sense light touch, texture at low frequencies; Pacinian corpuscles, very sensitive to high frequency vibration and deep pressure; Merkel discs, which sense fine spatial detail and sustained pressure; and Ruffini endings, which sense skin stretch and joint position. The muscle spindles, Golgi tendon organs, and the joint capsule receptors provide kinesthetic sensation, or the sense of the position of the limbs and their movement. Combined, these systems enable an expert surgeon to differentiate types of tissue, identify concealed structures, and control the amount of force used with great accuracy. Knowledge of the physiology of haptic perception is not just a theoretical concern among haptic interface designers. It has immediate engineering connotations. The frequency response of the Pacinian corpuscle determines the bandwidth needed by vibrotactile feedback. The density of the disc innervation of the fingertip determines the spatial resolution requirements of tactile display arrays. The just-noticeable difference in force is psychophysically dictated and is the rate at which force feedback is rendered stably, requiring about 7% or 10% of the reference force, which is the just-noticeable difference in force. It is a waste to design haptic systems that are below the perceptual threshold; it is a waste to design systems that are above the range of reliable human discrimination as well.

6.2.2 Types of haptic feedback

Force feedback: Kinesthetic feedback, also known as force feedback, is a type of feedback where controlled mechanical forces are applied to the arm, hand, or fingers of the operator to mimic the resistance, compliance, and weight of objects in the remote or virtual location. The simple one-degree-of-freedom resistance mechanisms are force feedback devices, but on the other extreme, there are more intricate multiarm exoskeletons that can provide six-axis force and torque. In surgical robotics, position-force bilateral teleoperation is the most common approach to force feedback, in which the measured contact forces of the slave robot are scaled and fed back to its master controller held by the surgeon. The engineering challenge is to have this rendering with enough bandwidth, stability, and transparency such that the surgeon sees the remote environment and not the control system.

Vibrotactile feedback: Controlled mechanical vibrations, which are normally provided by piezoelectric actuators, eccentric rotating mass motors, or linear resonant actuators, are used in vibrotactile feedback devices to stimulate skin mechanoreceptors but do not necessarily provide sustained forces. Due to the small size, lightweight, and power efficiency of vibrotactile actuators, they would be a good fit in wearable haptic feedback systems and surgical gloves, as well as miniaturized tools. Medical vibrotactile feedback has been applied to detect catheter contact in catheter navigation, to provide texture in palpation, and to indicate when a surgeon is facing an abnormal force situation through the use of a specified vibration pattern.

Multimodal and thermal feedback: Thermal haptic feedback uses the thermoreceptors of the skin to transmit information about the temperature of the tissue, which can have diagnostic importance, as in burn wound assessment or in operative tumor detection. The technical difficulties of thermal feedback are great: the thermal time constant of skin tissue is slow compared to clinical event rates, and there are limits on the temperature range that are limited by safety considerations, especially in a situation of sterilizable medical equipment. Multimodal feedback systems making use of two or more haptic modalities, usually a combination of force and vibrotactile, or thermal and force, seek to provide a more informative and descriptive picture of touch than a single modality can do. Studies continue to indicate that multimodal feedback is better than unimodal feedback in task performance as well as perceptual realism, and that the hardware and control are complex in multiple actuator systems, which have to be handled with care (Table 6.1).

6.2.3 Hardware components of haptic systems

A haptic system of a medical robot does not constitute any single device but is a combination of sensing, actuation, and computation components that have to collaborate on submillisecond accuracy. On the sensing end, the most common data source to be used in haptic rendering is force and torque sensors at the instrument tip, but tactile sensor arrays, the means to resolve the distribution of contact pressure over a surface, but not its integral, are becoming more of a feature of sophisticated surgical tools. The Fiber Bragg grating sensors that detect strain changes by optical wavelength are receiving popularity in minimally invasive surgical tools since they are not affected by electromagnetic noise, inherently sterilizable, and can be high-density (spatially) resolved.

Haptic feedback actuators have a very rigorous set of requirements: they need to provide forces across the full range of perceptual bandwidth, be able to work well in the presence of sterilization agents and moisture, use low energy, have no hazardous failure modes, and preferably be noiseless enough not to induce interference with the clinical environment. A single actuator technology that can meet all of these requirements at the same time does not exist, and this explains why the actuator technology selection is commonly the main design decision in a haptic interface development project. The system is topped off by controllers and signal processing pipelines, often at update rates of 1 kHz or more, with force feedback, special-purpose signal conditioning processors, control law execution processors, and communication with the robotic system.

6.3 Medical robotics and human–robot interaction

6.3.1 Overview of medical robotic systems

Medical robotic systems offer a broad spectrum of application areas, ranging from the dramatic high-stakes market of minimally invasive surgery to the more mundane yet no

Table 6.1 Comparison of haptic feedback modalities in medical applications.

Feedback modality	Mechanism	Sensing range	Medical application	Key advantage	Limitation
Force feedback	Mechanical actuators, motors	0.01 N–100 N	Robotic surgery, needle insertion	High fidelity tissue interaction	Bulky hardware, latency
Vibrotactile feedback	Piezoelectric/ ERM actuators	10–300 Hz	Endoscopy, catheter guidance	Compact, fast response	Limited force resolution
Thermal feedback	Peltier elements, heat resistors	15–45°C	Burn assessment, tissue diagnosis	Tissue characterization	Slow response, skin damage risk
Electrotactile feedback	Electrical stimulation of skin	0.1–10 mA	Prosthetics, neurorehabilitation	Bypasses damaged nerves	Discomfort, safety thresholds
Pneumatic/ tactile arrays	Air bladders, shape memory alloys	0–500 kPa	Laparoscopy, training simulators	Distributed contact sensation	Complex pneumatic control
Multimodal (Combined)	Fusion of 2+ modalities	Context-dependent	Advanced surgical platforms	Comprehensive sensory feedback	Integration complexity, cost

less vital markets of rehabilitation, orthotics, and diagnostics. The most technologically challenging and high-profile use of haptic interfaces is in the use of surgical robots, such as laparoscopic, cardiac, orthopedic, and neurosurgical robots. Although a massive commercial success, the da Vinci platform is also characterized by the lack of haptic feedback in its standard form, a choice of design that is accurate to the real-world engineering difficulties of providing stable force feedback to a flexible multidegree-of-freedom kinematic chain, but which has been repeatedly criticized by clinicians who were trained on robotic surgery before its introduction (Ueda et al., 2024). Haptic feedback is also used in rehabilitation robotics, such as end-effector and upper and lower limb rehabilitation exoskeleton systems, in which meaningful therapeutic interaction is generated. Several research platforms, such as the Lokomat gait rehabilitation system, the ARMIN arm rehabilitation robot, and many others, employ force-position control strategies to offer error-augmenting, resistive haptic, or assistive haptic guidance to patients performing motor tasks. There are significant differences between the design problems here and in the case of surgical robotics: the patient, not the clinician, determines the safety constraints, the therapeutic objective necessitates a careful balance of challenge and assistance in these sessions and levels of progression (Baronchelli et al., 2021).

6.3.2 Role of haptics in robotic surgery

The argument in support of the use of haptic feedback in robotic surgery is based on theoretical assumptions as well as empirical findings. In theory, the loss of force feedback can be replaced, and this restores the sensorimotor loop that evolution had been optimizing to achieve manual control, which surgeons can then use the same neural mechanisms that enable expert manual surgeons to be as effective. Experimentally, controlled experiments based on robotic surgery systems with and without haptic feedback have shown that peak applied force is reduced, knot tying is more consistent, and time to complete tasks is less with haptic feedback. These are not necessarily large effect sizes, and the transfer of bench-top experiments to a clinical outcome is not perfect, but the evidence trend is always the same way. Situational awareness. The haptic feedback also provides the surgeon with situational awareness to supplement visual information. The information that cannot be provided by the camera is that provided by a surgeon who is able to detect the hardness of a mass beneath a peritoneal layer. Only a surgeon who is able to sense the tension in a suture in the process of tightening it can check the sufficiency of tissue approximation, which cannot be achieved through visual examination. They are particularly necessary in processes that are complex and multi-step, with already high cognitive load and severe impact of errors.

6.3.3 Challenges in clinical integration

However, regardless of strong arguments supporting the usefulness of haptic feedback, clinical acceptance has a number of recalcitrant issues. Latency is, perhaps, the most basic.

Human haptic perception is time sensitive to the range of a few milliseconds, and delays longer than about 10 to 15 ms are also unstable to force feedback control loops and result in oscillation or chatter, which is unpleasant at best and may be dangerous at worst. Network delays in teleoperated surgery, especially across intercontinental connections, can often surpass this limit, and wave variable or model-based control measures are necessary to keep the system stable at the price of transparency (Ueda et al., 2024). The actuator technologies, electronics, and materials are highly constrained by sterilization requirements. A lot of commercially available tactile sensors and actuators are not engineered to be subjected to steam autoclaving, gamma radiation, or even high-level disinfection agents (Batty et al., 2022). Design and handling of haptic hardware that is sterilization-compatible is a current and significant area of research. Dependability and usability make up the clinical challenge set: a haptic interface that adds time to the current procedure, must be trained on by special operators, or fails in manners that jeopardize patient control, will not be implemented despite its technical functionality in controlled environments.

6.4 Artificial intelligence-driven haptic interface design

6.4.1 Machine learning for adaptive feedback

Implementation of machine learning in haptics interface design has led to possibilities that could never have been realized under the classical methods of control. Predictive force rendering is one of the most practical ones. In teleoperated surgery, the round-trip communication delay between master and slave robot implies that force feedback will always be an indication of what the state of the environment was somewhere in the past. The predictive models, trained on large datasets of tissue forces when subjected to representative surgical tasks, can predict the force to be experienced at the instrument end a few milliseconds in advance, and the haptic rendering system can then provide feedback, based on the present condition of the environment, not the delayed condition (Li & Hannaford, 2017). This well-calibrated approach can be used to push the stability limit of teleoperated haptic systems to much higher communication delays. Another machine learning field that has contributed to data-driven tissue modeling is the latter. Traditional methods of tissue simulation, as applied to haptic rendering, used finite element models wherein mechanical properties were manually defined and were laborious to build, and usually did not capture the heterogeneity of the biological tissue. Models based on neural networks can be trained on tissue behavior by using direct force-displacement measurements made either in an ex vivo experiment or intraoperative measurements to yield faster forward models, which are also generally more accurate than physics-based ones. Gaussian process regression is of particular value in this regard, not only to give point estimates of the tissue force response, but also to provide uncertainty quantification, which can be exploited to modify control strategy in areas where the model is ill-posed (Ou & Tavakoli, 2023).

6.4.2 Intelligent control algorithms

The idea of reinforcement learning has become an interesting solution to the task of controlling haptics in dynamic medical settings, where the optimal control policy is context-dependent in ways that cannot be characterized analytically. A reinforcement learning agent, that is, trained to control a surgical haptic interface, can learn by interacting with high-fidelity simulators how to control bandwidth, signal filtering, and feedback gain in response to varying surgical conditions, communication quality, and operator state without necessarily having a mathematical model of the relationships among the factors. Sim-to-real transfer problem. The transfer problem of simulation to real hardware and real tissue makes policies trained on simulators behave well on real hardware and with real tissue is an important research problem, but one (widely) currently being addressed using domain randomization and adaptive fine-tuning techniques (Kamelle, 2025). Control strategies based on neural networks, such as long short-term memory networks to process time-series force data, convolutional networks to process images of tactile sensors, are being used in tissue-type classification, detection of abnormal force events, and adaptive control of haptic feedback parameters. These systems are capable of handling information rates at competitive rates against classical digital signal processing, and they are able to introduce nonlinear relationships into the information that would otherwise have been missed by the linear methods. Close consideration of computational latency, model interpretability, and robustness to distribution shift is important in the practical implementation of such systems in clinical devices, and is a subject of current effort.

6.4.3 Personalization and context awareness

Personalization is one of the least studied areas of haptic interface design. The surgeons vary significantly in their haptic sensitivity, motor control, their preferred feedback intensity, and reaction to the various feedback modalities. The haptic rendering advantage that is natural and informative to one surgeon can be overwhelming or intrusive to the other. Adaptive systems, which develop models of each surgeon's haptic preferences, either through explicit feedback or through deduced signals such as task performance measures and physiological measures, can be used to adjust the interface to the perceptual attributes and working style of a particular user. Context awareness extends beyond interuser control of haptic feedback by modulating the haptic feedback dynamically in a procedure depending on how the surgical context evolves (Torabi et al., 2021). When the dissection is being done in a sensitive area around a big blood vessel, the system may become more sensitive to force feedback and have a lower rendering threshold to contact events. It may decrease the sensitivity and expand the repertoire of rendered forces during a vigorous suturing exercise. These context-based adaptations demand that the system keep a model of the current state of the surgical procedure, which in turn demands combining visual, kinematic, and force information using approaches that rely on signal processing and machine learning alike.

6.5 Design principles for medical haptic interfaces

6.5.1 Ergonomic design considerations

Medical haptic interfaces do not consider ergonomics in an incidental manner. It is one of the main engineering requirements defining almost all the physical design. Surgeons can be using a haptic interface for many hours at a time during complicated surgeries, and even minor ergonomic deficiencies will become fatigue of the hand, loss of motor control, and higher error rates toward the end of the long cases. The haptic controller grip geometry must be similar to the natural hand position during fine manipulation, which in most human beings is a pinch grip, not a power grip. The center of mass must be placed near to the hand as possible to minimize the torque in the wrist. Mechanical advantage and counterbalancing should be used to reduce the amount of finger force necessary (Fiveable, 2024). In addition to the physical interface, cognitive ergonomics is also important. A haptic display that requires conscious interpretation, making the surgeon actively decipher what the feedback signal entails as opposed to being able to intuitively perceive it as a natural expansion of their own sensory system, increases cognitive load and does not decrease it. The gold standard of haptic interface design is perceptual transparency, the property of the feeling that you are touching the remote environment and not interacting with a machine, and it is necessary to pay attention to the entire signal path from sensing to rendering (Zhang et al., 2025).

6.5.2 Safety and reliability requirements

The safety of medical devices is not a compromise, and haptic interfaces that impose forces actively on clinical users should be based on well-thought-out fail-safe behavior. The failure modes of a haptic device may consist of actuator saturation, sensor dropout, computational overrun, communication failure, and power supply interruption, and the failure modes need to be either avoided by design or dealt with in such a way that the system gets to a safe state. In the case of force feedback devices, failure is normally a passive, back-drivable condition where the device has no active forces on the operator. Sensory redundancy is a common practice in safety-critical systems, and medical haptic interfaces enjoy the advantages of sensor redundancy at a number of levels. Cross-validated dual force sensors, watchdog timeouts to computational processes, hardware current restrictors on actuator drives, and separately-operating safety processors that check system state and have the ability to stop power flow to actuators in response to abnormal state are components of a solid safety architecture (IxDF - Interaction Design Foundation, 2016). Reporting of failure mode and effects analysis (FMEA) required by medical device regulatory authorities compels design teams to think systematically through failure modes and their clinical implications, which is also a useful design discipline.

6.5.3 User-centered design approaches

Human factors and user-centered design approaches have become part and parcel of the medical device development process that has been progressively embraced by the medical community over the last 20 years. These approaches, in which clinicians, patients, and other final users are engaged in the entire design procedure (ronroll, 2011), as opposed to merely at the validation phase, have continued to yield safer, easier-to-use, and more acceptable devices that are modified to be used in clinical practice. In the case of haptic interfaces, a user-centered design is especially significant since the experience of the device in the sense of perception is subjective in the first place and differs in each user to a level that engineering requirements are inadequate to define it. One such area of user-centered design is clinical workflow integration, which is of particular concern to haptic interfaces. A device that interferes with the workflow that can be considered normal in an operating room, has an unusual setup process, and places new coordination demands among members of the surgical team, will be resisted in adoption, whether or not it performs well. The workflow context within which a device will be operated, which ought to be determined through observation studies, task analysis, and early prototype testing, in real simulated environments, should be a normal part of the development process.

6.5.4 Interface evaluation metrics

To assess the use of a haptic interface in medicine, there is a need to have a multidimensional assessment system that is not merely force-accuracy measurements. Technical metrics, particularly precision and responsiveness, are intrinsic to the concept of accuracy in how the device reacts to the forces it ought to facilitate, and with enough bandwidth to qualify it to execute the tasks it is created to support. Sensitivity to different impedance conditions, measured by passivity analysis and transparency indices, is also of significant concern, especially in teleoperated systems where the environmental impedance changes in unpredictable ways. The user satisfaction, which is generally determined with validated questionnaire scales like the system usability scale and the NASA Task Load Index, can give vital data on the clinical acceptability of the device, not available by technical metrics ((Abiri et al., 2019; Heney, 2019). The objective performance measures based on the data of execution of the task, such as time to complete, maximum applied force, force integral, and error rate, are useful in cases when they can be compared with the reference standards or baseline conditions in a meaningful way. Setting proper baseline criteria of comparative evaluation is a methodological difficulty in itself, as the corresponding comparisons can

Table 6.2 TabDesign criteria and evaluation metrics for medical haptic interface development.

Design dimension	Key criteria	Evaluation metric	Clinical relevance	Benchmark standard
Precision and fidelity	Accuracy of force reproduction, spatial resolution	Mean force error (N), Weber fraction	Tissue differentiation, palpation	< 0.5 N force error
Latency and responsiveness	Signal delay from sensor to actuator	End-to-end latency (ms), update rate (Hz)	Teleoperated and real-time surgery	< 10 ms for real-time; < 1 ms ideal
Safety and reliability	Fail-safe behavior, redundancy, sterilizability	MTBF, failure mode analysis, ISO compliance	Patient and clinician safety	ISO 13485/IEC 60601 standards
Ergonomics and comfort	Weight, grip, wrist fatigue, adjustability	NASA-TLX workload score, grip force (N)	Prolonged surgical procedures	NASA-TLX < 50; grip < 15 N
Stability and control	Passivity, Z-width, control loop robustness	Ziegler-Nichols stability, transparency index	Safe force display without oscillation	Passivity margin > 0; no limit cycles
User satisfaction	Intuitiveness, learning curve, perceived realism	SUS score, task completion rate, Likert scale	Training effectiveness, clinical adoption	SUS > 80 (excellent usability)
Regulatory compliance	Risk classification, documentation, testing	FDA 510(k), CE marking, MDR audit trail	Market approval, liability management	FDA Class II/III; MDR Annex IX

be between manual performance, between robotic performance without haptic feedback, or between other haptic interface designs, and each comparison addresses a different question (Table 6.2).

6.6 Applications of haptic interfaces in medicine

6.6.1 Robotic-assisted surgery

The operating room is the most critical place where a haptic interface can be applied, and it is here that the engineering need is the most severe and where the overall clinical benefit could be most significant. Haptics plays a safety and a performance role in minimally invasive robotic surgery. Regarding safety, a restoration of tissue force perception lowers the rate of accidental tissue damage, which continues to be a contributing factor to surgical complications in robotic operations. Haptic feedback can be used on the performance side to allow operations of an objectively challenging or impossible nature, such as vascular anastomosis (Bethea et al., 2004), where suture tension must be carefully regulated, and blunt dissection, where a plane boundary must be determined manually. The remote and teleoperated surgery procedure, such as the newly developing surgical telemedicine, is another challenge due to the physical distance between the surgeon and the patient, which creates communication latency and cybersecurity that domestic robotic surgery lacks. Haptic-driven telesurgery systems have been shown to work over ranges of thousands of kilometers in research settings, although the clinical implementation of these systems needs regulatory provisions and communication infrastructure, only now developing.

6.6.2 Medical training and simulation

Haptic feedback in surgery simulation has become one of the most successful uses of haptic technology in the medical field, and an increasing body of evidence has documented the transfer of skills learned in simulation to actual performance during an operation. Laparoscopic surgery, endoscopy, regional anesthesia, and interventional cardiology simulation virtual reality simulators with haptic feedback enable learners to practice over and over again without harming a patient, get quantitative feedback on their performance, and, more importantly, learn the haptic intuition they would need years of clinical practice to develop in real clinical practice (Team DHR-ICMR AI Cell, 2023). The skill evaluation platforms that are based on haptic and kinematic information on simulation give a new aspect of training evaluation. Instead of depending solely on the judgment of the supervisor, such systems may be used to numerically measure the quality of the procedural movements, monitor the performance patterns within the training programs, and detect particular technical flaws that the targeted instruction could be used to overcome. The information produced by haptic simulators also opens up the way to research in learning analytics that can potentially radically alter the conceptualization and presentation of surgical training programs.

6.6.3 Rehabilitation and assistive technologies

Haptic feedback plays a critical role in rehabilitation medicine, motor learning, and neural plasticity. Motor control and learning are facilitated by haptic devices, which provide mechanistic support for adaptive changes in the nervous system. Through sensorimotor coupling, the system acquires movement patterns by associating motor commands with the sensory consequences they generate. This continual pairing strengthens neural pathways, promoting both skill acquisition and plasticity. Haptic feedback provided by rehabilitation robots is the exact kind of consequential sensory information. Stroke rehabilitation studies have shown that robots that give error-augmenting haptic feedback, that is, reinforcing other errors, can lead to better motor learning than learning with assistance that gives error-correcting haptic feedback. The technology of prosthetic hands has significantly improved over recent years, and haptic feedback is becoming the most important technology in the control of natural prosthetic hands. In the lack of tactile information, the user of a prosthetic hand has to depend purely on visual feedback of the grip force, cognitively straining and incapable of functioning in visually impaired conditions. Sensory substitution strategies, whereby the sensors on the prosthetic fingertip are fed into the intact portions of the skin on a residual limb or other body parts, have demonstrated the ability to provide more natural, attention-free use of prosthetics, especially in fine manipulation tasks.

6.7 Governance, ethics, and regulatory considerations

6.7.1 Ethical implications of artificial intelligence-enhanced haptics

Application of artificial intelligence (AI) to haptic interfaces in medical applications also has some ethical implications that are not comprehensively covered in the current medical device paradigms. When an autonomous haptic system makes autonomous decisions about the presentation of feedback to a surgeon, the way such decisions are made, who is responsible in case of errors such decisions lead to, and how system-generated decisions can be comprehended and challenged by the clinician who depends on them are all literally open questions. The concept of clinical accountability, which posits that a qualified clinician should be capable of explaining and justifying each clinical ruling, is awkward with the AI systems, the decision-making of which can not be easily interpreted. The notion of transparency in AI-enhanced haptic systems entails a minimum of having the operating parameters of the system, provenance of the training data, common modes of failure, and performance limits reviewed and made available to clinical users and regulatory authorities (Singh & Keche, 2025). In addition to documentation, meaningful human control of AI-powered clinical decision support continues to be demanded, which in the haptic context entails the ability to make surgeons aware of it, modify it, and override its adaptive behavior as opposed to being an inactive receiver of AI-mediated sensory data (Abujaber & Nashwan, 2024).

6.7.2 Data privacy and security

Haptic interfaces within the medical field create rich data regarding the patient and the clinician. The profiles of force-displacement throughout surgery processes hold data related to tissues that are potentially deemed as classified health information when taken alongside other clinical records. The credentialing, professional assessment, and insurance have clear implications on surgeon performance data obtained through haptic systems during training or practice. The management of these data streams, such as the ownership, storage, access, secondary use, and international transfer, is an issue that should be given due care by the developers of the device, the healthcare institutions, as well as regulators. The related and more urgent concern is cybersecurity. Medical robotic systems do not have immunity against the interception or manipulation of haptic control signals that cross the networks. The impacts of a cyberattack on a working surgical haptic system can be devastating and may include disruptive deterioration of the quality of feedback, as well as the addition of bad force signals. The networked haptic medical device should also have security-by-design features, such as encrypted communication channels, authenticated command interfaces, and anomaly detection on control signals as the norm.

6.7.3 Regulatory frameworks

Medical haptic interfaces are medical equipment, and have to negotiate the regulatory systems that cover such equipment in every jurisdiction where they are sold and utilized. In the United States, the FDA has correlated devices into different classes in terms of risk, and the haptic interfaces used in surgery would be classified as either Class II or Class III, depending on the intended use and the type of clinical risk they pose. Some haptic devices can use the 510(k) clearance pathway, under which substantial equivalence must be shown to a predicate device, but novel AI-enhanced systems with no predicate may need the more rigorous premarket approval. The Medical Device Regulation, enacted in Europe in 2021, which supersedes the Medical Device Directive, contains more stringent clinical evidence, postmarket surveillance, and traceability requirements than the former. In the case of AI-enabled medical devices, both the FDA and the European Commission have published documents guiding the unique conditions of the software-based devices whose behavior can vary over time as they react to new data, which regulators call adaptive or learning algorithms. One of the key product development issues in this area is maneuvering these requirements in a way that does not compromise the adaptability that has enabled AI-enhanced haptic systems to be of clinical value.

6.8 Technical challenges and research gaps

Even though the medical haptic interface design has made considerable advances, there still exist a set of technical problems that persistently hinder clinical integration as well as restrict the performance of the systems. Latency and stability have become the most fundamental constraints: stability limits are imposed by the physics of force feedback control, which are increasingly more difficult to meet with higher communication delay, environmental impedance is diverse, and the dynamics of the haptic device itself are often an interacting component in the control loop. Advancements in this direction need improvement of both the control theory and the description of the environments where medical haptic systems are used. The complexity of the AI algorithms that can be integrated into the haptic control loop is limited by the real-time processing constraints. Models that take hundreds of milliseconds to generate outputs cannot be implemented directly in a 1 kHz force control loop, and in any case, more rapid models need to be written, or predictive control strategies need to be implemented, where the AI operation is carried out outside the loop instead of inside it. The research priority is underserved because it remains underdeveloped to develop efficient neural network systems that are particularly designed to fulfill the latency constraints of haptic control.

The haptic interface community is mostly not standardized, and that hinders the possibility of researchers and developers exchanging findings across platforms, reusing software, and building upon the efforts of others. This is due to the absence of standardized tissue phantom materials for haptic testing, standardized performance measurements of medical haptic interfaces, and standardization of communication between haptic devices and robots, all of which have a disjointed ecosystem, slowing scientific development and raising development costs. Coordinated actions of professional societies, standards bodies, and regulatory bodies in making similar frameworks would benefit immensely. Scalability and cost of barriers are also major barriers to clinical adoption. High-performance force feedback components such as precision force sensors, backdrivable actuators, and low-latency computational hardware are costly, and system integration issues of integrating the components with sterilized packaging and regulatory-compliant software are cost-additive. Methods based on commodity hardware, simple sensor modalities, or software-defined haptic rendering to enable lowering the cost-performance trade-off are also significant areas of research to open haptic-enabled medical robotics access.

6.9 Future directions and emerging trends

A closer fusion of multimodal sensory fusion, smarter adaptive control, and stronger correlation with digital representations of the surgical world is likely to typify the near-term future of medical haptic interfaces. Multimodal sensory fusion (the combination of tactile, force, proprioceptive, visual, and thermal information into a set of perceptual representations) is like how the human brain intuitively integrates sensory channels to allow the manipulation of objects dexterously but offers haptic experiences, which are qualitatively richer than even the best single-modal method can provide. Studies of the cross-modal interaction of perceptions, such as the effect that visual information has on the perception of a haptic input and vice versa, are yielding clues that will be used to develop more coherent multimodal interfaces.

Surgical simulator digital twins are a revolutionary opportunity. A haptic rendering system based on a computational substrate such as a digital twin of the anatomy of a particular patient, built by preoperative imaging, and updated intraoperatively by registration to real-time imaging, may be able to provide an anatomically accurate force presentation even in those parts of the body that have not yet been contacted by the instrument tip in use. This predictive nature, which is based on an ever-refined model of tissue properties and anatomical structure of the particular patient, has the potential to change the informational quality of haptic feedback to be anticipatory rather than reactive. AI-enhanced telemedicine is also expanding access to specialist surgical services not only to underserved communities and remote areas, but also haptic-enabled telesurgery represents one of the enabling technologies to fulfill this vision. The technical and regulatory issues are very significant, yet the trends of communication technology, latency-resistant control algorithms, and miniaturized haptic hardware indicate that dependable telesurgical haptic feedback across intercontinental ranges is within the domain of possibility within the next 10 years. The design space is growing quite broad with soft robotics and wearable haptic devices: soft actuators and sensors that can be fitted on complex anatomical surfaces, flexible electronics that may be integrated into surgical gloves, and wearable haptic feedback devices that may be used at home, not just in the clinical setting.

6.10 Case studies and practical implementations

A number of clinical applications of haptic capable robots systems offer useful lessons of both the potential and the real-world issues of medical haptic interface design. One of the first commercially developed robotic surgery systems to provide haptic feedback as a standard feature was the Senhance Surgical System, which was developed by TransEnterix and later acquired by Ascensus Surgical, featuring a proprietary force-sensing and rendering system that could provide surgeons with haptic feedback of the force between instruments and tissue. Surgical trials of the Senhance system have shown that the haptic feedback was useful to surgeons, especially with suture work, where they found better suture positioning and more intuitively responsive force control than with the other robotic systems they had used. Adoption has been constrained by the availability of platforms and training needs, but the system offers a real-world dataset on assessing the clinical effects of haptic feedback in minimally invasive surgery. In rehabilitation robotics, a large-scale example of integrating haptic technology in clinical rehabilitation is the use of the Hocoma Armeo series of upper limb rehabilitation exoskeletons in the stroke rehabilitation programs in Europe and North America. The patient outcome analysis in the context of Armeo-based rehabilitation programs showed a significant motor functional improvement, and it has been demonstrated that the haptic assistance mode and resistance mode of the device provide various patterns of motor learning, which could be exploited strategically by rehabilitation therapists. The workflow and training issues that go hand-in-hand with the implementation of clinical haptic devices are also reflected in the integration of these devices into clinical practice. The lessons learned during these and other deployments can be condensed into several common themes. The clinical user requirements in the development of haptic interfaces are very high and cannot be underestimated during the development of the device. The correlation between objective measures of performance and subjective measures of user acceptance is not always as expected or negative. And the journey between an exciting research prototype and a clinically deployed device is highly engineered, regulatory, and institutional, as it is challenging.

6.11 Discussion

Culminating the strands of this chapter, what is created is a scenario of a field that has accomplished extraordinary technical advances, but now suffers due to recurring difficulties in applying the advances to clinical reality. The merging of AI and haptic technology is truly thrilling, as it allows possibilities, such as predictive force rendering, customized adaptive feedback, and so on, which were simply not possible without the classical methods. However, such abilities are only useful when they can be implemented in a device that is safe, reliable, usable, and more affordable than a routine clinical use, and which meets the regulatory requirements that will safeguard the patient against unproven technology.

The connection between design concepts and clinical requirements is not necessarily simple. Professionals who make devices with optimization toward transparency, stability, and bandwidth might end up giving out technically spectacular yet clinically cumbersome devices. Current technology might prove inadequate to support the expectations of clinicians who are interested in more natural tactile feedback. Patients whose treatment is influenced by said technologies have interests that are not necessarily reflected in either the technical or the clinical discourse. To manage these tensions, the manner of interdisciplinary collaboration that is yet to be routinely exercised in the process of medical device development, such as engineers, clinicians, human factors experts, ethicists, patients, and regulators, must be introduced at the earliest design phases. The moral aspects of the haptic systems with AI improvement should be given more focus than they have usually been given. The next question of accountability, transparency, and meaningful human control that these autonomous decisions of AI involve when it generates sensory information to clinicians introduced to it is not just a matter of academic concern. They also have real-life implications on the design, validation, deployment, and governance of these systems and solutions to these problems, which will provide precedent on the integration of AI in healthcare environments more generally.

6.12 Conclusion

The design of haptic interfaces in the medical sector is at a crossroads. The technical infrastructure has been laid out, the clinical case in support of the usefulness of haptic-feedback continues to develop, and the artificial intelligence tools have been offering new opportunities to adaptable, personalized, and context-aware haptic presentation. The real work of integrating the systems, clinical validation, navigating regulations, and adopting and implementing them at the institutions into a standard clinical practice remains. This chapter has tried to present a topological landscape: starting with the software and hardware basis of haptic sensing, AI-based adaptation, design process, clinical uses, governance issues, and future perspectives. The two identified tables, which are provided below, are useful reference materials to the researchers and developers in this space, summarizing the comparative characteristics of the feedback mode and the multidimensional paradigm of analyzing haptic interface designs. Perhaps the most significant contributions of this chapter include the argument for a more human-centered approach to haptic interface design in the medical domain, the discovery of personalization and context-awareness as the two dimensions of high potential, and the invitation to take a more serious attitude toward the ethical and governance aspects of the AI-enhanced haptic systems. The solution to the latency, standardization, cost, and clinical validation issues highlighted herein will guide future research in the field to systems of haptic interfaces that can indeed be integrated into clinical practice, but not remain technologies of the future exemplified in the laboratory. The robots do not necessarily have to have no trace of the hand of the surgeon, the decades of trained judgment in the sense of touch. With continued investment in research, thoughtful design, and responsible governance, haptic interface technology can restore what robotic surgery has too long suppressed, not just as a technical feature, but as a genuine clinical capability that improves the safety, quality, and human character of medical care.

AI disclosure

During the preparation of this work, the author(s) used Grammarly to check the grammatical mistakes. After using this tool/service, the author(s) reviewed and edited the content as needed and take(s) full responsibility for the content of the publication.

References

Abiri, A., Juo, Y. Y., Tao, A., Askari, S. J., Pensa, J., Bisley, J. W., Dutson, E. P., & Grundfest, W. S. (2019). Artificial palpation in robotic surgery using haptic feedback. *Surgical Endoscopy, 33*, 1252–1259. https://doi.org/10.1007/s00464-018-6405-8.

Abujaber, A. A., & Nashwan, A. J. (2024). Ethical framework for artificial intelligence in healthcare research: A path to integrity. *World Journal of Methodology, 14*(3). https://doi.org/10.5662/wjm.v14.i3.94071.

Baronchelli, F., Zucchella, C., Serrao, M., Intiso, D., & Bartolo, M. (2021). The effect of robotic assisted gait training with lokomat® on balance control after stroke: systematic review and meta-analysis. *Frontiers in Neurology, 12*. https://doi.org/10.3389/fneur.2021.661815.

Batty, T., Ehrampoosh, A., Shirinzadeh, B., Zhong, Y., & Smith, J. (2022). A transparent teleoperated robotic surgical system with predictive haptic feedback and force modelling. *Sensors, 22*(24), 9770. https://doi.org/10.3390/s22249770.

Bethea, B. T., Okamura, A. M., Kitagawa, M., Fitton, T. P., Cattaneo, S. M., Gott, V. L., Baumgartner, W. A., & Yuh, D. D. (2004). Application of haptic feedback to robotic surgery. *Journal of Laparoendoscopic and Advanced Surgical Techniques - Part A, 14*(3), 191–195. https://doi.org/10.1089/1092642041255441.

Fiveable. (2024). 11.1. Design principles for haptic interfaces and systems – Haptic interfaces and telerobotics. Retrieved from https://fiveable.me/haptic-interfaces-and-telerobotics/unit-11/design-principles-haptic-interfaces-systems/study-guide/8KibF6KS0f9ao3AC. Accessed: August 2024.

Heney, P. (2019). Challenges of building haptic feedback for surgical robots. The Robot Report. Retrieved from https://www.therobotreport.com/haptic-feedback-design-challenges-surgical-robots.

IxDF - Interaction Design Foundation. (2016). What are haptic interfaces?. IxDF - Interaction Design Foundation. Retrieved from https://www.interaction-design.org/literature/topics/haptic-interfaces. Accessed: 19 December 2016.

Kamelle, S. (2025). Haptic feedback and machine learning in advanced robotic surgery. Retrieved from https://drscottkamelle.org/robotic-surgery/haptic-feedback-and-machine-learning-in-advanced-robotic-surgery/. Accessed: 20 January 2025.

Li, Y., & Hannaford, B. (2017). Gaussian process regression for sensorless grip force estimation of cable-driven elongated surgical instruments. *IEEE Robotics and Automation Letters, 2*(3), 1312–1319. https://doi.org/10.1109/LRA.2017.2666420.

Ou, Y., & Tavakoli, M. (2023). Sim-to-real surgical robot learning and autonomous planning for internal tissue points manipulation using reinforcement learning. *IEEE Robotics and Automation Letters, 8*(5), 2502–2509. https://doi.org/10.1109/lra.2023.3254860.

Ronroll. Da Vinci surgical system discussion [online forum comment]. 2011. Reddit. Retrieved from https://www.reddit.com/r/AskEngineers/comments/mzcc4/da_vinci_surgical_system/. Accessed: 05 June 2026.

Singh, M. P., & Keche, Y. N. (2025). Ethical integration of artificial intelligence in healthcare: Narrative review of global challenges and strategic solutions. *Cureus, 17*(5). https://doi.org/10.7759/cureus.84804.

Torabi, A., Nazari, A. A., Conrad-Baldwin, E., Zareinia, K., & Tavakoli, M. (2021). Kinematic design of linkage-based haptic interfaces for medical applications: A review. *Progress in Biomedical Engineering, 3*(2). https://doi.org/10.1088/2516-1091/abee66.

Team DHR-ICMR AI Cell. Ethical guidelines for application of artificial intelligence in biomedical research and healthcare. New Delhi: ICMR.

Ueda, Y., Miyahara, S. o, Tokuishi, K., Nakajima, H., Waseda, R., Shiraishi, T., & Sato, T. (2024). First clinical application of a surgical robot with haptic force feedback function for thoracic surgery: A case report. *Chest, 8*, 16. https://doi.org/10.21037/shc-24-6.

Zhang, L., Huang, S., Zuo, J., Gu, S., Xu, L., Zhang, Y., & Jiang, T. (2025). An innovative master haptic interface employing magnetorheological fluids for endovascular catheterization. *Sensors, 25*(24), 7450. https://doi.org/10.3390/s25247450.

SECTION II

Intelligent control and system architecture

7. Haptic feedback control in medical robotics 125
8. Artificial intelligence frameworks for intelligent haptic control 147
9. Design frameworks for multimodal haptic medical robots 169
10. Energy efficiency in haptic medical devices 181

CHAPTER 7

Haptic feedback control in medical robotics

Nasrulla K. Khan, Simna Shanavas and Jitha Janardhanan
Department of Computer Applications, Presidency College (Autonomous), Bangalore, Karnataka, India

7.1 Introduction

The Greek words "haptikos" (the feeling of touch) and "haptesthai" (to touch or establish a connection) are essentially the origins of the term "haptic" (Arnab et al., 2011). Haptics technology is a force/tactile feedback technology growing in disciplines linked to human–computer interaction (Mahmmod et al., 2024). Haptics is the research of how to interact with computer applications through touch (tactile) sensation and control. A haptic device provides users with a tactile sensation in computer-generated settings, making virtual objects that can be touched seem genuine and substantial. Haptics enables a user to touch, sense, manipulate, create, and/or modify simulated three-dimensional objects within a virtual space (Arnab et al., 2011). Haptics is used to train and improve skills in areas like surgery, robotics, and other tasks requiring precision and dexterity. Haptic medical systems are advanced healthcare technologies with mechanisms of tactile review integrated into them to improve various medical uses, including training, surgery, diagnostics, and rehabilitation. Force feedback vibration and tactile response are applied by the systems to provide an extremely realistic simulation of medical procedures, allowing medical professionals to work with virtual or robotic environments in the same manner as they would with real tissue or medical tools. Through the provision of real-time physical feedback, haptic technology allows precision dexterity and confidence to be built among medical professionals and students before they perform procedures on real patients. The technology is especially valuable in robotic surgery, where the surgeon applies AI-assisted haptic feedback to make controlled and accurate movements, thus reducing surgical complications and improving patient outcomes. Pillars of haptic medical systems are sensors, actuators, and artificial intelligence (AI) (Wang et al., 2023), which work together to render a review that is highly flexible and responsive. Sensors measure force, pressure, and motion, and actuators create physical sensations that imitate actual-world interactions with tools and tissue. AI interprets the information in real time, thus enhancing accuracy, flexibility, and the ability to learn. These systems are central in robot-assisted interventions as haptic review allows surgeons to perceive tissue resistance and adjust techniques accordingly. In prosthetics and rehabilitation, AI-driven haptic devices

Haptic Feedback Control in Medical Robotics
DOI: https://doi.org/10.1016/B978-0-443-27777-1.00001-9

help patients recover motor functions by delivering customized and responsive therapy. Telemedicine applications also employ haptic technology to enable remote surgeries and diagnostics, thus widening access to specialist medical care. As AI and haptic systems continue to adapt, they stand to innovate medical education, precision medicine, and patient care, thus improving the efficiency, security, and accessibility of healthcare.

7.2 Fundamentals of haptic technology

Haptics is a new age technology that emulates the sense of touch through the application of forces, vibrations, or motions toward the user. Haptics technology improves the experience with a more realistic and immersive one by producing tactile sensations that mirror real-world interactions (Arnab et al., 2011). Haptics is prevalent in a wide range of new-age devices and applications, both to improve user experience and functionality. For example, smartphones employ haptic feedback to produce tactile feedback while typing on a virtual keyboard, whereas game controllers deliver physical feedback that is aligned with in-game activity, such as producing a vibration when a character is hit. In virtual reality (VR) environments, haptics enhances immersion significantly by enabling users to feel virtual objects, thereby making the experience much more realistic. In the same vein, in medicine, haptic-enabled simulators are a critical tool in training healthcare professionals by mimicking the sense of touching or manipulating human tissue, thereby allowing realistic practice without risk. Haptic feedback is typically divided into three broad categories, each serving a different purpose in enhancing the sensory experience of the user. Tactile feedback addresses the skin and defines sensations that are perceived on the surface level, such as vibrations, pressure, or textures. A good example of this is the slight vibration of a phone alert, which informs the user of an event without the use of visual or auditory warnings (Mallipeddi et al., 2017). Force feedback provides a more intense experience by providing resistance to the movement of the user, thus simulating physical forces. For example, in racing games or driving simulators, the steering wheel can provide a backward force to the user's hands, simulating the resistance felt while driving a real car. Physical feedback addresses the sense of movement of the body and the application of force through muscles and joints. This form of feedback is generally applied in VR training environments where users interact with virtual objects, such as grasping a box or moving a limb, so that they can feel the associated weight or resistance. Together, these forms of review significantly improve realism and increase the user interface on a range of digital platforms.

7.2.1 Components of a haptic system

A haptic system is a sophisticated device that is aimed at improving user interaction with virtual or distant environments through the provision of tactile feedback via a network of

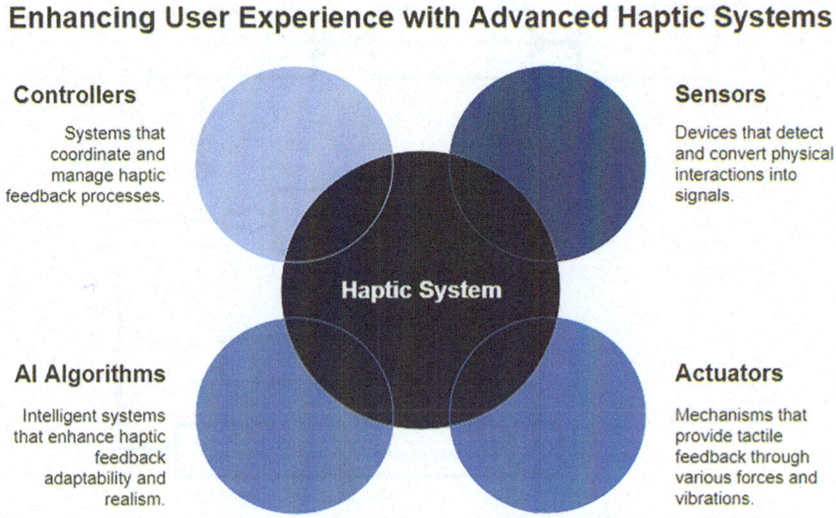

Figure 7.1 Components of the haptic system.

connected components. The fundamental elements of the system are sensors, actuators, AI, and controllers, each of which plays a critical role. Sensors, force motion touch, and temperature sensors detect and transform physical stimuli into signals compatible with processing. The sensors monitor user movement, pressure gesture, or even heat interaction, which is forwarded for feedback processing. Actuators are used to transform processed signals into felt sensations. The process can involve the creation of vibrations by actuators like eccentric rotating mass motors and linear resonant actuators, or the creation of forces and resistances by force feedback actuators. Sophisticated actuators like piezoelectric devices or electroactive polymers also give feedback by providing high-frequency responses or soft, compliant sensations, adding to the reality of the virtual environment. AI algorithms are integrated with the hardware components (Her et al., 2002) to provide the system with a layer of intelligence and flexibility. Machine learning (ML) models predict user goals, and deep learning (DL) enhances the system's ability to identify and respond to touch in real time, making user interaction better. Haptic rendering algorithms simulate real-world forces and textures, giving a more realistic feel by mimicking actual environmental feedback. Predictive algorithms further customize the experience by modifying the feedback in real time based on the user's conduct and context (Khamaj & Ali, 2024). The controllers, including microcontrollers, digital signal processors, and embedded control systems, are the main management units processing the sensor input and managing the interaction between actuators and algorithms. They cause the system to reply in real time with low latency and high performance, giving a seamless and dynamic haptic experience to the user (Figs. 7.1–7.6).

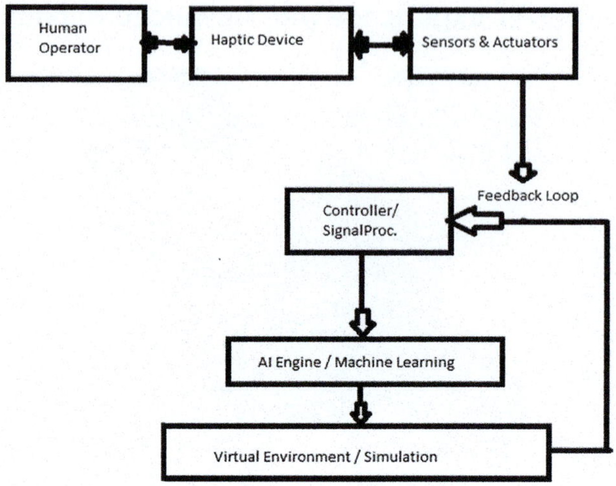

Figure 7.2 Block diagram of haptic system.

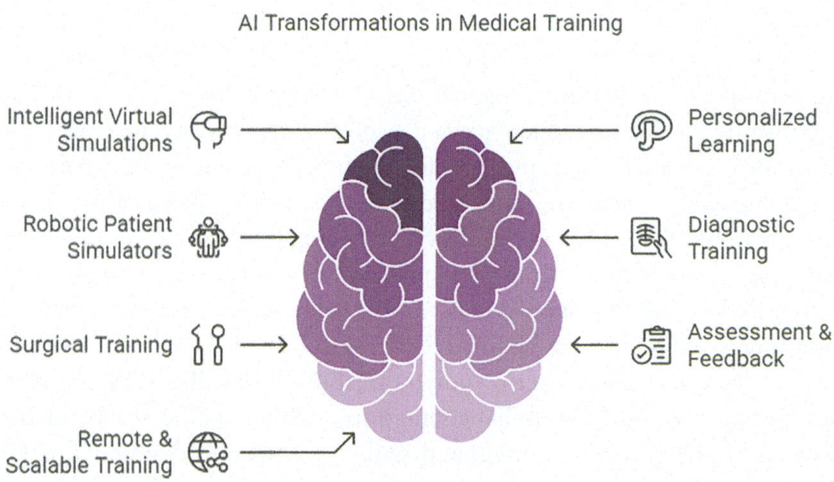

Figure 7.3 AI transformations in medical training.

7.2.2 Role of machine learning and deep learning in improving haptic responses

ML and DL are transforming haptic feedback systems to be more adaptive, real-time, and sensitive to user interactions. The most significant contribution is enhancing the accuracy of tactile feedback. The ML algorithms have the capability to interpret sensor data in real time and enhance the reproduction of the sensation of force, texture, and vibration (Khamaj & Ali, 2024). DL methods, particularly those learned from large touch interaction datasets, greatly enhance haptic rendering quality. AI also makes it possible

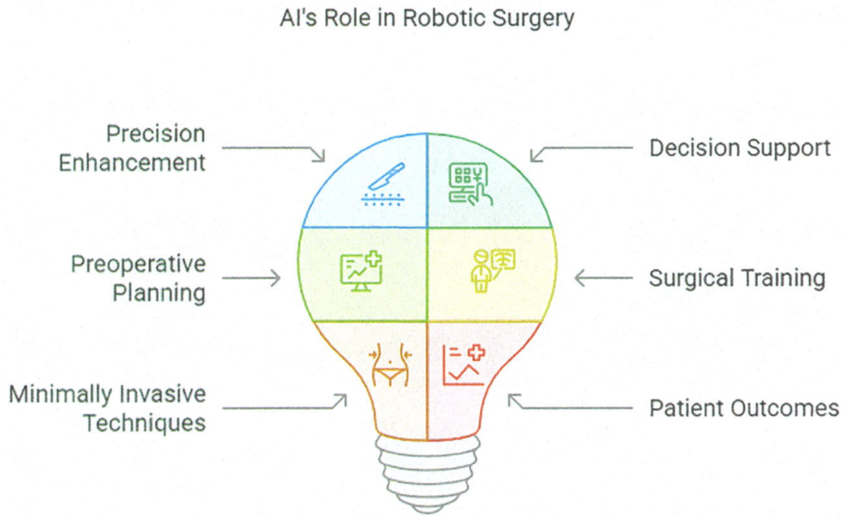

Figure 7.4 AI's role in robotic surgery.

Figure 7.5 Prioritizing challenges and ethical considerations in AI-driven haptic systems.

to customize haptic sensations by controlling feedback with user-specific preferences, history, and context. Adaptive algorithms can adapt the timing, texture, and intensity of vibrations to suit each user's own needs. DL models can reproduce realistic surface and texture by learning from diverse surfaces and materials. Generative models such

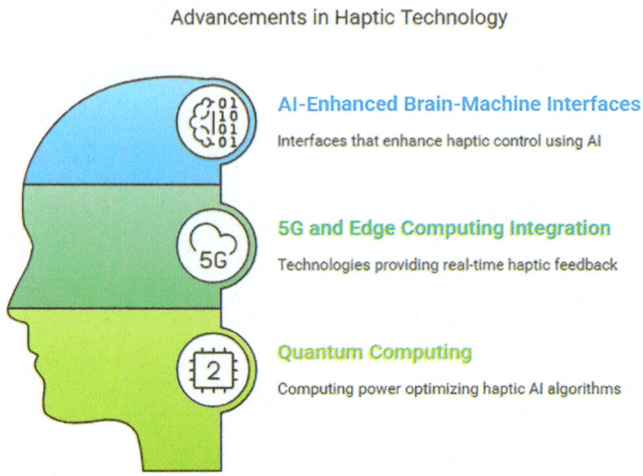

Figure 7.6 Advancements in haptic technology.

as generative adversarial networks can be employed to generate realistic virtual touch sensations. ML also supports predictive and adaptive haptics by anticipating the user's goal and adjusting the review accordingly, which helps reduce latency and improve overall responsiveness. These developments carry over into areas like artificial limbs and robotics, where AI improves grip and tactile sensing in prosthetic limbs. In augmented and VR, gloves and suits with ML capabilities mimic real-world tactile feedback, like temperature and resistance, to produce more realistic and immersive experiences. In medicine, AI enhances the precision of surgical robots and haptic-based training equipment for specialists (Iftikhar et al., 2024). Gaming and user interface experiences are improved as well through AI-driven haptic systems, wherein intelligent vibration patterns increase realism and touchscreen response, delivering a more dynamic and enjoyable user experience.

7.3 Artificial intelligence-driven applications in haptic medical systems

AI-based applications in haptic medical systems are revolutionizing the methods of learning, practicing, and performing medical procedures through the integration of AI with haptic feedback technology. Haptic systems simulate the sensation of touch, and the user can feel resistance, texture, or force like an actual medical procedure. The use of AI in these systems improves their performance significantly by providing intelligent, adaptive, and real-time feedback, thereby making training sessions more realistic and accurate.

Human operator and haptic device: The system is controlled by the user via a haptic interface such as a joystick, glove, or stylus. Sensors and actuators are used to capture movements and forces.

Sensors and actuators: Sensors measure position x(t), velocity v(t), and user-applied force f user(t). Actuators provide force feedback fenv(t) from the system.

Controller signal processor: Decodes sensor input and generates control signals. It is responsible for prompt response and stability in the haptic feedback loop.

AI engine ML: The AI component may learn user behavior, anticipate behavior, dynamically tune force feedback, or mimic more realistic object behavior based on training data. It may, for instance, use reinforcement learning (RL) to learn force feedback or neural networks to learn gestures.

Virtual environment simulation: The interaction with virtual objects is imitated using a model that is based on data or physics. By modifying force models or physical attributes, the AI can also contribute to improved realism.

7.3.1 Artificial intelligence in medical training and simulation

AI is greatly impacting the field of medical education and simulation, thus making the learning experience more realistic, accurate, and learner-centered. One of the most significant developments in this sector is the use of intelligent virtual simulations wherein students and medical professionals can carry out hands-on training in a virtual, risk-free environment (Mirchi et al., 2020). These AI-powered systems simulate the responses of actual patients based on actual physiological information, thus enhancing learners to address scenarios like trauma, cardiac arrest, or emergency surgery. Simulators like Fundamental VR and Osso VR blend virtual and augmented reality with AI to provide optimal realism and instant feedback, permitting trainees' skills and clinical findings to be improved without an actual patient's presence. AI has an important role in providing customized learning. Through monitoring individual performance and real-time analysis, AI systems generate customized learning pathways based on users' strengths and weaknesses. If a learner is having trouble with suturing or interpreting lab findings, the AI can recommend specific lectures and adaptive exercises to develop those skills. AI-powered robotic patient simulators provide realistic reviews, such as breathing, bleeding, or seizures, which are calibrated according to the actions of the learner. This degree of realism enables medical professionals to practice real-life emergencies, allowing them to not just learn technical skills but also critical thinking and decision-making abilities as well. AI supports surgical and diagnostic training by tracking hand movement accuracy and speed during procedures such as laparoscopy or robotic surgery. AI also helps students comprehend radiological images and pathology slides and compare their diagnosis with AI-driven results. In review and assessment, AI can analyze clinical reasoning time and communication skills using natural language processing (NLP), providing detailed feedback beneficial for board exams and certification. Through remote access, AI-based programs enable medical education to be scalable globally, allowing students from remote areas to study in a collaborative style (Ejaz et al., 2022). In the years ahead, AI is likely to incorporate real-time patient data, offer natural language lecturing, and facilitate interdisciplinary training among doctors, nurses, and paramedics, making medical education in the future smarter and more inclusive.

7.3.2 Artificial intelligence in robotic surgery

The use of AI is innovating the field of robotic surgery, making procedures increasingly precise, secure, and effective. Through integration with robotic platforms, AI enhances a surgeon's capacity to start complex operations with greater precision through the analysis of large data sets, instrument guidance, and the ability to learn from previous surgical procedures. The application of AI in robotic surgery reduces human error, enhances patient outcomes, and facilitates the acceptance of minimally invasive techniques, resulting in faster recovery times and shorter hospital stays (Iftikhar et al., 2024). One of the greatest advances attributed to AI is the contribution of real-time support for decision-making. In the case of surgery, AI systems can examine imaging data, such as CT or MRI scans, identify major anatomical structures, and direct robotic devices accordingly (Iftikhar et al., 2024). There are systems with the capacity to anticipate possible complications by examining patient history or intraoperative information. AI is involved in automated suturing, accurate cuts, and tissue type distinction, thereby allowing surgeons to undertake complex procedures wherein accuracy is a requirement. Examples include technologies such as the da Vinci Surgical System, which is being augmented by AI capabilities in order to improve control, forecast movement, and review outcomes.

AI is practically innovating preoperative planning and surgical training processes. Surgeons can now exercise skills on AI platforms that model individual patient anatomy using previous surgery data, along with predictive analytics capabilities. Robotic systems powered by AI can examine vast numbers of prior surgical histories and provide recommendations or insights during actual surgeries to surgeons. Further development in computer vision and ML can possibly enable increasingly independent capabilities in robotic surgery, allowing for an increasingly improved efficiency pace and global access to surgical practice.

7.3.3 Artificial intelligence in telemedicine and remote surgery

AI-powered haptic gloves and tele-controlled surgical robots are converting the practice of surgical procedures through the provision of highly accurate, immersive, and remotely controlled surgeries. AI-powered haptic gloves offer surgeons realistic haptic feedback, allowing them to feel the texture, pressure, and resistance of tissues during virtual or remote surgery (Ejaz et al., 2022). Combined with AI-powered robotic surgical systems, the gloves provide unimpeded remote control of robotic limbs with the AI system translating hand motion, compensating for tremors, and offering precise response of instruments. This combination of technologies enables surgeons to perform highly complex surgical procedures with greater accuracy and less risk, even from thousands of miles away, thus making advanced surgical care more accessible and reliable in remote or unserved areas.

One of the key barriers in the area of remote robotic surgery is latency, or the time between a surgeon's input and the outcome of the robotic device. Delays of just milliseconds can have a significant effect on surgical precision, particularly during complicated surgeries that demand swift, precise control. Latency may be due to delays in network transmission, signal processing time, or hardware capabilities, and is especially demanding when dealing with remote telesurgery, where the possibility of high-speed, uninterrupted internet connections cannot be assured. Another challenge involves providing real-time force feedback or haptic feedback. Surgeons must feel the resistance, tension, or texture of tissues during robotic surgery, which enables them to use the right amount of pressure during actions like cutting, suturing, or grasping. The remote simulation of these sensations and their real-time provision through haptic gloves or interfaces is a grand technical challenge. This requires high-fidelity sensors, fast data exchange, and highly advanced AI algorithms to accurately interpret and simulate force sensations. Any inaccuracy or latency in such review could lead to undue or inadequate application of force, compromising tissue integrity or causing surgical errors. Beating these challenges requires continuous innovation in AI predictive models' sensor technology and low-latency communication networks, such as 5G and next-generation internet infrastructure. AI is critical in the mitigation of signal delay experienced during remote surgery through the application of predictive approaches and real-time analysis of data to enable seamless and precise movement of robotic instruments. The principal technique applied is motion prediction, where AI forecasts the subsequent move of the surgeon based on past patterns of input and hand movements. AI provides predictions in advance, a few milliseconds to allow the robotic system to make the movements in real time, in spite of possible minor delays within the network. This develops an illusion of constant control and thereby reduces the possibility of delay that may influence the accuracy of surgery. AI can also eliminate micro latency jitters and provide robotic arm movement stabilization, especially in complex procedures like cutting or suturing tissue. ML algorithms implemented based on vast databases of surgical motion datasets have the capability to identify and correct deviations caused by delayed inputs, thus giving smooth transitions and avoiding the likelihood of errors. In advanced systems, AI can give priority to critical commands, thus enhancing data transfer and reducing the effects of high-latency environments. AI considerably improves the reliability and safety of telesurgery through effectively solving the errors related to signal delay, thus enabling surgeons to be in control and confident even at long distances.

A robotic surgical system is a cutting-edge technological innovation intended to help surgeons perform minimally invasive procedures with enhanced precision, flexibility, and control. Robotic surgical systems are usually made up of robotic arms, a surgeon's console, and a series of specialty instruments and imaging systems. The main features of robotic surgical systems are

- Surgeon console: The surgeon sits in a console, most often in the same room as the patient, and manipulates the robotic arms and instruments. The console provides high control over the instruments, often with superior 3D visualization magnification and natural motion.
- Robotic arms: These robots have the capacity to manipulate surgical tools with precise accuracy. The surgeon controls the arms through a console, which allows for the performance of precise movements that human hands would not be able to carry out on their own. The arms also incorporate cameras that provide a real-time view of the operating area.
- Endoscopic camera: This camera shows high-definition 3D images of the operating area, usually in real time, allowing the surgeon to have a far more precise view of internal organs and structures than is possible with older techniques.
- Accuracy and minimal invasion: Robotic platforms can execute severe, minute cuts, thus avoiding damage to the surrounding tissues and generally resulting in smaller recovery periods, minimal blood loss, and smaller scarring. Greater accuracy also results in better success rates for surgeries and allows more complex procedures to be executed safely.

7.3.4 High-end robotic surgery systems

The da Vinci Surgical System is one of the best-known systems for robotic surgery and is used for the performance of a variety of procedures, including urologic, gynecologic, and cardiothoracic surgery.

MAKO: Orthopedic robot surgery with a focus on joint replacement, in which the robot delivers high accuracy for preoperative and surgical processes.

Robot operating system (ROS): It is implemented in the area of research for a more programmatic and more flexible approach in robotic surgical practice; nevertheless, its clinical applications are limited by now.

7.3.5 Advantages of robotic surgery

- Higher precision: Robotic systems offer more precise movement and precision than human hands, which can be priceless in delicate procedures.
- Minimally invasive: Smaller cuts reduce body trauma, which leads to less pain, quicker recovery, and less hospital time.
- Increased visualization: High-definition 3D cameras provide the surgeon with a better and magnified view of the field of operation.
- Less human errors: Robots can help in executing repetitive work with precision, which can lower the risk of error.

In addition to these benefits, robotic surgery can be costly and necessitate specialized training for the surgeons and operating room personnel. Furthermore, the procedures depend on sophisticated technology, which can sometimes be challenging when it comes to maintenance and troubleshooting.

7.3.6 Artificial intelligence in prosthetics and rehabilitation

Haptic AI prosthetic limbs are redefining the assistive technology sector by enabling amputees to make more natural, intuitive, and responsive movements. Prosthetic limbs use AI to interpret the neural or muscular signals from the user's body and then interpret them into smooth, natural movements (Nayak & Das, 2020). ML algorithms continuously monitor the user's movement patterns and configure the functionality of the limb to the individual's gait, grip force, or activity level. As a result, the prosthetic device becomes an extension of the body and not a mechanical device. Haptic feedback integrated systems take this technology forward by allowing users to feel touch from the external world. Sensors in prosthetic limbs or fingers sense such attributes as pressure, texture, and temperature and feed the signals into the nervous system through vibrations or direct neural interfaces. AI converts these sensory inputs in real time and offers the users timely and precise feedback, which is essential for tasks such as handling delicate objects or traversing uneven ground. These prosthetics not only restore lost mobility but also significantly enhance the quality of life, autonomy, and confidence of the users by allowing human goals and robot responses to drift smoothly into one another. AI-powered smart rehabilitation systems are redefining patient recovery from injury, surgery, or neurological disorders by giving customized adaptive therapy based on each patient's progress and requirements. Sensors, motion tracking sensors, and wearable technologies are used to track a patient's movement, muscle activity, and biofeedback in real time. AI algorithms interpret this information to know how the patient is reacting to therapy, whether fatigue, compensatory movements, or mobility enhancements, and automatically adapt the difficulty intensity or exercise type accordingly.

For example, when a recovering stroke patient is having problems with specific arm movements, the system can reduce resistance, offer guided support, or fall back to a simpler version of the exercise. As the patient recovers, the AI increases the challenge to support further recovery, increasing participation and continued progress. Some systems also include game-based learning and virtual environment elements, engaging the patient more while also tracking detailed rehabilitation metrics. These sophisticated feedback-driven systems not only improve rehabilitation outcomes but also reduce therapist burdens by offering remote monitoring and real-time adaptation, making therapy more reachable and effective for home and clinic patients. AI is transforming the healthcare industry by taking a central role in predicting patient trajectories and outcomes. AI technologies have the ability to interpret large amounts of data, such as electronic health records, laboratory tests, imaging tests, and wearable device data in a bid to find patterns and coming health threats. ML algorithms possess the capability to detect the earliest signs of disease progression, predict hospital readmission probabilities, and estimate recovery intervals. This type of information allows healthcare practitioners to intervene early, update treatment approaches in advance, and optimize resource utilization. When dealing with chronic diseases, AI can continuously track a patient's vital signs and notify healthcare workers if medical conditions start deteriorating, thus facilitating faster and more accurate interventions. AI facilitates highly customized therapeutic interventions by matching treatment plans to the unique condition and response patterns of a single patient. Instead of relying on general advice, AI can customize therapies by considering an individual's genetic profile, medical history, lifestyle, and past reviews of treatment. For physical rehabilitation, for instance, AI can individualize the intensity of exercise and create routines from real-time feedback gathered from motion sensors and progress reports (Nayak & Das, 2020). For mental health, AI-driven chatbots and mobile applications can individualize the topic and timing of therapy based on a user's sentimental response and behavior. This level of personalization leads to improved treatment efficacy, greater patient connection, and overall higher satisfaction throughout the healthcare process.

7.3.7 Artificial intelligence in diagnosis and tactile sensing

AI-based haptic sensors are revolutionizing diagnostic techniques by enhancing extremely sensitive and noninvasive detection of abnormalities such as tumors and muscle rigidity. The sensors simulate the human sense organ of touch, but with much greater accuracy, using advanced tactile feedback mechanisms to detect variations in tissue texture, firmness level, and resistance (Xu et al., 2025). In combination with AI, the sensors can interpret data obtained through physical examination to identify patterns that indicate the presence of tumors or abnormal tissue density. Compared to manual examination, which is highly reliant on a physician's judgment, AI-based haptic sensors provide quantitative and valid outputs, thus improving the accuracy of early detection and diagnosis of tumors. In cancer diagnosis, these intelligent sensors can be embedded in robotic surgery instruments or portable diagnostic instruments. As the sensor scans a focused area of the body, it is collecting tactile data such as force deformation and elasticity (Won et al., 2021). The AI software interprets this data against huge databases of established tumor properties to distinguish between cancerous and benign tumors. This real-time assessment not only assists earlier diagnosis but also allows surgeons to detect and classify tumors intraoperatively, thus making cancerous tissue excision safer and more precise. In muscle stiffness testing, AI-driven haptic systems play an important role in assessing neuromuscular patients with spasticity secondary to a stroke or patients with chronic pain syndromes. These sophisticated sensors measure muscle responses to force, thus defining areas of rigidity or abnormal resistance. The AI system continuously processes patient information, making ever-refined decisions to monitor improvement and decline over time. This feature allows clinicians and therapists to tailor treatment protocols, monitor rehabilitation, and modulate therapy severity. AI-driven haptic sensors represent a significant advance in customized data-driven medical care, combining diagnostic accuracy with a more participatory and responsive clinical setting. AI-based robot palpation systems are reinventing clinical practice by conducting a precise, objective, and reproducible physical examination in identifying early signs of disease. Robot palpation systems consist of robotic arms mounted with highly sensitive tactile sensors that simulate human touch to probe tissue within the human body. Because robotic systems do not have the reliance on the clinician's skill and the latter's individual judgment, they provide quantitative information in the form of pressure deformation and stiffness of tissues. AI software then processes all this information on a real-time basis to detect abnormalities like lumps, nodules, or stiffening of tissues as indications of potentially critical conditions like tumors, cysts, and fibrotic changes.

One of the major benefits of AI in such systems is that it can precisely differentiate between normal and abnormal patterns of tissue. AI can detect subtle variations in tissue responses that are likely to be overlooked by humans. For instance, in breast cancer screening, AI-assisted robotic palpation can detect subtle or deeply seated tumors in the early stages of growth, even in dense breast tissue where conventional techniques such as manual examination or imaging may be unsuccessful. Uniform and repetitive measurements are another benefit of such systems, thereby reducing the risk of missed diagnoses due to human fatigue or error. Aside from cancer detection, robotic palpation systems also find use in the diagnosis of musculoskeletal disease, gastrointestinal disease, and liver fibrosis, where changes in tissue stiffness are early signs of disease. The AI adapts palpation force and technique in real time according to patient anatomy and feedback, both for safety and comfort (Ribeiro et al., 2017). Robotic palpation systems also enable remote and automatic screening, broadening access to early diagnosis in remote or unserved communities where specialist clinicians are not available. By taking advantage of the sensitivity of advanced haptic sensors and the intelligence of AI, robotic palpation systems are enabling more accurate, earlier, and more accessible detection of disease.

7.4 AI techniques used in haptic medical systems

ML has been extensively implemented in haptic medical systems to identify patterns from information gathered during surgery simulations or robot-assisted operations. These systems gather data through sensors like force, position, and movement data, and ML algorithms analyze the same to discern patterns in users' behavior (Minopoulos et al., 2023). For instance, ML can tell an expert from a novice based on how these users manipulate the surgical tools or press down on instruments. This enables the system to give immediate feedback and direct best practice guidance to users and aid in the analysis and improvement of their performance over time. DL, a strong branch of ML, enhances the capability of haptic systems to learn and react sensibly. With the help of neural networks like convolutional neural networks and recurrent neural networks, DL models are able to handle large and complex data like 3D medical images or dynamic touch feedback. DL models assist in creating realistic simulations of tissues, organs, and surgical environments through simulating their texture density and resistance. Such realism not only enhances training quality but also helps in the creation of robotic systems that are capable of conducting complex surgeries with higher accuracy and flexibility. AI-driven adaptation methods enable haptic systems to learn from the user's action (Minopoulos et al., 2023) and progress and adapt feedback accordingly. As the system learns through repeated use, it can tailor training scenarios with increasing difficulty or provide focused guidance where necessary. This customized approach ensures that users are trained at their skill level and learning speed. During robotic surgery, such an adaptation assists the system in adjusting its control processes in real time and thus renders the procedure more efficient and safer. DL and ML impart intelligence and flexibility to haptic medical systems, immensely developing their educational as well as clinical applications.

Computer vision is also very important in improving real-time haptic feedback for robot-assisted surgery by allowing precise visual examination of the surgical site. With high-definition cameras and sophisticated image processing techniques, computer vision systems can identify and monitor surgical tools, tissue motion, and alterations in anatomy in real time. This visual data is then used to determine the proper haptic feedback, such as resistance, texture, or force feedback, which is returned to the surgeon's console. This enables the surgeon to "feel" the tissue interaction even when out of physical contact with the patient, making it more accurate and less likely to cause damage to sensitive structures. Computer vision also facilitates the development of dynamic 3D models of the surgical field so that the system can mimic tissue or organ deformation when tools are interacting with them. The system, using visual information together with AI algorithms, can then update the haptic feedback in real time as a function of the visual processing of tissue dynamics, pressure locations, and tool location. This integration of vision and touch is a more realistic and accurate surgery experience, particularly in minimally invasive or robotic surgery, in which direct tactile feedback is minimal. Enhanced reality by computer vision increases the realism and safety of robotic surgery through smarter and more sensitive real-time haptic feedback. NLP is being used more by AI-driven training systems in the healthcare industry to give more dynamic and interesting learning environments. Through the use of spoken or written language, users can interact naturally with haptic medical training simulators thanks to NLP. Real-time feedback or clarification requests can be made by trainees. The NLP-powered system can understand the context of the questions and provide relevant responses to trainers and automated systems, especially in situations requiring individualized coaching or distant learning settings (Zhou et al., 2024). NLP is also necessary for clinical diagnosis to manage large volumes of unstructured medical data, such as radiology reports, laboratory results, clinical narratives, and patient histories. AI systems with NLP capabilities can extract important information, spot trends, and suggest possible diagnoses or treatment plans based on the data. This speeds up the diagnosis process and reduces mistakes brought on by attention to detail or information overload. NLP can help healthcare professionals by highlighting important information, offering summaries, and even helping with decision-making in challenging situations (Zhou et al., 2024). The accuracy and efficacy of AI-supported training and diagnosis systems in modern healthcare are thereby increased by NLP.

A powerful AI technique called RL can assist haptic systems in automatically improving their responses through trial and error. Through repeated interactions, RL allows haptic medical systems to learn how to effectively give tactile feedback. The haptic system is rewarded or penalized based on how realistic, precise, or useful the haptic sensations it provides are. Finer and more effective haptic feedback is the consequence of the system's behavior being optimized over time to increase favorable outputs, such as a tighter correspondence to a surgeon's expectations and a more precise simulation of tissue textures. In robotic-assisted surgery and surgical training simulators, where real-time corrections are crucial, this continuous learning mechanism is especially useful. When the user performs surgery or manipulates virtual tissue, the RL algorithm detects how well the haptic feedback matches the user's actions and makes the necessary adjustments. The system rewards responses if it discovers that certain feedback improves accuracy or reduces errors. Constant adaptation ensures that the haptic system is smarter and more responsive over time, which improves the realism of the user experience and ultimately improves therapeutic results.

7.5 Challenges and ethical considerations

Haptic feedback systems powered by AI are important for precision in remote procedures, robot-assisted surgery, and medical simulations. Keeping accuracy and dependability is still the most difficult task. Inaccurate haptic feedback can have serious consequences, especially when it comes to diagnosis or surgery. To imitate real-world human perception, the AI models must be updated often with high-quality data, which is a complex and time-consuming process. Reliability is a top priority for engineers and medical practitioners because any delay, error, or breakdown in the review process would compromise patient safety. Data security and privacy are also major issues. These systems' learning and decision-making processes mostly rely on patient data. Unauthorized access and security breaches must be prevented for sensitive health data. Cyberattacks can target AI systems, and inappropriate data processing might result in patient confidentiality violations. Regulations about health data, like the General Data Protection Regulation (GDPR) and the Health Insurance Portability and Accountability Act (HIPAA), must be in effect. To preserve public trust, developers and healthcare institutions must guarantee robust encryption, frequent audits, and publicly accessible data management rules. Ethics are also raised by the availability and cost of AI-driven haptic devices, particularly in undeveloped countries or medical facilities with inadequate infrastructure. Their availability may be restricted by high development and maintenance costs, resulting in unequal access to advanced medical care. Furthermore, ethical concerns are brought up in remote treatments and robotic surgery, where AI plays a role in decision-making. Issues of patient autonomy, informed consent, and accountability become complex. In high-stakes healthcare settings, it's critical to strike a balance between automation and human intervention so that these advanced tools complement human judgment rather than replace it. The application of AI in haptic medical systems is confronted with a range of challenges, primarily because of the intricacies of combining sensory feedback with ML systems. Haptic feedback designed to mimic the sense of touch needs the close synchronization of the hardware components with AI algorithms to accurately simulate tactile sensations present in the real world. One of the basic challenges is the inconsistency of human touch and perception. Various patients or medical practitioners will interpret tactile feedback differently based on variations in skin sensitivity, pressure tolerance, or overall physical condition. Another challenge is the development of AI systems that can adjust to these variations in real time without introducing delays or inaccuracies. This requirement entails the development of large training sets that are diverse and representative of the infinite range of situations a haptic system can be exposed to, which can be resource-intensive to develop.

Another limitation comes from the hardware limitations inherent in haptic technology. AI algorithms need high-fidelity sensory feedback from haptic devices; however, the technology generally cannot provide sufficient resolution or sensitivity in tactile feedback. Such insufficiency could further result in poor simulation of subtle forces or textures that play pivotal roles in particular medical procedures, especially those related to minimally invasive surgery. Besides, haptic devices are generally large, expensive, or lacking in delicacy for fine procedures. Placing AI in such systems needs overcoming such hardware limitations while being safe and reliable for medical applications. The requirement for such systems to work well in realistic clinical settings, where they must aid a range of tasks, poses a significant challenge.

7.5.1 Privacy and security for healthcare providers

7.5.1.1 Health Insurance Portability and Accountability Act (HIPAA)

The Health Insurance Portability and Accountability Act (HIPAA) has established various ethical and legal responsibilities for healthcare providers to protect patients' privacy and confidentiality. The digitization of healthcare has significantly benefited both medical professionals and patients; however, digital services in the sector have also increased the range of vulnerabilities.

HIPAA primarily emphasizes the privacy and security of health data in the U.S. It applies to organizations such as healthcare providers, insurance firms, and their business partners, as well as third-party vendors that manage protected health information (PHI) (Lee et al., 2023). For haptic medical systems that may include the gathering and relay of sensitive health information (such as patient vitals or therapy details), adherence to HIPAA is crucial when

- Health information is being gathered, saved, managed, or communicated.
- The system is utilized within a healthcare environment or in relation to U.S. healthcare suppliers

Key privacy and security measures under HIPAA that would apply to haptic medical systems:

Privacy rule

- Patient consent: Patients are required to give clear consent for their health information to be gathered, distributed, and utilized. This encompasses information from haptic medical systems.
- Use and disclosure: PHI must be utilized solely for permitted reasons, like delivering care or for operational activities, with limited sharing.

Security rule

- Access control: Haptic systems need to guarantee that only permitted users can reach confidential health information.
- Data encryption: All health data that is stored or sent must be encrypted to guard against unauthorized access, particularly when transmitted via networks.
- Audit trails: Every action related to patient data (e.g., access, modification) must be recorded for accountability.
- Data integrity: The system must guarantee that the data remains accurate and protected against unauthorized changes.
- Breach notification: If a data breach occurs that involves health information, HIPAA requires prompt notification to impacted individuals and the U.S. Department of Health and Human Services (HHS).

7.5.2 General Data Protection Regulation (GDPR)

The General Data Protection Regulation (GDPR) is a significant regulation that outlines the right of individuals to access their personal data. Numerous cryptographic methods and centralized systems have been suggested to aid in data sharing (Lee et al., 2023). A blockchain-oriented approach can address numerous issues related to data sharing. Blockchain-based platforms help to address the key challenges of data sharing and traceability with transparency of the data transactions.

GDPR: EU regulation for data protection and privacy

GDPR is an extensive regulation for privacy and data protection for individuals in the European Union (EU) and the European Economic Area (EEA). It pertains to any entity that handles the personal data of EU citizens, irrespective of the organization's location. This encompasses health information gathered by haptic medical systems

Key privacy and security measures under GDPR that would apply to haptic medical systems:

Data privacy

- *Informed consent*: Haptic systems are required to secure explicit, informed consent from individuals prior to gathering and handling their health information. The agreement needs to be precise, knowledgeable, clear, and offered voluntarily.
- *Data minimization*: The system must gather only the essential data needed to achieve its medical objective.
- *Access right*: Patients can request their health data gathered by the system.
- *Right to deletion:* Under specific circumstances, individuals possess the right to ask for the removal of their data (the "right to be forgotten").
- *Transparency*: Users must be clearly informed by the system regarding how their data will be utilized, stored, and shared.

Data security

- Privacy by design and default: Haptic systems must embed privacy features into their design from the beginning (e.g., encryption, anonymization) and guarantee that privacy is preserved by default (i.e., by reducing data collection).
- *Data protection impact assessments (DPIAs)*: For data processing activities that are considered high risk (like handling health information), it is essential to perform a DPIA to evaluate and reduce risks.
- *Encryption*: Similar to HIPAA, GDPR mandates that health information be encrypted to safeguard it from unauthorized access or exposure.
- *Access management:* Only permitted personnel must access health information, and the system must guarantee appropriate user verification.
- *Cross-border data transfers*: Should the haptic system require data to be sent beyond the EU (e.g., to India or the U.S.), GDPR enforces stringent regulations on cross-border data transfers. The system must guarantee that sufficient safeguards exist, like Standard Contractual Clauses (SCCs), or that the receiving country has sufficient protection established.

7.5.3 How HIPAA and GDPR apply to haptic medical systems

Haptic medical systems gather, analyze, or relay health information to allow the system to engage with patients or healthcare professionals (e.g., tactile responses for treatment or surgical procedures). These systems must adhere to privacy and security protocols to safeguard sensitive health information, and they must guarantee that their use aligns with regulations based on the patients' or users' geographical locations.

Key areas for haptic medical systems:

- *Data gathering*: Prior approval is required from patients before collecting any data. Information from haptic systems, particularly sensitive details such as health conditions or medical backgrounds, needs to be handled with caution.
- *Data transfer*: All health data transmissions (particularly across networks) must be encrypted and safeguarded against unauthorized access, in accordance with HIPAA and GDPR.
- *User privacy*: Both HIPAA and GDPR focus on honoring patient/user privacy entitlements. Haptic systems need to guarantee that patient information is not disclosed without appropriate consent and that users are aware of how their information will be utilized.
- *Data security*: Both regulations highlight the importance of strong security protocols to guard against breaches and unauthorized access, including encryption, user authentication, and audit trails.
- *Cross-border operations*: If the system functions across various jurisdictions (such as U.S. patients engaging with an Indian system), it must adhere to the applicable regulations, ensuring that data protection protocols are established for the involved countries.

7.6 Future directions and innovations

Creating haptic control with AI-driven brain–machine interfaces is among the most intriguing possible applications. BMIs convert neurological impulses into precise commands for haptic devices, enabling more efficient communication between robotics and human brains (Kim et al., 2017). Using AI, these systems have the ability to learn and adapt to the unique neural patterns of each individual and provide more natural and intuitive control. The capability to control surgical robots or prostheses with their minds only would greatly boost rehabilitation and autonomy for individuals with motor impairments. Another cutting-edge technology is combining 5G and edge computing to provide real-time haptic feedback. 5G provides very low latency and high connectivity to provide real-time haptic data transfer, particularly in telemedicine or remote control. By processing nearly at the point of origin, edge computing minimizes latency and reliance on far-off servers. These technologies allow haptic devices to produce realistic and immediate touch sensations across long distances, opening up new possibilities for remote diagnosis, training, and teamwork. Future haptic AI systems could be redefined by quantum computing, which can optimize complex algorithms at previously unheard-of speeds. Large amounts of sensory input must be processed in real time for haptic feedback, and current computer systems are often unable to handle the volume and speed required. Quantum computing has the potential to greatly improve AI models' efficacy by enabling them to more closely mimic human touch and respond more skillfully to unexpected circumstances. By helping to overcome computational limitations, haptic systems may become more intelligent, responsive, and scalable for a variety of medical applications as quantum technology advances.

7.7 Conclusion

In summary, haptic medical devices powered by AI represent a significant advancement in medical technology. By providing precise tactile input, they enhance precision and realism in surgical simulations and remote treatments. However, they also present significant obstacles, including the need for high accuracy, dependability concerns about data security and privacy, and problems with accessibility and cost. To ensure patient safety, informed consent, and a prudent balance between machine intelligence and human decision-making, ethical concerns in robotic and remote surgeries must be handled carefully. The performance and functionality of haptic systems will be improved in the future by technologies like AI-enabled brain-machine interfaces, 5G, edge computing, and the potential application of quantum computing. Globally, these innovations will continue to result in healthcare services that are more responsive, accurate, and reasonably priced. In haptic technology, AI is about more than just making machines smarter; it's about increasing healthcare workers' skills and helping patients get better results. AI in haptic systems will be essential in advancing a new generation of intelligent patient-centered care as medical practice continues to change.

AI disclosure

During the preparation of this work, the author(s) used Napkin AI and Grammarly to generate the images and Grammarly to correct grammatical errors. After using this tool/service, the author(s) reviewed and edited the content as needed and take(s) full responsibility for the content of the publication.

References

Arnab, S., Petridis, P., Dunwell, I., & de Freitas, S. (2011). *Enhancing learning in distributed virtual worlds through touch: A browser-based architecture for haptic interaction* (pp. 149–167). Springer Science and Business Media LLC. https://doi.org/10.1007/978-1-4471-2161-9_9.

Ballantyne, G. H. (2002). Robotic surgery, telerobotic surgery, telepresence, and telementoring. *Surgical Endoscopy, 16*(10), 1389–1402. https://doi.org/10.1007/s00464-001-8283-7.

Bolopion, A., Ni, Z., Agnus, J., Benosman, R., & Regnier, S. (2012). Stable haptic feedback based on a dynamic vision sensor for microrobotics. In *2012 IEEE international conference on intelligent robots and systems* (pp. 3203–3208). IEEE. https://doi.org/10.1109/IROS.2012.6385557.

Ejaz, H., McGrath, H., Wong, B. L. H, Guise, A., Vercauteren, T., & Shapey, J. (2022). Artificial intelligence and medical education: A global mixed-methods study of medical students' perspectives. *Digital Health, 8*, 205520762210890. https://doi.org/10.1177/20552076221089099.

Her, M.-G., Hsu, K.-S., & Yu, W.-S. (2002). Analysis and design of a haptic control system: Virtual reality approach. *The International Journal of Advanced Manufacturing Technology, 19*(10), 743–751. https://doi.org/10.1007/s001700200085.

Iftikhar, M., Saqib, M., Zareen, M., & Mumtaz, H. (2024). Artificial intelligence: Revolutionizing robotic surgery: Review. *Annals of Medicine & Surgery, 86*(9), 5401–5409. https://doi.org/10.1097/ms9.0000000000002426.

Khamaj, A., & Ali, A. M. (2024). Adapting user experience with reinforcement learning: Personalizing interfaces based on user behavior analysis in real-time. *Alexandria Engineering Journal, 95*, 164–173. https://doi.org/10.1016/j.aej.2024.03.045.

Kim, M., Jeon, C., & Kim, J. (2017). A study on immersion and presence of a portable hand haptic system for immersive virtual reality. *Sensors, 17*(5), 1141. https://doi.org/10.3390/s17051141.

Lee, T. F., Chang, I. P., & Su, G. J. (2023). Compliance with HIPAA and GDPR in certificateless-based authenticated key agreement using extended chaotic maps. *Electronics (Switzerland), 12*(5). https://doi.org/10.3390/electronics12051108.

Li, Y., Raison, N., Ourselin, S., Mahmoodi, T., Dasgupta, P., & Granados, A. (2024). AI solutions for overcoming delays in telesurgery and telementoring to enhance surgical practice and education. *Journal of Robotic Surgery, 18*(1), 1863–2491. https://doi.org/10.1007/s11701-024-02153-9.

Lin, J., Ma, Z., Gomez, R., Nakamura, K., He, B., & Li, G. (2020). A review on interactive reinforcement learning from human social feedback. *IEEE Access, 8*, 120757–120765. https://doi.org/10.1109/ACCESS.2020.3006254.

Mahmmod, B. M., Naser, M. A., Al-Sudani, A. H. S., Alsabah, M., Mohammed, H. J., Alaskar, H., Almarshad, F., Hussain, A., & Abdulhussain, S. H. (2024). Patient monitoring system based on Internet of Things: A review and related challenges with open research issues. *IEEE Access, 12*, 132444–132479. https://doi.org/10.1109/ACCESS.2024.3455900.

Mallipeddi, S. R., Goda, D. R., Yerram, S. R., Varghese, A., & Ande, J. R. P. K. (2017). Telemedicine and beyond: Navigating the frontier of medical technology. *Technology and Management Review., 2*(1), 37–50.

Minopoulos, G. M., Memos, V. A., Stergiou, K. D., Stergiou, C. L., & Psannis, K. E. (2023). A medical image visualization technique assisted with AI-based haptic feedback for robotic surgery and healthcare. *Applied Sciences (Switzerland), 13*(6). https://doi.org/10.3390/app13063592.

Mirchi, N., Bissonnette, V., Yilmaz, R., Ledwos, N., Winkler-Schwartz, A., Del Maestro, R. F., & Pławiak, P. (2020). The virtual operative assistant: An explainable artificial intelligence tool for simulation-based training in surgery and medicine. *PLOS ONE, 15*(2), e0229596. https://doi.org/10.1371/journal.pone.0229596.

Nayak, S., Das, R. K., Application of artificial intelligence (AI) in prosthetic and orthotic rehabilitation. IntechOpen, (2020), https://doi.org/10.5772/intechopen.93903.

Ribeiro, M. L., Lederman, H. M., Elias, S., & Nunes, F. T. L. S. (2017). Techniques and devices used in palpation simulation with haptic feedback. *Computing Surveys, 49*(3). https://doi.org/10.1145/2962723.

Wang, C., He, T., Zhou, H., Zhang, Z., & Lee, C. (2023). Artificial intelligence enhanced sensors-enabling technologies to next-generation healthcare and biomedical platform. *Bioelectronic Medicine, 9*(1), 17. https://doi.org/10.1186/s42234-023-00118-1.

Won, C. H., Lee, J. H., & Saleheen, F. (2021). Tactile sensing systems for tumor characterization: A review. *IEEE Sensors Journal, 21*(11), 12578–12588. https://doi.org/10.1109/JSEN.2021.3078369.

Xu, J., Chen, W., Qian, H., Wu, D., & Chen, R. (2025). ThinTact: Thin vision-based tactile sensor by Lensless imaging. *IEEE Transactions on Robotics, 41*, 1139–1154. https://doi.org/10.1109/tro.2025.3530319.

Zhou, B., Yang, G., Shi, Z., & Ma, S. (2024). Natural language processing for smart healthcare. *IEEE Reviews in Biomedical Engineering, 17*, 4–18. https://doi.org/10.1109/RBME.2022.3210270.

Zhu, M., Sun, Z., Zhang, Z., Shi, Q., He, T., Liu, H., Chen, T., & Lee, C. (2020). Haptic-feedback smart glove as a creative human-machine interface (HMI) for virtual/augmented reality applications. *Science Advances, 6*(19), 2375–2548. https://doi.org/10.1126/sciadv.aaz8693.

CHAPTER 8

Artificial intelligence frameworks for intelligent haptic control

Deepika Roselind Johnson[a], Srivats Raghavan[a], Kalyanasundaram Venkataraman[a], AJ Keerthi[a], Utkarsh Mishra[a] and Aaryan Gupta[b]
[a]School of Computer Science and Engineering, Vellore Institute of Technology, Chennai, Tamil Nadu, India
[b]Department of Artificial Intelligence and Machine Learning, Manipal University Jaipur, Jaipur, Rajasthan, India

8.1 Fundamentals of haptic medical systems

Haptic medical systems are becoming a more indispensable aspect of modern healthcare, offering precise and sensitive sensory perception that significantly enhances the abilities of medical professionals in clinical practice. By duplicating the feeling of touch, the systems enable professionals to interact with virtual worlds and actual human anatomy in ways previously unfathomable. This physical linkage between the real and virtual domains allows the patient to feel variation in tissue resistance, pressure, and texture critical to an accurate procedure in sensitive intervention. Tactile feedback technology is being developed in many different medical applications, including robotic surgery procedures, testing diagnostics, and therapy interventions, and leading toward a new accuracy, safety, and minimally invasive approach. With advancements in actuators, sensors, and real-time feedback algorithms, haptic systems are transforming traditional practices into highly optimized, data-driven processes. This chapter explores the building blocks of haptic systems, including tactile perception, force feedback, and their several clinical applications.

8.1.1 Tactile perception and force feedback mechanisms

At the heart of all haptic systems lies the mechanism of tactile perception—the innate human ability to sense, interpret, and respond to physical stimuli through the skin. The highly sensitive sense is facilitated by a complex network of special receptors, including mechanoreceptors, which detect pressure and vibration, and nociceptors, which respond to pain and temperature changes. Human sensitivity to touch is exceptionally great, with people able to tell apart infinitesimal differences in the texture, hardness, and heat transfer of surfaces. The nervous system relays the stimuli to the brain, which interprets and combines them into a cognitive representation of the world. This ability is especially applicable in the clinical setting, where sensory feedback enables clinicians to distinguish between healthy and ill tissues, recognize anatomical features such as nerves

Haptic Feedback Control in Medical Robotics
DOI: https://doi.org/10.1016/B978-0-443-27777-1.00017-2

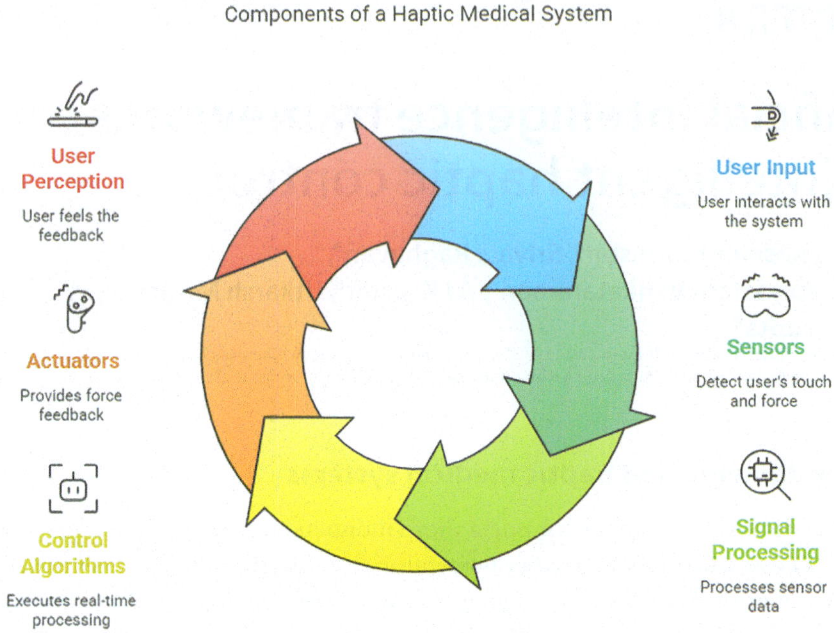

Figure 8.1 Components of a haptic medical systems.

and vasculature, and perform complex procedures with precision (Selim et al., 2024). To achieve the sensation of touch in robots, force feedback systems are employed. These systems give controlled mechanical responses—forces of resistance—that mimic human interaction. Actuators, electromechanical transducers, are also essential in replicating these forces. They are governed through feedback from a series of sensors that track user movement, contact force, and interaction dynamics in real-time. The feedback loop gives haptic systems the realistic sensations required to execute robotic surgery, simulation training, and remote diagnostics.

These sensors measure the forces exerted by the user and the virtual or real object with which they are operating. The output of these systems is typically sensed through the hands, fingers, or other appendages of the user, providing a direct interface between the system and the user. For optimal natural and effective interaction, the system must offer force accuracy—the ability to produce realistic force feedback sensations in accordance with user actions. Force accuracy is important in medical applications such as robotic surgery, where small variations in force have significant effects on patient safety (Steinbach et al., 2011). In addition, response latency, i.e., the time lag between user input and resultant feedback, is another factor that influences haptic system performance. Time delay can disrupt the flow of normal interaction, degrading accuracy and increasing the cognitive load of the user. Mitigating force accuracy issues and response latency is of utmost priority for enhancing the user experience and ensuring the reliability and safety of the system within the medical environment. Fig. 8.1 visualizes how sensory input, processing, and feedback generation happen in a haptic system.

8.1.2 Applications in surgery, diagnosis, and rehabilitation

Utilization of haptic feedback technology in healthcare systems has revolutionized the precision, safety, and effectiveness of operations in various fields most notably, surgery, diagnosis, and rehabilitation. In robot-assisted surgery, haptic feedback bridges the patient-surgeon sensory gap by offering the perception of touch, enabling surgeons to feel resistance, pressure, and texture during interventions. This is particularly important in minimally invasive treatments such as endoscopic and laparoscopic surgery, where there can be loss of direct contact sensations that impair precision. Interfaces like the Da Vinci Surgical System allow operators to detect variations in tissue density that permit discrimination of thin structures like nerves, fibrous tissues, or blood vessels. The result is increased surgical skill, reduced complication rates, and enhanced confidence, even in difficult or high-risk procedures (Torabi et al., 2021). In the clinical diagnosis field, haptic technologies are transforming medical training and diagnostic testing. Medical students and physicians can utilize haptic-enabled simulators to hone the art of diagnostic palpation, for instance, identifying tumors, cysts, or organ abnormalities, through realistic virtual environments (Mueller & Grunwald, 2023). These systems mimic the haptic nuances of real tissues, thus improving diagnostic accuracy in a controlled, ethical, and reproducible setting. Moreover, diagnostic tools like robotic ultrasound systems now include force feedback systems, enabling clinicians to assess the stiffness, elasticity, and texture of tissues in real-time. This capability aids in the diagnosis of pathological states such as cancer, fibrosis, or cysts at the initial stage that would otherwise not be detected using regular visual or auditory inspection.

In rehabilitation, haptic systems serve a vital purpose in helping patients recover from injuries, strokes, or amputations. Robotic exoskeletons, prosthetics, and assistive devices with haptic feedback enable patients to restore motor functions by mimicking natural motion and offering real-time sensory feedback. For instance, robotic rehabilitation platforms assist in helping stroke victims in relearning motor functions through guiding their movements and providing force feedback to impart a feeling of resistance and control (Shazhaev et al., 2023). They not only aid in physical recovery but also stimulate the brain's sensory and motor systems, which is critical in enhancing recovery and enhancing the patient's quality of life. In addition, virtual reality (VR) and augmented reality (AR) platforms with haptic feedback are being utilized to develop immersive rehabilitation experiences in which patients can interact with virtual spaces to practice motor functions in an effective and safe way. Fig. 8.2 shows the key medical areas where haptic technology is applied.

With improvements in haptic technology, it is being implemented in medicine increasingly more accurately, responsively, and intuitively (Shazhaev et al., 2023). These offer tremendous potential to enhance medical practice, training, and the rehabilitation

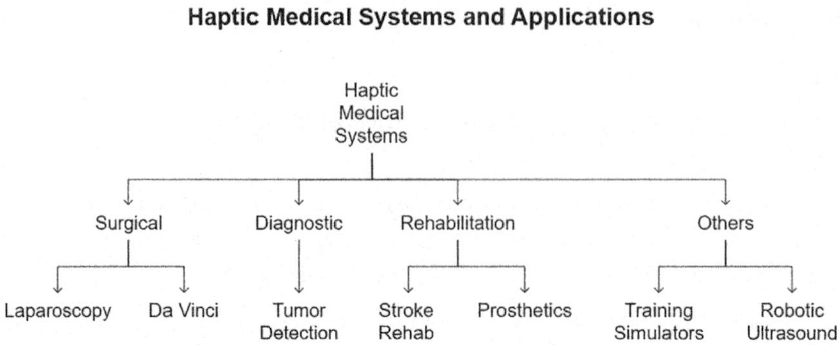

Figure 8.2 Applications of haptic systems in medicine.

of patients at a faster pace. Nevertheless, there are ongoing challenges such as enhancing force precision, minimizing the latency of the response, and increasing the system's ability to adapt. Therefore, future research and development will probably be aimed at overcoming these limitations so that haptic systems can become an even more integral component of contemporary healthcare.

8.2 Limitations in traditional haptic control systems

Classical haptics control systems have played an important role in moving medical robotics to the next level by giving tactile feedback to the surgeon and clinician. Nevertheless, such systems suffer from central issues, especially related to the accuracy of force acting on the tissue and the latencies associated with a robot's response to external stimuli, which can be translated directly into the effectiveness of any procedure and its safety.

8.2.1 Force accuracy and response latency constraints

Haptic systems play an essential role in medical robotics, allowing surgeons and medical practitioners to gain tactile feedback that would increase precision and control over a surgical operation. Nonetheless, conventional haptic control systems have serious constraints such as insufficient force, accuracy, and response time. These limitations can have a direct impact on safety, efficacy, and overall user experience in medical applications, including robotic-assisted surgeries, telemedicine, and rehabilitation systems.

8.2.1.1 Force accuracy constraints

Traditional haptic systems are a bit rigid and often fail to provide the operator with high-fidelity force feedback, causing challenges in discrimination between the experience difference in tissue tactile texture and tactile density (Fig. 8.3). This is particularly daunting in minimally invasive surgery, and surgeons depend on their haptic sense to

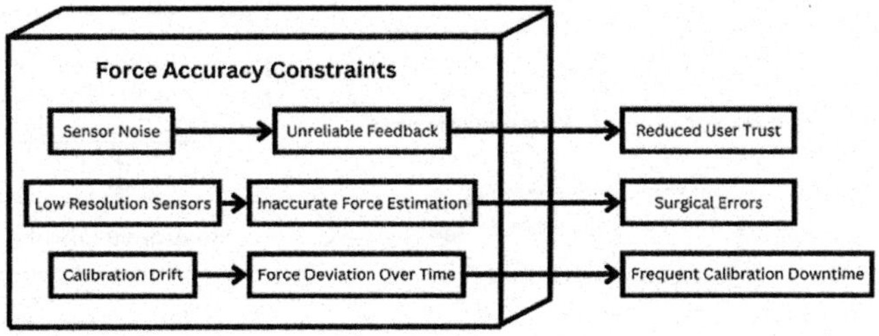

Figure 8.3 Cause-and-effect of force accuracy issues.

differentiate between various anatomic structures. During laparoscopic procedures, for example, insufficient feedback can prevent the surgeon from spotting a blood vessel, raising the risk of damage. Force feedback is measured and replicated using force sensors and actuators in haptic devices. With limited dynamic ranges of low-resolution sensors and actuators, the generated feedback and estimations for forces become inaccurate. Moreover, sensor noise and drift can further reduce accuracy, resulting in haptic sensations that are inconsistent or unreliable. For instance, in a robotic-assisted heart valve repair, slight inaccuracies in force feedback may compromise the integrity of the delicate tissue being sutured. Force sensors in haptic systems also drift over time, which creates a stack of errors. Due to the high precision requirement in medical procedures, the inexact force measurement and misalignment of actuation always yield poor performance. Every time it needs to be recalibrated, it adds complexity to the overall operation of the system, leading to a higher amount of downtime. An example of this in practice is found in robotic rehabilitation therapy, where variability in force feedback can affect patient recovery and therapy efficacy.

To understand force accuracy constraints, let us have a look at a few real-life scenarios where several challenges continue to remain in their practical application. In a laparoscopic hysterectomy, surgeons must separate the uterus from surrounding tissues. Though traditional haptic systems often achieve lower fidelity force feedback, they may struggle to adequately feel connective tissues or important structures like blood vessels, leading to increased strain or excessive force. Another example highlights the precision in pediatric surgeries that is much higher because the anatomy is smaller. Which in a case like the repair of a congenital heart defect, suturing very fine tissues is involved. The sensor drift could cause the wrong contact force to be applied, which could end up being incorrect suturing, either too much force leading to tissue damage or not enough leading to inadequate healing.

Haptic systems employed in next generation prosthetics system help restore a realistic sense of touch to amputees. But calibration error in the sensor due to time could create a difference between the force applied by a user and the force they felt back, leading to discomfort or inefficient usage of the prosthetic limb. Skills like precise haptic feedback

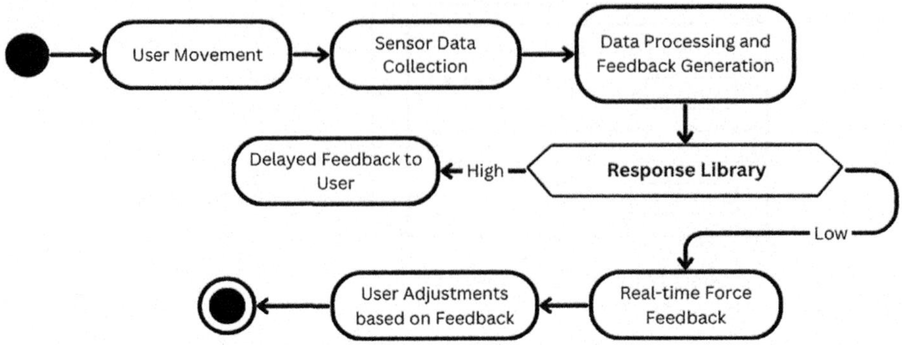

Figure 8.4 Feedback loop in haptic systems.

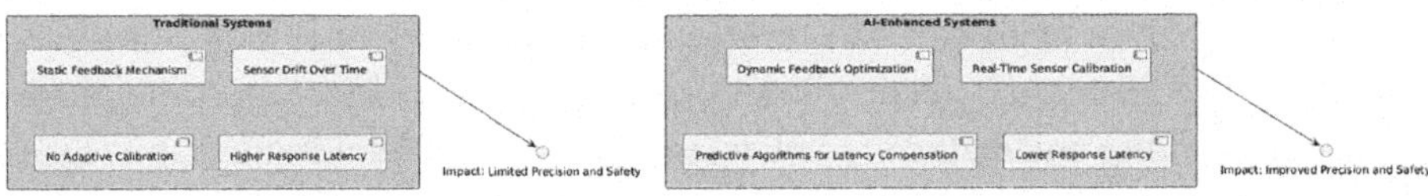

Figure 8.5 Comparison of traditional and AI systems.

are also critical to avoid damaging important parts of the human brain during minimally invasive neurosurgery. A baseline, preexisting strain in the network of nerves, brain, and other parts is maintained by making sure the patient remains stable and dormant. If there exists even a slight delay or deviation in the calibrating force feedback, a lot of sensitive neural areas may end up compressed with unnecessary, delicate but strenuous pressure, leading to heightened surgical risks. This situation only serves to worsen the patient's present condition.

8.2.1.2 Response latency constraints

In haptic systems (Fig. 8.4), there exists a bidirectional flow of force and motion data between the control system, actuators, and sensors. In traditional systems, there are delays during data transmission and processing, leading to response latency. In medical robotics, even low levels of latency can adversely affect the user's control precision, especially in time-sensitive scenarios. In robotic prostate surgery, for example, delayed feedback may impair the surgeon's ability to make fine adjustments, increasing procedural risk.

Traditional haptic (Fig. 8.5) systems rely on control algorithms that require a lot of computational power. If the hardware being used is of lower performance with the aid of complex algorithms, there can be further delays. Force feedback with high computational latency creates lag that makes real-time interaction difficult. Delayed response may affect the efficiency of skill acquisition and procedural learning, for instance, in simulated medical training systems. One of the key concepts is that of network-induced latency introduced from the distance between the operator and robotic system in teleoperated medical procedures. Existing haptic devices do not currently exploit advanced compensation strategies for fixed natural perceptual systems, creating a poor experience with sensitive feedback. This becomes a risk when performing remote surgeries where accuracy and timing are of the essence. For example, the Da Vinci surgical system is an actual robotic device that, while it has advanced capacities, still struggles with latency management in teleoperated surgeries.

Force accuracy constraints are critical in several real-life scenarios. When it comes to surgery on children, nowhere in the human body is greater precision needed than when performing surgery on children, whose anatomy is obviously smaller. In the repair of a congenital heart defect, for instance, the surgeon has to stitch together incredibly delicate tissues. Sensor drift can also lead to inaccurate force estimation, which means stitches might be either too tight and damage tissue, or too loose, risking poor healing. Advanced prosthetic devices utilize special haptic systems to simulate lifelike sensations for users. When the sensors drift in their calibration with time, the feedback becomes detached from the force applied by the user, which causes discomfort or inefficient control of the prosthetic limb. Haptics systems implemented in high-end prostheses try to give amputees as realistic tactile sensations as possible. However, over time, the calibration of the sensors and sensor drift from the point of calibration which causes a discrepancy between the force applied by the user and the feedback received, leading to discomfort or inefficient use of the prosthetic limb.

8.2.2 Handling dynamic and complex medical interventions

Inconsistent force feedback can impair precision for the surgeon in making fine adjustments, leading to greater chances of damage to surrounding tissue or procedural mistakes. In an area like neurosurgery, where small errors could have such a dramatic impact, these limitations in the accuracy of the force scan limit the efficacy of the surgery.

High latency detracts from the naturalness of interaction, leading to fatigue and diminishing trust in the system. Surgeons may now have an increased cognitive load, which can lead to longer operating times and decreased overall performance. Delay in sending force feedback on manipulation during vital tasks may lead to unintentional force exertion, threatening the well-being of the patient. In remote robotic-assisted trauma surgeries, latency may prevent a surgeon from reacting promptly to contingencies. To alleviate these drawbacks, advanced computational approaches need to be incorporated. Predictive algorithms, adaptive feedback mechanisms, and real-time data analysis offer promising solutions through AI. They address latency compensation, accuracy, and immersive haptic feedback with fundamental AI-based model development.

Such as using AI-driven predictive algorithms to determine the responses of a force based on previous experience, minimizing the perceivable latency. It also allows for dynamic calibration of sensors for accurate force feedback even in long-term operations with machine learning (ML) models. AI and ML algorithms for communication network optimization in telesurgery, providing a variety of healthcare services in real-time, require advanced networking capabilities and protocols that preserve high connectivity; however, in real-time telesurgery, introducing AI-based network optimization algorithms will enable the prediction of communication delays and maintain constant force feedback.

The accuracy of force and delay of response of the robotic system in the medical industry are the most important factors in order to guarantee a good outcome. In spinal fusions, for example, robotic-assisted surgery helps steer surgeons in screwing them into the vertebrae with exactitude. Yet, errors in force feedback and response time can result in a lack of screw alignment, leading to the potential for nerve damage or loose fusion. Likewise, in rehabilitation for stroke patients, the robot is developed to assist with motor function recovery. Unpredictable force feedback and response time lags in these systems can be interruptive to a therapeutic gesture flourishing during training sessions; they can lead to frustration for the patients and slow down their learning. Also, corneal transplants performed during ophthalmic robotic surgery require ultra-sensitive haptic feedback because corneal tissue is extremely fragile. Any delay or errors in force sensing will affect the tissue manipulation and may cause corneal injury or even procedure failure. Dealing with these two challenges of force accuracy and response latency is also central to the improvement of reliability and safety in robotic-assisted medical interventions.

8.3 Artificial intelligence frameworks for intelligent haptic control

8.3.1 Introduction

Haptic technology, which provides tactile feedback to users, has become an important component of many human–computer interactions (HCI), including surgical robots, VR, prosthetics, rehabilitation devices, and support systems for people with disabilities. The main goal of haptic systems is to replicate the touch feeling and enable users to cope with virtual or distant environments instinctively and engagingly. Nevertheless, traditional haptic control systems often encounter challenges related to delay, adaptability, and situational awareness. Including AI in your haptic system greatly improves response, flexibility, and accuracy. Haptic AI-equipped systems can learn from user interaction, adapt to specific people, and provide immediate feedback. This chapter examines the architectural framework of AI haptic systems and explores predictive analytics that optimize sensory processing, learning, adaptive processes, integration into ML, and real-time feedback mechanisms performance.

8.3.2 Architectural design: Sensory processing, learning adaptation

Intelligent haptic control systems usually consist of three important elements: Sensory processing for the acquisition and understanding of tactile data, developing knowledge to adapt and improve system performance, and developing adaptive techniques (Fig. 8.6).

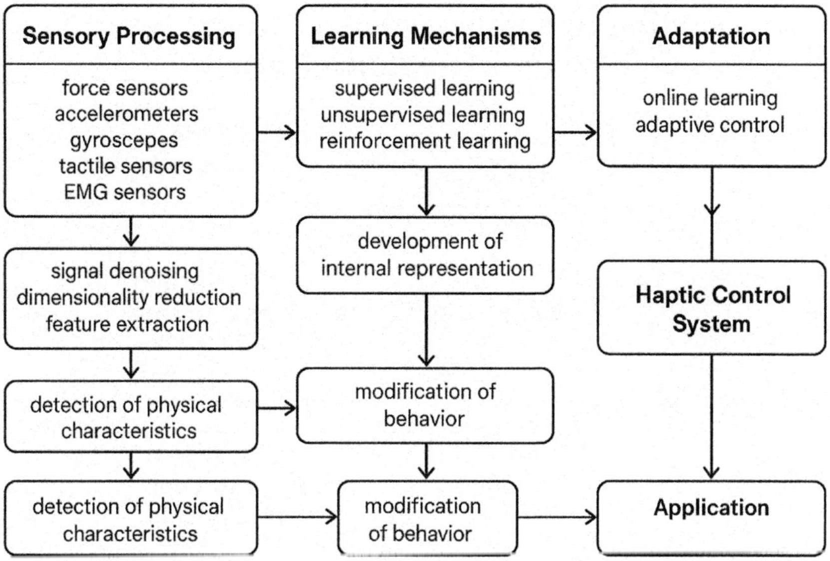

Figure 8.6 Architecture design with sensory processing, learning, and adaptation.

8.3.2.1 Sensory processing

Sensory processing is the first and most important stage of all tactile systems. These sensors provide data on force, torque, pressure, temperature, and movement. Transform raw sensing data into abstract representations using AI methods such as signal removal, dimension reduction, and characteristic extraction. This allows the system to recognize detailed physical properties such as the stiffness, surface roughness, and texture grains of the object (Ahmed et al., 2022). Highly developed signal processing technologies, such as wavelet transform and folding networks (CNNs), are often used for accurate and effective tactile data analysis. Immediate filtering and integration of various sensor data gives you a strong understanding of the environment. This is extremely important for changing applications such as robotic surgery and self-directed navigation.

8.3.2.2 Learning mechanism

Learning allows haptic systems to create internal representations of their environments and user actions. Surviving learning methods allow the system to classify different tactile sensations using salient data, but unattended learning helps to recognize patterns without direct annotation. Reinforcement learning (RL), in particular, deep reinforcement learning (DRL), is widely popular in tactile control tasks, characterized by sparse and delayed feedback. For example, a robot gripper with RL determines the best power to edit different vulnerable elements. Over time, the system improves its efficiency by optimizing reward features that are punished for slips and damage. Learning algorithms can also be used to present user preferences, task development, and interaction goals.

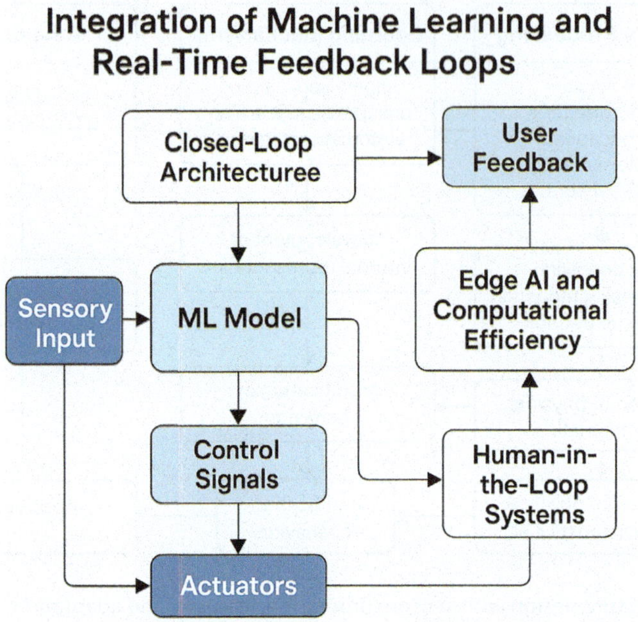

Figure 8.7 Integration of machine learning and real-time feedback loops.

8.3.2.3 Adaptation

Adaptation refers to the ability of a haptic system to change its behavior based on user interactions, environmental factors, or variations in dynamics. Online learning algorithms allow the system to gradually adapt the model without the need for full retraining. Transfer methods use findings from previous tasks to improve the learning pace for new but similar tasks. The adaptive controller changes actuator parameters, such as stiffness and damping coefficients, in real time, depending on sensory input and feedback. It is a known fact that haptic systems with feedback context usually adapt themselves according to the user's physiological needs or skills to improve generalization and to boost their performance (Srinivasan et al., 2023).

8.3.3 Integration of machine learning and real-time feedback loops

An intelligent haptic control system with a closed loop is an amalgamation of ML algorithms (Fig. 8.7). These algorithms ensure that the control system responds effectively to environmental stimuli, which in turn leads to improved accuracy and efficiency.

8.3.3.1 Closed-loop architecture

Initially, the user input is fed to a closed-loop haptic control system, the system analyses the control signals and information using a suitable ML model and actuator. The closed-loop feedback system adapts its action in real time to ensure system stability in uncertain circumstances. Deep learning architectures such as recurrent neural networks (RNNs), long short-term memory models (LSTMs), and transformers are increasingly used for time-based modeling for closed-loop control systems. These models pursue relationships at time intervals. That is, it is suitable for predicting future conditions and modifying control settings as needed (Garg et al., 2023).

8.3.3.2 Edge artificial intelligence and computational efficiency

A key challenge in implementing AI models in haptic systems is the delay caused by data transmission and processing, especially when using cloud-based solutions. Edge AI solves this problem by directly performing calculations on embedded devices such as microcontrollers and Edge GPUs. This reduces response time, saves bandwidth, and increases data protection. Edge-AI methods include model compression, quantization circumcision to implement deep neuronal networks in energy-efficient devices. Union learning methods are becoming increasingly popular, allowing a variety of haptic devices to train models together without sharing raw data (Okamura, 2022).

8.3.3.3 Human-in-the-loop systems

HITL systems (Human-in-the-Loop) focus on users as the core of the learning and management process. These systems collect user feedback consistently—directly (via assessment or input) or indirectly (via physiological sensors or behavioral signals)—using this data to adapt system parameters. Collaborative adaptive learning enables users and AI systems to develop together over time. For example, the exoskeleton of rehabilitation can change the level of resistance according to patient data and performance data as patients are increasingly qualified to operate the device. These interactive systems improve participation and adaptation security (Patel et al., 2024).

8.3.4 Predictive analytics for performance maximization

Predictive analytics employs AI models to foresee future scenarios, identify possible problems, and enhance system performance in advance. Within haptic systems, predictive analytics enhances interaction fluidity, lowers error rates, and boosts task results.

8.3.4.1 Forecasting tactile interactions

Models that predict future interactions can measure their nature and time using historical tactile data and user behavior. For example, in a virtual training setting, the system can predict when and where users will interact with the surface and prepare appropriate feedback in advance to ensure a smooth experience. This model of prediction is particularly useful in instances where temporal precision is crucial, such as in robotic relief and collaborative robotics. Using sequence modelling techniques combined with attention mechanisms, predictive models are able to capture and record fine-grained interaction signals (Rakhsha et al., 2023).

8.3.4.2 *Anomaly detection and fault prediction*

Tactile systems are also subject to the negative effects of sensor deterioration, sensor fatigue, and delayed communication. Anomaly detection enables pattern matching, which allows the system to see signals immediately and capture defined patterns where parts of the system have erred. Time series predictive models, like those for automobiles and statistical control diagram systems, are employed to determine deviations from expected behaviors and recognize the anomalies. AI employs forecast expectations in assessing the remaining useful life (RUL) of components while task planning, which reduces downtime and enhances the operational lifespan of the system (Wang & Chen, 2022).

8.3.4.3 *Performance optimization*

Strategies for AI optimization continuously adapt to system configuration to enhance performance along several criteria like accuracy, energy consumption, user satisfaction, and task completion time. Bayesian optimization, evolutionary algorithms, and multilens optimization are implemented to achieve the set objective. For instance, the intelligent haptic gloves can enhance strength, feedback details, and signal transmission to provide the best multiuse, efficient touch interface while preserving battery power (Levine et al., 2021)

Haptic control systems under AI mark a complete and radical change in machine interaction with the physical environment. It consists of programs contemplating advanced sensory processing, adaptive learning, instantaneous feedback, and predictive analytics to give a more precise, more efficient, and better user-oriented approach. Incorporation of edge computing and human-in-the-loop designs brings relevance to haptics in areas critical to healthcare, manufacturing, and immersive media. Currently, research is in the process of further refining the technology, with future prospects revolving around self-learning systems, explanations in haptics, cross-modal interaction modelling, and collaborative intelligence. The advancements in AI and haptics simultaneously would thus bring in a paradigm shift in education, therapy, entertainment, and much more.

8.4 Technical and ethical barriers in artificial intelligence-driven haptic systems

AI-haptic systems are one technology that tends to combine robotic, rehab medicine, and even VR applications most finely. An adaptable and engaging user experience is made in an attempt to provide AI with information processing and generation into meaningful motion depicting touch in the real-world from a virtual or remote interaction. The balance of this Haptic feedback system was suspended by those fathomless technological and ethical hurdles. Touches on the confidentiality of data as well as the ethics of computation optimization, which need to be addressed if AI-based haptic systems are to be secure, ethically sound, and trusted.

8.4.1 Ethical and data security concerns

Advances in haptic technology usher in new privacy and ethical concerns because they generate personal data that the haptic systems pick up and store. This track information is attached to bodily touch: e.g., force, pressure, and motion pattern, which will be manifested into a possible feature of the user or even for determining health status in inpatient rehabilitation conditions. For example, tremors may be implicit in haptic data in rehabilitation, or muscle reaction time, where protecting the privacy is most critical (Emami et al., 2024). Manipulated data would be used in activities such as targeted discrimination, illicit medical profiling, and identity theft that require hard security mandates.

The main ethical problem is in informed consent and educating the participants to the fullest extent. How much haptic-controlled systems conceal from the researchers the mechanism by which data processing, saving, or even reusing occurs (Steinbach et al., 2011) makes it challenging for participants to recognize the mechanism by which data are processed. This access becomes more maddening when haptic information is being used for some other goal other than the initial goal, i.e., for some goal such as behavior monitoring or emotion detection in VR (Abdossalami & Sirouspour, 2009). These finer nuances of grip force or latency of user touch can thus be reaped to predict degrees of stress, autonomy, and consent in question. Developers should thus accord the highest priority to open data practice provision and communication mechanisms under which users ought to be able to opt out or limit the use of data.

Bias in AI-driven haptic systems is also an ethical issue. The systems learn and react to user input from training data, but if the training data are representative, the resultant models may prefer some groups over others. Take the example of a haptic device for simulating surgical training: if training data are mainly characteristic of hand sizes or motor capabilities of a particular population, the system could fail for individuals with non-average body dimensions, undermining fairness and performance (Pacchierotti et al., 2017). Countermeasures involve cautious curation of data sets and inclusion of fairness-conscious algorithms to achieve equalized performance across multiple groups of users (Hui et al., 2021). If not addressed, this can enhance disparities in access to more advanced haptic technologies, especially in important applications such as the healthcare sector.

Data protection is a technical foundation for ethical haptic system design. Real-time capability, a characteristic of haptic feedback, must be achieved through rapid data transport, usually over publicly accessible networks of eavesdropping. With teleoperation

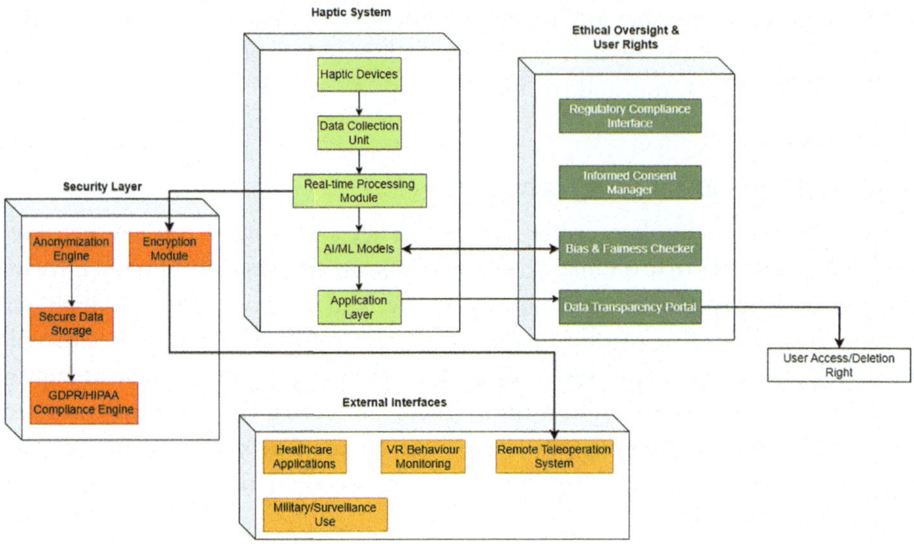

Figure 8.8 Architecture of ethical and data security concerns.

systems, where robot movement across remote locations is manipulated through haptic feedback, unencrypted streams of data would be susceptible to hijacking, stealing essential user input or system requests (Winfield et al., 2019). Haptic data long-term storage also poses additional danger, especially for personal data. To counteract these attacks, end-to-end encryption and anonymization of the data can be done by the developers, but these need to be done cautiously so as not to add latency that compromises real-time performance (Simsek et al., 2016). It also needs to adhere to legislation like the General Data Protection Regulation (GDPR) or the Health Insurance Portability and Accountability Act (HIPAA) and protect information and people's rights to access or delete their information (Chaudhariu et al., 2015).

Besides technical protection, broader ethical aspects of AI-based haptic systems are also of concern. Abuse potential for uses like military training or general surveillance gives ethical regulation a consideration worth pursuing (Rajkomar et al., 2018). As an example, haptic information can be employed to train autonomous systems with destructive motives, which has led to regulations to prevent such a situation. By facilitating interdisciplinary ethics teams and open development processes, haptic system development consistent with societal values, where benefits far outweigh the risks, can be facilitated. AI-based haptic system ethics and data protection need to be addressed in a holistic manner. Fig. 8.8 depicts the architecture of ethical and data security concerns, which means privacy for the users by secure dealing with data, consent in the instance of algorithmic transparency, a lack of bias by design participation, and vision regarding societal effects other than the immediate. All these must be achieved to establish trust and secure the integrity of haptic technology use.

8.4.2 Computational optimization

The achievement of AI-based haptic systems relies on such systems' capacity to deliver continuous, real-time haptic feedback, and it involves overcoming vast computational challenges. Processing high-dimensional haptic data in real time with force, position, and velocity sensing is problematic, but it is further compounded when AI algorithms are typically much more computationally intensive. The subsequent subsection assumes widespread optimizations of the nature of edge computing, data compression, low-energy ML structures, and adaptive algorithms in delivering a high-performing haptic experience. Edge computing has the potential to enhance latency in AI-based haptic systems. Computed on or close to the device (Okamura, 2004), as opposed to faraway cloud servers, edge computing reduces transmission delay, a major aspect of real-time responsiveness. With local processing of touch signals in a haptic teleoperation system, for instance, the user can get immediate feedback from a remote robot, enhancing accuracy in control (Hacker, 2018). But edge devices are not strong, so they require low-power consumption algorithms and low memory usage. It is technically challenging to satisfy these conditions under the demands of AI-based analysis.

Data handling is also of similar concern, particularly because haptic signals are high and in real-time. High-dimensional data will saturate processing pipes and lower system response. Dimensionality reduction, for example, by principal component analysis (PCA), can re-map haptic data onto its most relevant features, optimizing computational cost with preserved fidelity (Arkin, 2010). Similarly, new compression methods optimized for haptic signals—i.e., normative coding structures—enable efficient transmission across bandwidth-limited networks, i.e., for application in remote surgery or multiplayer VR. These approaches maintain high tactile content and keep computation loads low.

AI haptic systems must therefore further optimize their ML models for real-time execution. Deep learning architectures are very resource-heavy; therefore, they are not usually applicable for situations requiring low latency simply because of their respective resource requirements. There are lightweight systems that reduce model complexity, like pruned CNNs or quantized RNNs. Pruning zeros out redundant parameters while quantization lowers the precision of weight parameters, much in the same way, for significantly reduced computation time at almost no cost in terms of accuracy. Accelerated hardware could help achieve further performance: for haptic applications, the AI tasks could either be executed on a fast GPU or TPU platform. These further optimizations are crucial to maintain the instantaneous nature of tactile feedback.

On the other hand, adaptive algorithms offer the dynamic nature of optimization, which is usually an efficient manner of resource allocation, depending on the occasion.

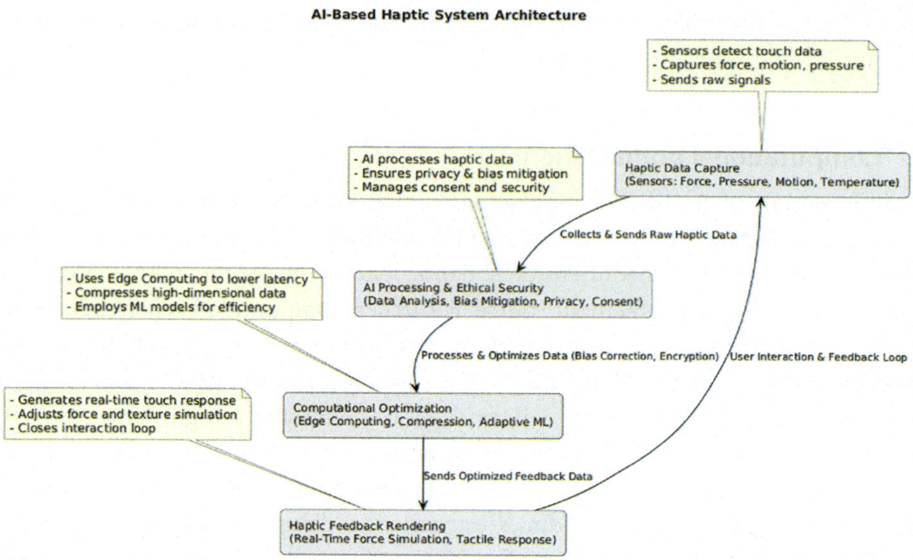

Figure 8.9 Architecture of haptic system.

For instance, during a passive scene, the system could cut down haptic rendering resolution for power saving in a virtual-reality game, whereas it ramps up accuracy during active interactions, such as when the user manipulates an object. Such adaptive optimizations guarantee performance on resource-constrained devices like wearables for the haptic glove while still ensuring immersion for users. However, the design and implementation of the adaptive process involve complex control logic capable of adjusting processing fast while ensuring that no user-perceptible interruptions occur.

Along with these years of progress, optimization has always had the trade-offs that it has to endure. These operations that typically compromise quality feedback, like model simplification or data compression, can also cause artifacts that influence the quality of touch. A case in point would be the type of aggressive compression that distorts the sense of texture in a virtual environment, thus shattering the immersion. Hybrid approaches that balance the efficiency of old-school control with the accuracy of AI are being explored by researchers to meet both the demands of efficiency and precision. Fig. 8.9 depicts the AI-based haptic system architecture. Contributing to the complexity of optimization attempts is also the necessity for compatibility on heterogeneous hardware platforms, from high-performance workstation computers to low-power ones that may include wearables, again demanding some level of scalability. It is the computational efficiency of AI-driven haptic systems that makes them deployable in real-time through the help of edge computing, data compression, optimized ML, and adaptive methods. Fig. 8.9 depicts the architecture of the haptic system; all these methods combined solve the computational complexity versus user experience trade-off to provide more immediate and accessible haptic technologies. Further research into these will be critical in an effort to address the growing demands being made by next-generation applications.

8.5 Conclusion and future work

AI-driven haptic systems have emerged quickly, but whoever has constructed even a basic haptic system has felt that ethical-moral implications can be intricate to tainted depending upon the broad spectrum of technologies from gamified VR to autonomous health care systems, thus a technology or design should complement with ethical-moral responsibility to liberate from the hype the final potential of such technologies. Modern AI models use advanced algorithms and best computation techniques for force fidelity and response latency limits' solutions that provide the gateway for haptic perception that almost feels real without consuming too much time in computation. In comparison to that, intelligent haptics being applied to autonomous medical devices is another option for a future where the robot would be able to conduct fine operations autonomously with matching sensing, feeling as much as doing things very skillfully. Robots can revolutionize nutritional procedures and patient outcomes by enabling precise, component-level distribution with zero wastage. But anthropocentric these technological advances and their ethical effect on data privacy, (Kuhail et al., 2024) informed consent and algorithmic discrimination must be carefully integrated with open action, secure security practice, while cosmetic application and the nature of collaborators must be directed to prevent actions which prioritize fairness and user freedom in transparent processes. Finally, in a time when haptic technology is coming of age, how much value it can create, introducing to the digital world a level of control and responsiveness even greater than that which we can provide today, will be a matter of overcoming computation barriers and creating techniques attuned to our social values, so that we can share more widely the richness of a touch-sensitive world without losing a sense of the potential pitfalls. Therefore, future endeavors must be spent refining these two pillars, technical superiority and ethical stewardship, by ongoing research, field testing, and actively regulated environments, to build the foundation for haptic systems to be the cornerstone of human-machine interface for the next decades.

8.5.1 Next-generation artificial intelligence models for haptic interaction

Next-generation models capable of addressing current limitations in force accuracy, response latency, and ethical considerations are pivotal to the progression of AI-based haptic systems (Fig. 8.10). Using the latest ML algorithms like RL and generative adversarial networks (GANs) to make haptics more realistic and configurable. For example, RL could allow haptic systems to adaptively tune feedback according to real-time user interactions, thus increasing the accuracy of use cases such as robotic surgery

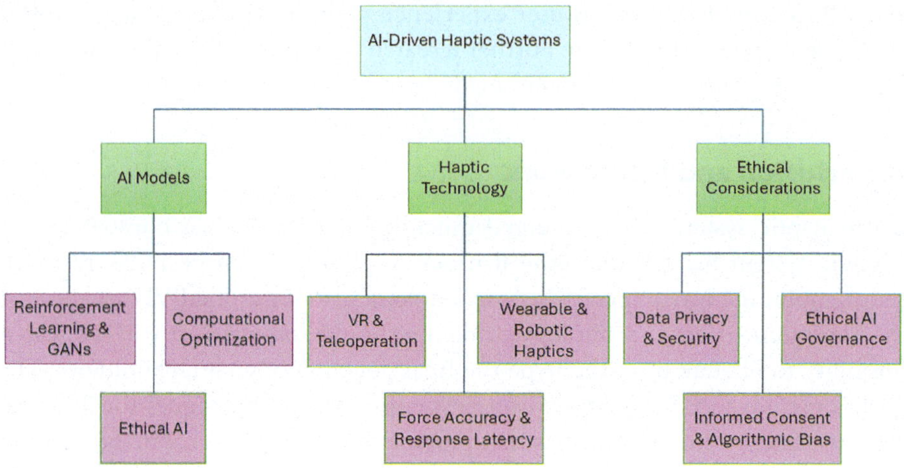

Figure 8.10 Conceptual framework of AI-driven haptic systems.

or care rehabilitation. At the same time, GANs could be used to synthesize high-fidelity haptic data from relatively scarce training inputs, allowing for wider training sets to be built, which would reduce bias in the trained models through simulation of different user profiles (Maheshwari et al., 2024). With the incorporation of such AI systems, haptic systems are moving toward a more intuitive and cohesive user interface, pushing the limits of VR and teleoperation.

The real-time computational requirement of next-gen AI models for haptic feedback computational optimization techniques, model pruning, quantization, GPU use, TPU use, and specialized hardware, just discussed in the computational optimization section, are hybrid approaches that will reduce latency while still maintaining accuracy. Such optimizations are critical for resource-limited devices such as wearable haptic gloves that are major bottlenecks in power and processing. Finally, another option would be to implement adaptive algorithms for rendering that allow for resolutions to vary based on the context of dropping fidelity during passive interactions without significantly reducing users' perception of the low-quality pixels. As the models themselves get better, they will still have to balance performance, as shown in Fig. 8.10, and scale to work for any platform from high-end workstations to low-power wearables.

Next-generation AI models in haptic interaction also need to consider transparency and fairness from an ethics perspective. This means building systems that clearly indicate when haptic data is processed and used, to counter previous discussion on consent challenges. User data that fits this model but comes with a caveat might be that the users would be able to control how the data is collected, for example, opting out of secondary usage analytics that could be used for behavioral analysis. Representing the target groups with diversity and inclusiveness, both in terms of hand size

(Qaisar et al., 2024), motoric abilities, and cultural styles for interacting with technology during training would be extremely important to minimize bias in the models built. Such systems could be designed to be compatible with social values through interdisciplinary collaboration between AI engineers and ethicists, along with end-users.

8.5.2 Expanding intelligent haptics for autonomous medical systems

The field of autonomous medical technologies empowered by intelligent haptic systems is a paradigm-changing new frontier in the delivery of healthcare (Mackenzie et al., 2022). These systems could enable robots to perform complex tasks, such as suturing or manipulating tissue, with minimal human supervision and the aid of AI-driven haptic feedback so they can feel and respond to their surroundings. For example, in robotic-assisted surgery, an autonomous system with sophisticated haptics could determine differences in tissue resistance and thus whether the tissue is healthy or diseased and act on that information. This kind of ability would be groundbreaking in terms of both precision and safety in life-or-death situations like neurosurgery or corneal transplants, where traditional systems struggle with both force fidelity and latency (see limitations section).

Effective security will be vital to ensure smart haptics are also integrated into autonomous medical systems, with ethical frameworks in place to enable their secure deployment. At the same time, due to the sensitive nature of the medical haptic data, such as muscle response patterns or tremor detection, these types of systems have to endure end-to-end encryption and anonymization, along with ensuring patient confidentiality in (Zhang et al., 2023) compliance with GDPR and HIPAA regulations. Oversight also prevents misuse, such as the making of AI systems that are autonomous and used to deploy harm to the extent of a military attack. As long as there are clear guidelines, and if stakeholders, such as medical professionals and patients, are involved in the design process, developers have greater hope of ensuring that new technologies deliver on patient well-being and positively influence society, rather than inflict unintended harm.

As intelligent haptics becomes a part of autonomous medicine, scalability toward such systems in the future will rely heavily on improvements made in edge computing and real-time data processing. Edge-based AI would enable these systems to function (Sheik et al., 2022) autonomously from cloud servers, minimizing latency and improving reliability in remote or resource-poor environments, including rural telemedicine. Also, the incorporation of predictive algorithms that predict tissue responses as per the existing knowledge base can also help mitigate delays or improve outcomes in critical time-sensitive procedures, which are in need of immediate interventions. As studies continue, pilot studies in controlled environments, like simulated operations or rehabilitation clinics, will be vital to affirm these systems and advance the path to larger clinical use, changing autonomous healthcare delivery.

References

Abdossalami, A., & Sirouspour, S. (2009). Adaptive control for improved transparency in haptic simulations. *IEEE Transactions on Haptics, 2*(1), 2–14. https://doi.org/10.1109/TOH.2008.18.

Ahmed, S., Xu, Y., & Li, Z. (2022). Predictive modelling of tactile feedback in tele-operated surgical systems. *IEEE Transactions on Haptics, 15*(4), 876–884.

Arkin, R. C. (2010). The case for ethical autonomy in unmanned systems. *Journal of Military Ethics, 9*(4), 332–341. https://doi.org/10.1080/15027570.2010.536402.

Chaudhariu, R., Schuwerk, C., Danaei, M., & Steinbach, E. (2015). Perceptual and bitrate-scalable coding of haptic surface texture signals. *Journal on Selected Topics in Signal Processing, 9*(3), 462–473. https://doi.org/10.1109/JSTSP.2014.2374574.

Delwar, T. S., Aras, U., Mukhopadhyay, S., Kumar, A., Kshirsagar, U., Lee, Y., et al. (2024). The intersection of machine learning and wireless sensor network securiity for cyber-attack detection: A detailed analysis. *Sensors, 24*, 1–44.

Emami, M., Bayat, A., Tafazolli, R., & Quddus, A. (2024). A survey on haptics: Communication, sensing and feedback. *IEEE Communications Surveys and Tutorials*, 1. https://doi.org/10.1109/COMST.2024.3444051.

Garg, R., Sharma, P., & Singh, A. (2023). Deep learning for closed-loop haptic feedback systems. *Journal of Intelligent & Robotic Systems, 110*(2), 221–239.

Hacker, P. (2018). Teaching fairness to artificial intelligence: Existing and novel strategies against algorithmic discrimination under EU law. *Common Market Law Review, 55*(4), 1143–1185. https://doi.org/10.54648/cola2018095.

Hui, A. T., Ahn, S. S., Lye, C. T., & Deng, J. (2021). Ethical challenges of artificial intelligence in health care: A narrative review. *Ethics in Biology, Engineering and Medicine: An International Journal, 12*(1), 55–71. https://doi.org/10.1615/ethicsbiologyengmed.2022041580.

Kuhail, M. A., Berengueres, J., Taher, F., & Al Kuwaiti, M. (2024). Challenges of designing haptic experiences. In *Advances, applications and the future of haptic technology* (pp. 43–53). Springer Science and Business Media LLC. https://doi.org/10.1007/978-3-031-70588-5_4.

Levine, S., Finn, C., Darrell, T., & Abbeel, P. (2021). End-to-end training of deep visuomotor policies. *Journal of Machine Learning Research, 17*(1), 1334–1373.

Mackenzie, C. F., Harris, T. E., Shipper, A. G., Elster, E., & Bowyer, M. W. (2022). Virtual reality and haptic interfaces for civilian and military open trauma surgery training: A systematic review. *Injury, 53*(11), 3575–3585. https://doi.org/10.1016/j.injury.2022.08.003.

Maheshwari, S., Makwana, H., & Lal, D. K. (2024). Real-time haptic rendering: Perception, optimization, and multi-modal integration. *Library of Progress-Library Science, 44*(3). https://doi.org/10.48165/bapas.2024.44.2.1.

Mueller, S. M., & Grunwald, M. (2023). Perceptual dimensions of the haptic system. In *Human touch in healthcare: Textbook for therapy, care and medicine* (pp. 1–41). Springer Berlin Heidelberg. https://doi.org/10.1007/9783662678602_1.

Okamura, A. M. (2022). Haptic feedback in robot-assisted minimally invasive surgery. *Current Opinion in Urology, 32*(1), 76–81.

Okamura, A. M. (2004). Methods for haptic feedback in teleoperated robot-assisted surgery. *Industrial Robot: An International Journal, 31*(6), 499–508. https://doi.org/10.1108/01439910410566362.

Pacchierotti, C., Sinclair, S., Solazzi, M., Frisoli, A., Hayward, V., & Prattichizzo, D. (2017). Wearable haptic systems for the fingertip and the hand: Taxonomy, review, and perspectives. *Transactions on Haptics, 10*(4), 580–600. https://doi.org/10.1109/TOH.2017.2689006.

Qaisar, T., Mumtaz, M., Raza, H., Ali, M. M., Khalid, K., & Bajwa, I. (2024). Emotional engagement with haptic feedback in virtual scenarios: A literature review. In *2024 Proceedings - 2024 international conference on engineering and computing, ICECT*. Institute of Electrical and Electronics Engineers Inc https://doi.org/10.1109/ICECT61618.2024.10581267.

Rajkomar, A., Hardt, M., Howell, M. D., Corrado, G., & Chin, M. H. (2018). Ensuring fairness in machine learning to advance health equity. *Annals of Internal Medicine, 169*(12), 866–872. https://doi.org/10.7326/M18-1990.

Selim, M., Dresscher, D., & Abayazid, M. (2024). A comprehensive review of haptic feedback in minimally invasive robotic liver surgery: Advancements and challenges. *The International Journal of Medical Robotics and Computer Assisted Surgery, 20*(1). https://doi.org/10.1002/rcs.2605.

Shazhaev, I., Mihaylov, D., & Shafeeg, A. (2023). A review of haptic technology applications in healthcare. *Open Journal of Applied Sciences, 13*(02), 163–174. https://doi.org/10.4236/ojapps.2023.132013.

Sheik, A. T., Atmaca, U. I., Maple, C., & Epiphaniou, G. (2022). Challenges in threat modelling of new space systems: A teleoperation use-case. *Advances in Space Research, 70*(8), 2208–2226. https://doi.org/10.1016/j.asr.2022.07.013.

Simsek, M., Aijaz, A., Dohler, M., Sachs, J., & Fettweis, G. (2016). 5G-enabled tactile internet. *IEEE Journal on Selected Areas in Communications, 34*(3), 460–473. https://doi.org/10.1109/JSAC.2016.2525398.

Srinivasan, M., Lee, S., & Kim, Y. (2023). Optimization in AI-based haptic systems: A review. *Robotics and Autonomous Systems, 162.*

Steinbach, E., Hirche, S., Kammerl, J., Vittorias, I., & Chaudhari, R. (2011). Haptic data compression and communication. *IEEE Signal Processing Magazine, 28*(1), 87–96. https://doi.org/10.1109/MSP.2010.938753.

Torabi, A., Nazari, A. A., Conrad-Baldwin, E., Zareinia, K., & Tavakoli, M. (2021). Kinematic design of linkage-based haptic interfaces for medical applications: A review. *Progress in Biomedical Engineering, 3*(2), 022005. https://doi.org/10.1088/2516-1091/abee66.

Mosqueira-Rey, E., Hernandez-Pereira, E., Alonso-Rios, D., Bobes-Bascaran, J., & Fernandez-Leal, A. (2022). Human-in-the-loop machine learning: A state of the art. *Artificial Intelligence Review, 56*, 3005–3054.

Winfield, A. F., Michael, K., Pitt, J., & Evers, V. (2019). Machine ethics: The design and governance of ethical AI and autonomous systems. *Proceedings of the IEEE, 107*(3), 509–517. https://doi.org/10.1109/jproc.2019.2900622.

Wu, X., Xiao, L., Sun, Y., Zhang, J., Ma, T., & He, L. (2022). A Survey of Human-in-the-loop for Machine Learning. *Future Generation Computer Systems, 135*, 364–3810.

Zhang, J., Shu, Y., & Yu, H. (2023). Fairness in design: A framework for facilitating ethical artificial intelligence designs. *International Journal of Crowd Science,* 7(1), 32–39. https://doi.org/10.26599/IJCS.2022.9100033.

CHAPTER 9

Design frameworks for multimodal haptic medical robots

Divya Pokar Patel and Nivedita Ravindra Dharne
School of Business Analytics, Woxsen University, Hyderabad, Telangana, India

9.1 Introduction

Healthcare facilities have encountered broad transformation since robots and automation technology have pervaded all areas of their operations. Medical technology innovations have transformed various operational aspects of clinical processes and enhanced levels of performance as well as patient security. **Haptic medical robots** now represent a transformative medical device that changes surgical and rehabilitative practices for clinicians. These robotic systems convey touch-associated information that allows medical operators to feel tissue contacts, strength changes, and resistance levels. The haptic sense simulation technology enables medical doctors to blend their instincts with machine precision using touch feedback in very demanding medical treatments that require outstanding accuracy and control, as per Iftikhar et al. (2024).

Advanced medical robots incorporating haptic features have wide applicability in many fields, such as **minimally invasive surgery, telemedicine, and rehabilitation treatment,** due to the fact that they effectively create replication of touch (Selim et al., 2024). The devices tap into modern-day **sensor–actuator** platforms to feel and create a sense of touch so that medical specialists are able to perform intricate interventions with greater accuracy, rapid response times, and added dexterity. The **da Vinci surgical system** delivers haptic feedback that gives resistance feedback to surgeons when they work with tissues, leading to sutures or cuts, ensuring surgeons do not injure fine tissues. Feedback mechanism allows surgeons to have instrument control while minimizing the risk of medical errors, finally providing improved treatment outcomes.

Haptic robots are determinative in **rehabilitation therapy** since they offer imperative sensory feedback to patients who regain their health following injuries and operations. The robots facilitate guided training through their ability to offer real-time tactile feedback, which encourages patients to restore movement skills in addition to strength and coordination by recurring therapeutic exercises (Okamura, 2009). Quick tactile feedback input using these devices enables patients to learn proper movement positions, which enables them to record improved recovery results and rehabilitation outcomes. This capability offers the important advantages to stroke patients and survivors

Haptic Feedback Control in Medical Robotics
DOI: https://doi.org/10.1016/B978-0-443-27777-1.00013-5

of spinal cord and orthopedic surgery due to its facilitation of the recovery of muscle functions and self-mobility.

Haptic medical robots make safer patient care with more accessible healthcare while reducing treatment cost levels. The technological advances are likely to guide healthcare innovation ahead through the development of more accurate and customized medical procedures that function with higher efficiency (Mühlbauer et al., 2024). Haptic medical robots will become ever more important amid healthcare changes due to technological development, which will create further opportunities to innovate patient care.

9.2 Literature review

Recent medical robotics research exhibits rapid progress due to haptic feedback systems as indispensable tools for more precise surgery. The ground-breaking paper *Haptic Feedback in Robot-Assisted Minimally Invasive Surgery* analyzes contemporary technical features of haptic feedback in surgical robots, even though it was published early. Physical phenomena and touch sensations that take place in surgical hands directly influence their surgical performance and reduce the likelihood of making mistakes. The authors extend the current haptic system capability to address latency problems and multiple feedback integration challenges in existing haptic technology. This review article examines the influence of artificial intelligence (AI) on robotic surgical systems, drawing on insights from the article "Revolutionizing Robotic Surgery." AI algorithms have presented three essential functions, such as predictive model prognostications, real-time decisions, and learning processes, which overcome robotic limitations. AI offers the prospect of enhancing robotic surgery efficacy along with precision, thereby lowering surgery risk as well as procedural variation, as per the conclusion of the research paper.

The paper critiques *Deep Learning-Based Haptic Guidance for Surgical Skills Transfer*, which discusses how deep learning (DL) replicates expert surgical maneuvers using haptic guidance to train novice surgeons. The writers prove in this paper how DL mechanisms advance skill training methods and reduce learning time, in addition to final result enhancement, especially in surgical areas. The study illustrates strategic connections between AI technology and haptic feedback systems as measures for advancing medical robotic sectors. The paper *Multimodal Haptic Feedback for Grip Force Reduction in Robotic Surgery: How Multiplex Haptic Feedback Can Improve Robotic Surgery Grip Force Control* presents evidence that excessive grip force can lead surgeons to cause tissue injury and compromise movement accuracy. By integrating multimodal feedback—combining tactile input with body-motion cues—the research illustrates how surgeons can achieve improved force-control skills and greater precision in robotic procedures. Several research studies verify that robotic surgery works more effectively with various types of sensory inputs, which are mixed to increase security levels and accuracy performance. The reduction of feedback latency and minimization of sensory integration require further research emphasis because these problems hinder effective operations. The research paper, published in: *The Benefits of Haptic Feedback in Robot-Assisted Surgery and Their Moderators* assesses several studies through meta-analysis to determine haptic feedback effects on the outcomes of surgery. The study confirms that haptic feedback provides three fundamental advantages that make up superior accuracy while enhancing tissue preservation and accelerating training time. The authors conducted a study of performance variables, which led them to identify findings related to haptic feedback type and surgical complexity levels alongside experiential conditions. System calibration protocols, user learning abilities, and adjustments define the performance effectiveness of haptic feedback for improving surgical performance. Existing haptic technology needs improvement to maximize its benefits to robotic-assisted surgery procedures.

9.3 Haptic medical robots

9.3.1 Definition and working principle

Paradigm medical robots apply force feedback technology to provide medical personnel with feelings of touch, such as pressure, along with resistance and texture during medical interventions (Zhou et al., 2017). This technology brings together forces from sensors, actuators, and control systems, which cooperate to replicate tactile feelings that provide accurate feedback in real time during medical procedures.

9.3.2 Components of haptic medical robots

- The **sensors** track different factors, such as force and pressure, in addition to movement, by which robots log their interactions with patients or medical equipment.
- By applying physical force, **actuators** assist users to feel immediate feedback, which comprises resistance and roughness.
- The **control algorithms** perform information obtained from sensors to drive the actuators, which alters the way the robot interacts with environmental factors and tasks allocated.
- **User interface** is an element of the system that offers healthcare practitioners two choices to control robot interactions, either by means of physical controllers or virtual interfaces, to ensure smooth and user-friendly operation.

9.3.3 Types of haptic medical robots

- The robotic surgery system provides accurate, minimally invasive **surgical procedures** that aid surgeons in the precise execution of complex procedures in small, hard-to-reach spaces.
- **Rehabilitation robots** assist patients by guiding therapy exercises and providing resistance feedback to help patients acquire controlled movement.

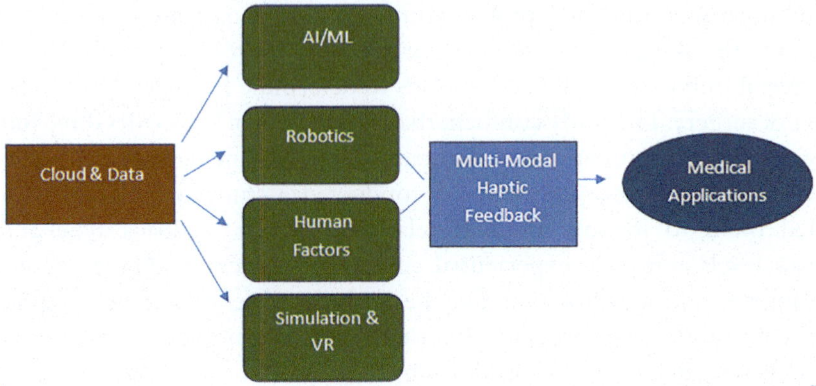

Figure 9.1 The given system for a multi-modal medical robot overcomes current system issues through integration with enhanced sensory functionality, AI-enabled decision systems, and cloud connectivity.

- These robots create real-life medical scenarios to enable medical practitioners and students to utilize **simulated training** facilities that mimic the actual diagnosis or surgery touch experiences.

The healthcare industry has been revolutionized by medical robots that offer precise control and simulation of the human touch. Robotic surgery systems with AI features, along with human touch abilities, have been found effective in successful surgery operations, remote entity control operations, patient care education, and physical therapy deliveries.

9.4 Applications, advantages, and limitations

In this section, we will discuss on the appliations, advantages, and limitations of heptic medical robots (Fig. 9.1 and Table 9.1).

9.5 Proposed multimodal medical robot

The given system for a multimodal medical robot overcomes current system issues through integration with enhanced sensory functionality, AI-enabled decision systems, and cloud connectivity.

9.5.1 Phase 1: Foundation—Cloud and data infrastructure

- *Data acquisition and storage*: A secure system retrieves pertinent medical information from multiple sources, such as patient history, surgical footage, and sensor data, to save them in a cloud-based system. The data obtained is the critical fuel AI/ML requires, and at the same time, serves as a critical data set used during the analysis process.

Table 9.1 Exploring the applications, benefits, and constraints of haptic medical robots.

Application	Description	Example	Advantages	Limitations
Minimally invasive surgery	Enables surgeons to perform surgeries through small incisions with enhanced control, thanks to tactile feedback.	da Vinci surgical system (prostate, gynecological surgeries)	Improved precision and control, reduced risk of human error, smaller incisions, quicker recovery, reduced scarring	High cost, complexity, and training requirements, limited haptic sensitivity in delicate procedures
Telemedicine and remote surgery	Allows surgeons to perform surgeries remotely, transmitting tactile feedback in real-time to enhance precision.	Remote surgery systems (telemedicine)	Enables access to expert surgeons in distant regions and reduces healthcare accessibility issues	Latency and feedback delays, complexity in transitioning from traditional systems
Rehabilitation therapy	Assists patients in recovering from injuries by providing real-time tactile feedback during physical therapy exercises.	MIT-Manus (stroke rehabilitation)	Accelerates recovery, improves patient outcomes, and provides real-time feedback for correct movement and position (Wang et al., 2024)	High cost, training requirements, limited range of tactile sensations in some therapies
Medical training and simulation	Provides realistic surgical simulations with tactile feedback, allowing students to practice without patient risk.	Touch surgery platform (virtual reality simulators)	Realistic practice enhances medical training and improves proficiency without patient risk	High cost of simulators, limited tactile range in some training modules
Surgical skill training	Provides a safe environment for medical professionals to practice surgical skills with feedback to simulate real situations.	SimSurgery haptic training system	Improves surgical technique, reduces training time, enhances learning efficiency	Requires extensive training, lacks full sensory fidelity compared to real-life surgeries

(*continued on next page*)

Table 9.1 Exploring the applications, benefits, and constraints of haptic medical robots—cont'd

Application	Description	Example	Advantages	Limitations
Robotic-assisted biopsy	Allows precise tissue sampling with tactile feedback, improving accuracy in targeting and extracting tissue.	Robotic biopsy systems (e.g., intuitive surgical's biopsy robotics)	Increased accuracy in locating and extracting samples, reduced patient trauma, improved diagnostic results	Complexity in system setup, high cost, integration with other diagnostic systems
Orthopedic surgery assistance	Provides support for joint replacements and bone procedures by offering tactile feedback to ensure correct placement.	RAS joint systems (orthopedic surgery robots)	Enhanced precision, reduces surgery time, improves post-surgical recovery	High cost, challenges in adapting to different body types, specialized training needed
Pain management	Delivers targeted haptic feedback to stimulate nerves and reduce pain in certain medical conditions (Fekri et al., 2021)	Haptic-based TENS units (transcutaneous electrical nerve stimulation)	Noninvasive pain management, quick relief, personalized treatment based on tactile response	Effectiveness varies per patient, need for continuous calibration, and limited evidence for chronic pain treatments
Neurorehabilitation	Offers real-time sensory feedback to aid in the rehabilitation of neurological injuries, improving motor function recovery.	Neofect smart glove	Enhances recovery for neurological disorders like stroke, and provides real-time feedback to optimize exercises	Limited effectiveness in severe neurological conditions requires extensive patient engagement and motivation
Patient monitoring and diagnostics	Provides real-time tactile feedback to assist in diagnosing conditions through touch-based data collection (Fekri et al., 2021)	Haptic sensors for diagnostic procedures	Improves diagnostic accuracy, allows for early detection of diseases, and enhances patient comfort	Data interpretation challenges, potential for misinterpretation of tactile signals

- *Platform development*: Creating cloud-based platforms for managing data, remote access, shared tools, and secure communication. The platform shall act as the overall support of the entire system.
- *Security and privacy*: A secure and reliable security infrastructure needs to be created to secure precious patient information and uphold privacy requirements.

9.5.2 Phase 2: Core technologies—Development and integration

This phase happens concurrently, as these technologies are interdependent:

- *2A. AI/ML development:*
    - *Algorithm selection and training*: The group adopts appropriate machine learning (ML) techniques (DL, reinforcement learning, and so forth) that process information during Phase 1.
    - *Sensor fusion and data processing:* AI models should be created that bring together a variety of sensors for force, vibration, and temperature measures that create collective surgical environment intelligence.
    - *Control algorithm development*: Auscultate required control algorithms for robots to process surgical inputs of surgeons with appropriate haptic feedback.
- *2B. Robotics development:*
    - The **hardware component design and selection** process involves robotic arms in conjunction with end-effectors, actuators, and sensors that work best for multimodal haptic feedback.
    - AI **haptic rendering research** focuses on transforming virtual and remote world information into tangible haptic sensations for end-users.
    - *Control system integration*: Merging the robotic hardware with the AI-driven control algorithms.
- *2C. Human factors engineering:*
    - **User interface (UI) design** development is aimed at creating simple user interfaces and easy-to-use designs for surgeons to control the robotic system while receiving different kinds of haptic feedback.
    - Medical user testing will assess system usability and performance through **usability testing** between professionals and the robotic system.
    - Surgeons will be provided with training sessions that instruct efficient robotic system use and correct interpretation of multimodal feedback through **protocol development**.
- *2D. Simulation and VR development:*
    - Healthcare volunteers would apply **virtual environment creation** to produce detailed virtual areas that replicate actual surgical procedures for simulation and training.
    - **Haptic rendering algorithms** programming allows the system to provide realistic touch sensations in simulation environments.

- There would be a **realistic training framework** created for medical personnel to utilize the haptic robotic system within safe simulated clinical environments.

9.5.3 Phase 3: Multimodal haptic system integration

- *Hardware and software integration*: Integration of the robotic hardware, AI-based control algorithms, and haptic rendering software into an integrated system.
- *Multimodal feedback calibration*: The system necessitates multimodal feedback calibration, which allows for accurate and reliable presentation of force, vibration, and temperature, and other tactile forms of feedback.
- *System testing and validation*: The integrated system is systematically tested through controlled simulated conditions for establishing its safety, along with reliability and efficacy.

9.5.4 Phase 4: Medical applications

- As far as **clinical trials** are concerned, we must assess how the multimodal haptic robotic system works through actual healthcare procedures.
- Clinical application and medical staff training for the use of the technology in actual medical procedures must be **implemented** in clinical settings.
- The users **offer feedback** to produce continuous updates of the system through clinical practice and by integrating technology updates into the process. The process loops back repeatedly to earlier phases, like AI retraining based on new data and hardware updates.

The systematic path traces the diagram pattern to illustrate the particular functions that each element performs in medical devices utilizing multimodal haptic technology. The process illustrates an iterative process since feedback and new information allow for improvement of outcomes in previous stages.

9.6 Artificial intelligence and machine learning in haptic medical robots

Haptic medical robots have been significantly improved by AI and ML in terms of their measurable properties, as well as their adaptability, responsiveness, and precision within healthcare environments (Wang et al., 2024). A real-time process of breaking down large datasets of sensory data allows robots to understand touch feedback while guiding healthcare staff with precise procedures. ML-based AI algorithms assist **haptic medical robots** in getting improved feedback through the analysis of force and pressure and robot motion data to enhance robotic movement. The system allows for better accuracy in taking care of sensitive types of treatment like robotic-assisted surgical procedures that need sensitive touch perception (Fekri et al., 2021). The robots' **adaptive learning ability** allows them to learn from past experiences, which results in changing reactions and constant performance improvement over time (Iftikhar et al., 2024). AI serves as a critical component for **error detection and prevention** systems (Zhou et al., 2017). Haptic robots' AI controllers can recognize upcoming hazards from excessive pressure forces and improper tool placement, causing automatic surgical safety adjustments. Robots acquire the capability to identify varied tissue types through ML analysis. This allows surgeons to adjust their grip strength and enhance instrument handling. A combination of force sensor feedback, motion tracking feedback, and computer vision feedback is made viable through AI and ML algorithms. Combining the above medical technologies is made easier by improving depth perception and granting instantaneous support to medical professionals. Space-based models adopted in AI-controlled haptic robots make it easier for them to predict the behavior of surgeons that results in more controlled and more streamlined surgery operations. AI and ML also help with **data processing and sensory integration** by merging force sensor inputs, motion tracking, and computer vision data. This integration improves depth perception and offers real-time help to medical experts. For example, AI-powered haptic robots can utilize predictive models to anticipate a surgeon's next action, resulting in smoother and more regulated procedures. AI-based haptic simulators in medical **skill training and improvement** allow medical students to acquire complex procedures within simulated, controlled environments. Simulations of training provide real-time visual feedback that enables medical personnel to hone their skills before treating real patients (Selim et al., 2024). The performance of trainees is evaluated by AI systems prior to presenting them with personalized recommendations for building their capabilities.

9.6.1 Integration of artificial intelligence and machine learning in haptic systems

- To implement AI and ML in haptic robots, multiple technological components, including computer vision coupled with DL and sensor fusion, must be interlinked. Some sensory interpretations come into robots from these technologies, which allows them to give exact and quick replies.
- The forces that are demanded for force control must be foresighted to minimize damage during procedures that are executed softly.
- The use of reinforcement learning helps haptic responses improve continuously until human interactions become more natural and efficient.

9.6.2 Impact on medical training and skill enhancement

AI-operated haptic simulation technology allows medical students to practice surgical procedures without endangering patients at all. Simulations provide real-time feedback, which allows students to develop flawless movement and precision irrespective of human patients. AI technology assesses performance metrics to identify training gaps so that it can personalize defined programs for specific students. The medical sector gains improved accuracy as well as safety when integrating AI and ML technology into haptic medical robots for training systems. Robot-assisted medical procedures are enhanced by these advancements that improve their operational efficiency and accessibility, according to Bergholz et al. (2023). The development of AI and ML technology has encouraged great improvement in haptic medical robots, using their precision control operations, minimization of errors, and enhancement of medical education techniques. Safety and operation efficiency of robot-assisted operations have been enhanced through recent improvements, which enhance medical therapy efficacy and availability.

9.7 Conclusion and future scope

Medical robots incorporating haptic systems will revolutionize medicine by providing greater surgical accuracy while enhancing rehab outcomes as well as remote medical operations. Haptic feedback integration into robotic-assisted surgeries is advantageous because it manages to both facilitate safer surgeries by minimizing errors and improve accuracy during complicated procedures. Robot learning is enhanced due to AI models because they provide on-demand progress of surgical techniques and adaptive automated responses during operations. Deep haptic guidance technology quickens learning in medical school by means of training protocols and maximizes performance in medical facilities (van der Meijden and Schijven, 2009). The new multimodal system advances haptics technology by integrating AI automation, tactile output, and remote medical records retrieval capabilities. These various improvements yield higher results.

9.8 Future directions

1. *5G connectivity for remote surgeries*: Remote surgical procedures are made possible through rapid responses using the integration of 5G technology that reduces latency. The rapid low-delay communication infrastructure allows surgeons to perform optimally with robotic instruments, which render telesurgery an effective medical treatment for far patients.
2. *Augmented reality (AR) and virtual reality (VR) integration*: Surgical training equipment and enhanced visualization of the operating field will be the outcomes of haptic robotic technology combined with AR and VR systems. The combination of essential patient data with procedural information and anatomical material within AR-augmented robotic surgery systems enables surgeons to have high accuracy while monitoring their operating space.
3. *Cost reduction strategies for wider adoption*: The purchase of haptic robotic systems is hindered by implementation challenges because their acquisition prices are still on the higher side. Development efforts should focus on low-cost haptic robot product

development and programming, and flexible manufacturing processes to make these systems reach global accessibility to healthcare facilities.

4. *AI-driven autonomous surgical assistance*: Medical robots will gain self-defined decision-making ability through advancements in AI that allow them to process clinical information and detect errors while creating robotic surgery protocols for surgeons. The system updates reduce human operator errors through thorough surgical precision that extends across all levels of medical proficiency.
5. *Personalized and adaptive haptic feedback systems*: Medical robots will have autonomous decision-making powers through AI technology that provides predictive clinical data, error identification, and automation of surgery procedures for surgeons. Other applications can be made of existing haptic feedback systems available on the current market. The conduct of intuitive surgery operations requires further development, including adaptive feedback systems that account for the surgeon's individual preferences and unique patient conditions.
6. *Enhanced security and cloud-based data sharing*: Some minimum requirements for cloud computing-based medical robots and AI models involve effective data security and privacy protection. Patient records, as well as recordings of surgeries and real-time telemetry data, must be protected by enforcing encryption systems that integrate with data transmission security mechanisms. Contemporary haptic medical robots are the building blocks of healthcare in the future due to ongoing innovations in robotics and AI alongside haptic feedback technologies, as explained by Wang et al. (2024). They will set new healthcare standards through addressing current constraints in current conditions, along with foreseen development approaches that enhance the quality of clinical practices, medical training, and remote surgery capabilities to increase global healthcare successes.

AI disclosure

During the preparation of this work, the author(s) used ChatGPT to improve the readability and language of the work. After using this tool/service, the author(s) reviewed and edited the content as needed and take(s) full responsibility for the content of the publication.

References

Bergholz, M., Ferle, M., & Weber, B. M. (2023). The benefits of haptic feedback in robot-assisted surgery and their moderators: A meta-analysis. *Scientific Reports, 13*(1). https://doi.org/10.1038/s41598-023-46641-8.

Fekri, P., Dargahi, J., & Zadeh, M. (2021). Deep learning-based haptic guidance for surgical skills transfer. *Frontiers in Robotics and AI,* 7. https://doi.org/10.3389/frobt.2020.586707.

Iftikhar, M., Saqib, M., Zareen, M., & Mumtaz, H. (2024). Artificial intelligence: Revolutionizing robotic surgery: review. *Annals of Medicine & Surgery, 86*(9), 5401–5409. https://doi.org/10.1097/ms9.0000000000002426.

Mühlbauer, M., Hulin, T., Weber, B., Calinon, S., Stulp, F., Albu-Schäffer, A., & Silvério, J. (2024). A probabilistic approach to multi-modal adaptive virtual fixtures. *IEEE Robotics and Automation Letters, 9*(6), 5298–5305. https://doi.org/10.1109/lra.2024.3384759.

Okamura, A. M. (2009). Haptic feedback in robot-assisted minimally invasive surgery. *Current Opinion in Urology, 19*(1), 102–107. https://doi.org/10.1097/MOU.0b013e32831a478c.

Selim, M., Desscher, D., & Abayazid, M. (2024). A comprehensive review of haptic feedback in minimally invasive robotic liver surgery: Advancements and challenges. *The International Journal of Medical Robotics and Computer Assisted Surgery, 20*(1). https://doi.org/10.1002/rcs.2605.

van der Meijden, O. A. J., & Schijven, M. P. (2009). The value of haptic feedback in conventional and robot-assisted minimal invasive surgery and virtual reality training: A current review. *Surgical Endoscopy, 23*(6), 1180–1190. https://doi.org/10.1007/s00464-008-0298-x.

Wang, T., Zheng, P., Li, S., & Wang, L. (2024). Multimodal human–robot interaction for human-centric smart manufacturing: A survey. *Advanced Intelligent Systems, 6*(3). https://doi.org/10.1002/aisy.202300359.

Zhou, J., Li, Y., Zhang, Y., Liu, G., Tan, H., Hu, Y., Xiao, J., & Shi, H. (2017). Solitary ground-glass opacity nodules of stage IA pulmonary adenocarcinoma: Combination of 18F-FDG PET/CT and high-resolution computed tomography features to predict invasive adenocarcinoma. *Oncotarget, 8*(14), 23312–23321. https://doi.org/10.18632/oncotarget.15577.

CHAPTER 10

Energy efficiency in haptic medical devices

Nikhil Sharma[a], Pooja Choudhary[b] and Jofia Jose Prakash[c]
[a]Business Analyst, DP World, Hyderabad, Telangana, India
[b]School of Business Analytics, Woxsen University, Hyderabad, Telangana, India
[c]Computer Society, Aldie, VA, United States

10.1 Introduction

In the early 2000s, the medical field began implementing touch-based technology into surgical robotics, allowing for remote sensor-based procedures. This technology provided surgeons the ability to remotely assist in surgical processes by gaining the ability to sense the resistance or compliance of tissues that lie beneath the skin, and differentiate between structures, during open surgery, enabling greater precision and control of tissues and structures, while also assisting in the identification of, and addressing surgical challenges in the target organs and structures, and thereby improving surgical outcomes (Okamura, 2009). Rehabilitation robotics and patient assistive technologies employ a technology called force compliance control, in which a rehabilitative robotic device applies a control force to a patient during the patient's voluntary movements. This technology adjusts the level of support offered by the robotic device, thereby enabling the patient to relearn movements. Elastic resistance or counter-resistance to patient movements. A similar telemedicine application uses virtual touch technologies to enable clinicians to perform remotely accessible physical examinations. Temporal digital tactile simulation allows clinicians to perform remote physical examinations. Digital touch technologies restore tactile sensation to automated, remote care processes.

Integrating haptic technology with real-world medical applications faces a unique challenge pertaining to power usage. Unlike games or virtual reality technologies, apparatus used in hospitals must be subject to stringent rules regarding energy consumption. There are mobile units in which battery size is one limitation, excessive heat can be another, as it can be detrimental to a patient. Surgeries can last up to 6 h and require continuous tactile feedback signals (Rosen et al., 2006). Similarly, recovery devices will require sufficient power to be operational all day, which may aggravate or push against the limbs of the patient. Each of these factors creates the need for engineers to design devices with more considerations. There can be more emphasis on power-saving measures than in other applications, as it will dictate the design in its entirety.

Haptic Feedback Control in Medical Robotics
DOI: https://doi.org/10.1016/B978-0-443-27777-1.00011-1

What is the challenge? Advanced touch feedback technology will require rapid motor response, rapid cycling of the controlling processor, and strong actuation, all of which consume a lot of power. More electricity will be consumed as the overall system is designed with more precision and the individual components are optimized. The tradeoffs will be the most pronounced when designing battery-powered medical devices.

The primary focus of this work revolves around the examination of the utilization of power in tactile medical instruments. To begin with, some of the most rudimentary mechanisms that these instruments utilize to produce energy and motion come to the forefront. Following this, the greatest amount of energy consumption becomes evident, especially with some reasonable scrutiny. Recent techniques that have been developed to reduce this energy consumption come next in order. Strategies that focus on the efficient prioritization of energy consumption follow closely. The utilization of energy from the environment becomes more pronounced later in the conversation. Empirical evidence collected from medical and rehabilitative therapeutic institutions serves to bolster the argument further. The criteria for satisfactory outcomes seem to be delineated toward the conclusion. The voids that remain in the research context present the greatest challenges. The range of potential future directions for this work is somewhat more confined. The concluding remarks integrate the remaining disparate comments in order.

10.2 Fundamentals of haptic medical devices

10.2.1 Haptic feedback mechanisms

Tiny vibrations travel through fingers at intervals in the region of 10–1000 times/s to discern texture. Such phenomena are present in medical education, where, to illustrate, what is the surface of an organ, or the interface of a tool to a user through a vibrating alert (Prattichizzo et al., 2012). On the other end, arms and legs detect a sense of motion and exertion. This is because internal organs have a mechanism to sense muscle elongation and the angular position of the joints (Tzafestas, 2003). Machines provide the illusion of touch, where tissues are firm to the touch, and just as guiding touch through force, recovery exercises are conducted where the affected limbs have to work against a predetermined resistance. While touch is concerned with details, motion is focused on the overall weight and the direction of movement.

The main actuators in an active haptic system are motors, and with movement, the physical interface "feels" real. This is used in virtual environments, as well as in rehabilitation. These systems are, on the other hand, energy-hungry. They are responsive, and can be designed in a way to provide energy-glutton status, low for minimal energy consumption. In contrast, fully passive systems, as opposed to a motor system, are used, which may be impeding any form of feedback to the user through applied effort, but paradoxically are designed in a way to work without the application of energy to the system. More effort is required to implement both to create a hybrid system, where, in this case, the effort is minimal. The function may be achieved to reduce energy consumption and maintain active feedback.

10.2.2 Actuation technologies

Most medical devices utilize electric motors since they are small and have predictable and safe performance (Van der Linde et al., 2002). Despite brushless direct current (DC) motors and permanent magnet synchronous motors achieving 80%–90% efficiency, they heat up during continuous operation. Voice coils are good for fast tactile feedback; however, they are only about 50% efficient and have poor energy feedback. Newer designs, such as coreless and slotless designs, achieve better output/lower resistance during motion, thus conserving energy during critical areas.

Need to lift heavy things without using large components? Pneumatics and hydraulics are ideal for therapy devices and surgical instruments because they provide power in a very compact package. A little motion, such as with air-based systems, can provide a cushioning effect that is very useful in machines that operate close to people. However, air compressors are often very audible, are energy wasters, and leaks are common, slowing operation. Furthermore, refined adjustments are unachievable. While fluid pressure systems provide a large amount of force for a given volume of fluid, large pumps are required, which poses problems if the fluid is spilled near a person on a surgical table. When considering fluid dynamics, as well as the energy that is lost in the conversion of energy forms and the resistance to flow, both systems fall short of electric motors that effectively convey energy directly via wires.

They are coming out of the lab; these smart materials are taking the spotlight as new alternatives for moving parts without typical motors. Piezoelectric devices are defined by quick movements—high voltage is a big constraint for piezoelectric devices, as they are only able to create quick, low-distance movements; they can create quick "taps" but are not able to create "sweeping" type movements. Some more "liquid" type materials that can stretch are also slowed by heat (think of a liquid as a foam that can stretch and only uses a minimal current to thicken). This is good for dials that need a slight push backup, while overheating risks loom, the foam can also "shift" without warning during use. Some more "liquid" type materials that can stretch are also slowed by heat (think of a liquid as a foam that can stretch and only uses a minimal current to thicken). This is good for dials that need a slight push backup, while overheating risks loom, the foam can also "shift" without warning during use (Table 10.1).

10.2.3 Sensing and control architecture

The utilization of feedback loops that incorporate a variety of sensors is necessary to obtain precise touch feedback. Rather than relying on a single type of sensor, a system

Table 10.1 Comparison of actuation technologies for haptic medical devices.

Technology	Efficiency (%)	Power density (W/kg)	Response time (ms)	Typical power (W)	Best application
Electric motor (Brushless DC)	80–90	500–1000	5–15	10–100	General-purpose haptics
Pneumatic actuator	40–50	800–1500	30–100	50–200	Compliant interaction
Hydraulic actuator	55–65	2000–3000	20–50	100–500	High-force applications
Piezoelectric	50–60	400–800	0.5–2	1–20	High-freq tactile feedback
Shape memory alloy	30–40	300–600	200–1000	5–50	Large displacement tasks
Magnetorheological fluid brake	65–75	200–400	2–10	2–15	Semiactive control

that combines multiple inputs is more likely to provide accurate feedback. Strain gauges and capacitive sensors provide force and torque feedback that indicate the intensity of a user's push on the machine. Based on the feedback received, the machines will vary the level of resistance provided. In isolation, this type of sensor feedback system is insufficient. The movement feedback that gets processed in the system through the use of optical encoders or magnetic resolvers that track the motion of the system's components down to extremely small (sub-micrometer) distances. Accuracy in tracking is critical as small tracking errors tend to accumulate substantially during high velocity operations. Sensor operation tends to go unnoticed; however, the consumption of electrical power invariably increases in proportion to the number of sensors activated. Even a fully operational six-axis force sensor will consume between 2 and 5 watts of power, surprisingly, surprisingly, surprisingly. The operational consumption of power increases even more with the addition of multiple position sensors, where each minutely position detector is also operationally consuming 1–3 watts. Therefore, the combination of operating multiple channels simultaneously turns minor operational consumption of each system into a significant operational consumption. The operational performance of the system tends to either operationally consume significant power or operationally consume no power. With the wear and operational consumption of the system, design parameters for the system are influenced.

With haptic systems, multiple attributes should work simultaneously, such as keeping the system stable, maintaining a natural feel to the user, and conserving power. Rather than just tracking motion, impedance systems control how a user perceives a force in relation to their motion. This is useful for creating a digital system that simulates stiff or rigid objects. Conversely, admittance systems shift the point of contact when a user pushes the object and work well for larger systems (Van der Linde et al., 2002). Impedance haptic devices must maintain a constant viscosity, which requires the processor to do a lot of computational work to reduce latency. Recently, many designers have integrated small computers that perform specialized number-crunching directly into the haptic devices. Resource-conserving, high-speed computing has led to systems on chips being used in haptic devices.

10.3 Energy consumption in haptic medical systems

10.3.1 Sources of energy loss

Heating occurs in touch-feedback machines due to the energy losses in a myriad of ways. Not all the energy consumed in electric motors gets converted to movement; in fact, a good quality motor will still lose about 10% to just heating wire and metal components. When the system exerts power in a push, and the system does not translate that to movement, that power gets wasted. When a system has to move a certain distance, that distance creates friction in the system due to the resistance from the system's parts. That energy, lost from the system, is converted to heat. The movement is not perfectly transferred through the machines; there will always be energy loss from the system. When motion is transmitted through a gearbox, some energy is lost due to friction among the components. A substantial amount of energetic loss occurs due to friction in the components, and surgical robots prefer to use cables to drive the mechanism rather than gears because it is more preferable. However, those cables rub against guide wheels, and there is always a loss of energy to friction at the pivots and rods of the mechanism when the parts bend due to stress. When a feedback system is required to touch the tissue, the system is designed to have a weak mechanism. Thus, the system requires larger motors to make it more powerful to counter the losses. As a result, the power requirements increase.

As systems require more computing strength, the power consumed by the controls and signals keeps increasing. Operating today's haptic controllers requires the simulation of environments, the detection of collisions, and the recognition of touch inputs across multiple senses. Some small control units constantly consume power in the range of 5–15 watts, the same as the motors they drive when they are under light load. The various tasks associated with cleaning sensor data, digitizing it, and moving data create a significant process load. Even small power draws accumulate quickly. Collectively, in modern systems, the electronic components consume a substantial portion, typically 20%–40%, of the power.

10.3.2 Power constraints in medical applications

When using batteries for running a device, a number of restrictions are in place, which include the total operational run time, the size of the device, and the estimated time of usage for the device. Consider a rehab suit you would have to wear, which would need to operate for about 6–8 h and would need to weigh no more than 2 or 3 kg. That weight limit determines everything. Each lithium-ion battery typically has an energy density of 150–250 watt-hours per kilogram, which means the entire system can only have about 300–750 watt-hours. If an average draw of 50 watts is expected from the system, then the operational time will no longer be optimal and will be limited to short periods of time. At this point, every portion of the battery is severely limited and tied to an inflexible power draw. Models that are designed to be plugged in bypass the battery concerns entirely. However, even though the plug systems do not have to deal with the battery issues, the cords can become tangled, limit the patient's range of motion, and become hazardous if the patient moves into a different area of the treatment zone. When devising with a person, heat rules impose additional power limits. If a device contacts a user's skin, its outer surfaces must remain below 41°C, or else a skin burn hazard exists from extended use. Components that generate motion or carry out processing functions can generate heat quickly and become quite hot. If no active cooling is available, temperatures can reach these dangerous levels very quickly. With completely passive cooling based on the device, each 100 cm^2 surface area can dissipate 10–20 watts before it becomes too hot. Integrating fans or liquid cooling systems adds complexity and cost to a device while also using additional power and generating noise. Battery heat rules also prevent a device from operating at maximum power continuously, even if the battery has the capacity to power the device at that level.

Due to regulatory requirements, restrictions from end-user hardware, and the clinical needs of modular medical devices, power management is especially challenging for design. For example, medical devices frequently require embedded safety features and backup modes, which restrict the flow of current and necessitate greater complexity regarding the power handled. Additionally, battery charging circuitry must comply with medical standards, which often results in the use of fused circuits with lower charging rates than those found in consumer electronics. Lastly, due to the risks associated with electromagnetic interference, regulatory noise mitigation often necessitates the use of energy sinks, which can be counterintuitively antiproductive. For example, the testing requirements for medical devices often require hospitals to perform extensive testing of the medical devices, with the expectation of obtaining the same results after multiple uses, which results in engineers favoring conservative design principles and ultimately sacrificing high power or short circuit designs. Predictability is valued over design best practices (Fig. 10.1).

10.4 Energy-efficient actuation strategies

10.4.1 Low-power actuator design

The way you choose and design actuators determines how haptic devices are optimized for power savings. Motors using high-quality permanent magnets and optimized wire and magnetic materials are often greater than 90% efficient. In frameless designs, omitting the

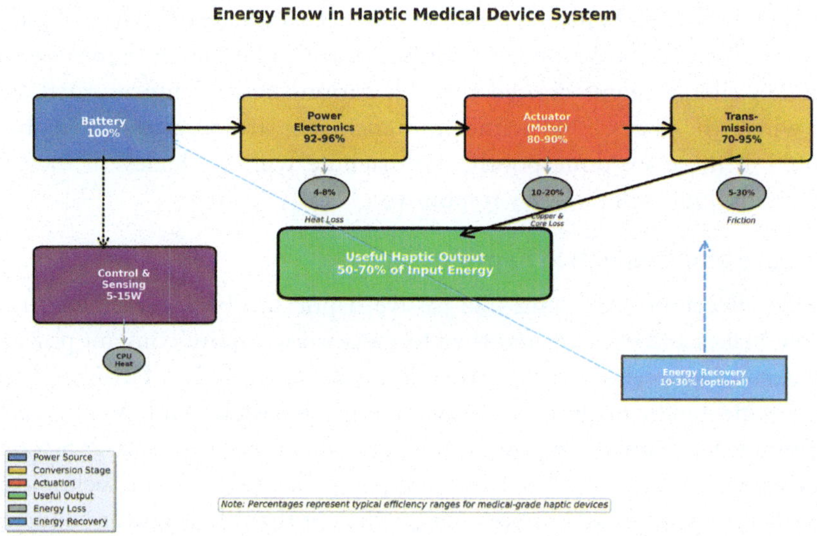

Figure 10.1 This figure illustrates the flow of energy through a haptic medical device system, showing how power moves from the source through actuators, sensors, and control electronics, and where energy losses occur at each stage.

outer case and supporting components reduces the motion loss, contributes to 5%–10% increased performance, and reduces weight. Direct coupling of the motor to the load eliminates gears and the associated power loss, but requires larger motors with greater torque. The optimum solution is a function of the type of push-and-move profile required for the application. For example, steady and high force requirements usually lead to geared solutions despite some power losses, while short and fast requirements are well served by direct coupled configurations.

Not all motor systems use the same amount of energy under the same loading conditions. A few of them, however, are better from an engineering perspective. Consider self-adjusting resistance systems. They separate the load from the resistance. Springs, either located beside or inside the drive path, capture energy instead of dissipating it (Pratt & Williamson, 1995). In repeating motion cycles, the spring will store the energy from the load instead of letting it be converted to heat through the resistance. One therapy machine with this mechanism reduces energy consumption by almost 50%. Other systems use multiple components: more parts mean more weight and more complicated systems. The energy savings will be apparent on the energy meter, but the machine will visually become quite large.

Designing a responsive feel for a user has typically meant designing energy storage elsewhere. However, splitting this paradigm with high responsiveness and effective energy storage has been achieved with recent designs of cables, levers, and split motion gears. A good example is capstans: With the right design, they can transmit almost all of the force applied to them while also being very responsive to applied forces. However, there are several hidden energy drains created through various design choices, such as bearing selection, where the wires bend within the enclosing structure, and how the tension is distributed. If these are done poorly, the machine will feel lifeless; if they are done correctly, the machine will feel very responsive.

10.4.2 Passive and semiactive haptics

Based on the description, it seems like passive haptic feedback technology makes use of adjustable brakes and clutches and therefore does not require constant power. Rather than using force, these technologies create resistance. The viscosity of magnetorheological (MR) fluid is affected by magnetic fields, which means MR braking systems can provide variable resistance. Creating MR fluid is energy efficient because only 2–5 watts are required to create the magnetic field. Therefore, passive haptic feedback technology breaks magnets can be more than 90% more energy efficient than their motor counterparts.

Most designs of both particle clutches and electromagnetic brakes are energy efficient, but they can behave very differently from one another. While one design can create a large amount of torque, it does have a problem when it comes to quickly adjusting. The latter uses more energy to engage and can spring into action very quickly. What design is required for a system really comes down to what the system is required to do. Surgical simulation training tools may require systems that use electromagnetic brakes to provide rapid engagement. In contrast, tools that are designed for rehabilitation use MR brakes that provide a constant resistance.

As people push and pull on any haptics-enabled device, some of the energy spent usually dissipates as heat. In contrast, some devices capture that motion and convert it into electricity, because motors generate voltage when spun in reverse (as happens with the motor when it is spun backward). That electricity is stored in energy storage devices, such as batteries and supercapacitors. Energy recovery systems in rehabilitation robots capture approximately 25% of the energy used in repetitive motion in the same direction. The reduction in energy consumption allows for extended training sessions without the need to recharge. The amount of energy recovered is largely dependent on the operator's motion—recovery is optimized in push/pull repetitive cycles as opposed to linear motion. The addition of recovery circuitry increases cost and design complexity; however, for devices that operate autonomously, the additional effort is justified (Fig. 10.2).

10.5 Control strategies for energy efficiency

10.5.1 Impedance and admittance control

When the device moves without touching any object, a lower resistance allows for energy saving, but otherwise still maintains the object. Depending on the position of the device,

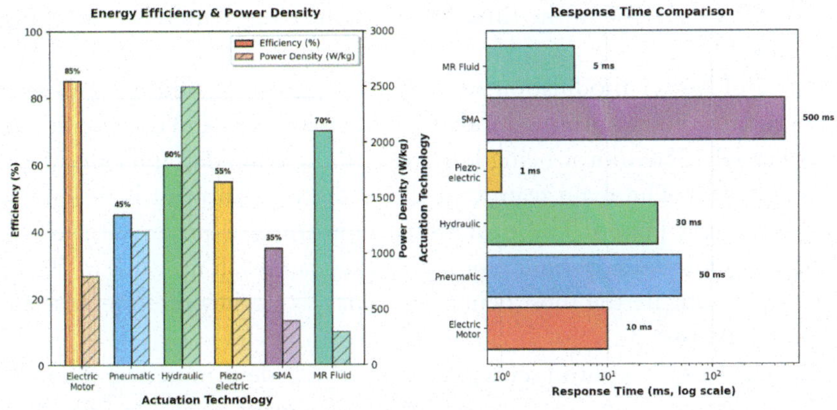

Figure 10.2 This figure compares key performance metrics, including energy efficiency, response time, force output, and weight across different actuation technologies used in haptic medical devices, such as electric motors, pneumatic systems, and smart materials.

power demands change if the device is in free motion or in contact position. The system will tighten to mimic contact. This type of situational adjustment to response can cut energy expenditure from 25%–40% in surgeries where contact is made and unmade repeatedly. What is the challenge? Switching to different modes can result in disruptions. Therefore, making these modes shift without friction is the primary goal.

Impedance control is largely dictated by designing a consistent response or one that saves energy. Responding to fast changes of the system requires a stiff response, which in turn necessitates a strong control effort, resulting in the system requiring more power to respond and more power to counter interference. Some approaches achieve stability by ensuring some intentional loss of energy in the system, thus preventing the system from ever operating at its full potential (Colgate & Brown, 1994). New approaches control the system to prevent instability and permit higher control efforts without the risk of instability, maintaining a balance of energy expenditure and system stability (Hannaford & Ryu, 2002).

10.5.2 Event-based and adaptive control

When actual changes take place, changes can be made in 2013; there's no need to constantly recalculate. Rather than the repeated ticking of a clock, a system only responds when the measured movements cross a specified limit. By maintaining a set pressure, a motor does not change its output and can even go long stretches of time without making any changes. Still, conventional arrangements issue new commands every millisecond. Smarter systems, by closely monitoring the important parameters, determine when changes need to be made on their own. In most surgical procedures, the processors are only active for less than 15% of the time. Likewise, the energy used is also proportionate to the runtime.

In the case of power modulation of an actuator, on-off control patterns are used. Rather than running continuously, a motor is active in intervals and is constant in position, or is moving slowly. The motor being less active in the extended intervals results in less energy consumption. The main feature of this is the regulation of how long and how strongly the motor is pulsed. The operational uniformity is preserved and is without visible shaking. In idle-like configurations, the losses can reach upwards of 50%. Devices used for assistive work are the most beneficial from this, as they often require a constant push for several minutes.

With changes to tasks, control settings change as well, determined by the exact action being performed by the user. Fast movements need to exert less force and subsequently lower the energy used. As the user's hand approaches a target, surface, or goal, the control responsiveness increases to allow for better touch clarity. Artifacting behavior in users' movements is detected by intelligent algorithms, and as a result, preemptive adjustments are made to control feel to increase efficiency and eliminate power wastage. Overall energy consumption is reduced by over 20%, while the focus of the control settings remains clear and adaptable to changes.

10.5.3 Model predictive and optimal control

Rather than using conventional approaches, model predictive control conceptualizes tactile feedback as an energy puzzle simulating touch feedback within boundaries where energy saving is as important as precision. By anticipating the possible actions of both the control system and its operator, the controller selects actions that will minimize both errors and energy consumption. This type of intelligence, such as predictive control, operates on a continual basis, with each time step projecting the next state of the system and adjusting the control set appropriately. The engineer designing the control system uses the adjustable parameters in the control system to determine the trade-off between realism and energy savings. When these parameters are adjusted in a real surgical process using a digital tactile feedback system, the control system can perform these predictions and adjustments as many as 1000 times per second and achieve an energy savings of 25% energy savings in the control system.

Understanding how the control system operates in an energy-efficient manner generates important concepts for designing tactile feedback systems. The model predictions cited in the research state that energy saving is maximized through gentle movements rather than sharp discontinuities. In cases where an acceptable time interval is present, the movements are planned to reduce the energy consumption by 50% compared to a system that operates without any regard for energy consumption. However, the human factors research indicates that if the movements are too slow, the feedback system will lose its realistic feel and will not be effective. This has been identified as the trade-off for an energy-efficient touch feedback system.

First, the control designs must incorporate the medical safety regulations clearly to perform optimally. When the maximum force limits, speed limits, or movement zones are implemented, risks are mitigated, although in some cases, power consumption can increase if the limits are breached. Rather than simply setting limits, newer control methodologies are designed to encourage and guide control actions to the proximity of the boundary, improving both efficiency and safety. In medical applications, the control systems must conserve energy and incorporate the safety physiological limits based on the tissue/body shape/boundary behavior, which are unpredictable.

10.6 Power management and energy harvesting

10.6.1 Power electronics and system-level optimization

Power switches connect batteries to motors, and the energy lost in the switches affects the overall performance of the system. With the use of field-oriented control with smart switching strategies, the motors can achieve between 92% and 96% efficiency over a wide range of operating points. In bridge circuits with MOSFETs, synchronized switching of the active semiconductor devices eliminates the conduction voltage drops present in standard diodes, which results in an additional 3%–5% improvement in efficiency when the load is light. The optimal selection of control strategies to achieve the desired performance results and operating efficiency can be particularly beneficial. When the components of a system operate in a coordinated fashion, it is possible to achieve a more efficient use of energy. Components such as the motors, sensors, and computers are provided with just the right voltage so that energy is not wasted. In the case of sensors, when they are not in use, their operation can be suspended, and the system can be adjusted, including slowing down the main processor. A radio can be configured to transmit data only at certain intervals instead of doing so continuously. Through proper design, the operational power consumption of a system can be significantly reduced from the order of 20–2 watts, which is a notable improvement when a system is powered by batteries.

As circuits become more dense, the importance of halting the flow of wasted energy becomes greater, and this can be accomplished by turning off the power to inactive regions of the chip. Some contemporary chips can turn off and on individual regions of the chip with precise control at a time scale of microseconds, allowing the delay to be obscured. Thanks to these intelligent power-down methods, the vibrating wearables of today can be expected to have lifetimes extended by a further 30%–50%.

10.6.2 Energy harvesting opportunities

As users manipulate components of a haptic device, the motor goes into generator mode, which is how energy is retrieved. Motion studies indicate how frequent hand movements occur for surgeons in normal practice. Because these movements occur quickly and repeat themselves, there are opportunities for on-the-fly energy retrieval. Surgical assistance devices, through the use of the described energy reciprocity, can potentially reduce energy consumption by 10%–20% due to the described energy reciprocity. Rapidly captured and released electricity is ideal for supercapacitors, which temporarily store the captured energy and release it to motors or sensing systems as needed. The energy generated by human–robot contact encompasses more than the recovery of movement. Worn devices transform the energy generated by the human body into electricity through the use of temperature differential materials during the process of human–robot contact. Although each powering element generates between 10 and 50 uW/cm^2, the entire frame generates between 1 and 3 watts. This output is sufficient for power monitoring and low-level computation devices (without batteries). Also, sunlight contributes, owing to the presence of solar cells on the outer contours of the device, particularly when the integrated circuit is designed to illuminate the interior. Light is sporadic, so the solar cells use less than 5% of the total power requirement of the device.

Minuscule quantities of energy are generated from pressure-sensitive touch sensors through their piezoelectric components, which convert the physical push of a user's finger into energy. While one component produces trivial amounts of energy, many components placed over a surface that will experience a lot of finger movement will result in a cumulative increase in energy production. Robotic limbs drive small, spinning energy generators. Energy is produced when the joints of the flexible or wearable frame bend. However, the added resistance from these new components could decrease the range of motion of the joints or the speed and ease of movement of the whole garment. If the movement of the user suffers too much, the amount of usable energy that the system generates may be too much. The final value of these systems in the real world is a result of the trade-off between the usable generated energy and the user's functionality.

10.7 Case studies in medical haptic systems

10.7.1 Surgical robots

An example of these power demands can be seen in the da Vinci Surgical System. Here, the likes of high-output DC motors are used, which can power each of the movements of the axes at the surgeon's console. They consume power in the range of 150–200 watts and will subsequently be operating at maximum capacity for the duration of the surgical procedure. Since the machine draws continuous power, the machine can be left plugged in at all times to avoid battery management, and subsequent heat management must be considered. Surgeons can anticipate heat at the grips of the models where the motors are located, which can be addressed by keeping the grips cool to 35°C. This is accomplished by the use of a forced air cooling system, which consumes 10–15 watts.

Of the smaller touch surgical instruments, power is a limitation. Tape instruments are on the smaller end of the 8–12 mm diameter range and need sensors to be located in multiple axes to sense forces. The challenge of transmitting power and data wirelessly is compounded by the use of magnetic fields. In body tissue, magnetic fields will use >50% of the energy as waste. Every component in a system must be designed to consume as little energy as possible. The most current models utilize a unique design in which a single small current (sip) amplifier is used for each strain gauge, which is only activated to send a signal when necessary. This brings the total consumption of the instrument below 100 mW and is sufficiently enough for the duration of the entire surgery.

When a doctor performs a surgery using robotic arms, the delay during the procedure affects the energy consumption of the system. In a network, a computer has to make predictions to compensate for the time delays in signal transmission, which uses more of the computer's processing resources. In robotic arms, these predictions can result in erratic and imprecise arm movements, which can lead to excessive energy consumption. Engineers have discovered smarter ways to predict robotic arm motions, ensuring that the arms remain stable without overworking the servo motors, even when communication delays temporarily occur. In experimental robotic surgery, the engineers were able to achieve a delay of 0.3 s, and using their techniques, the energy consumption was reduced by 25%.

10.7.2 Rehabilitation and assistive devices

Bulky batteries have an adverse effect on the usability of rehabilitation devices. For example, one of the arm devices requires 40–80 watts worth of power, which makes the device inefficient. Arm devices assist at 30 NM at the shoulder and 15 NM at the elbow (Cappello et al., 2018). The runtime of a 2 kg battery pack is 3–5 h, which makes long sessions rather difficult. The addition of bulk is a clear solution to the issue of having a pack and battery with a long runtime, which is to ensure that the therapy devices assist in rehabilitation. In fact, pack and battery systems have been designed for rehabilitation devices that last 8 h to complete the rehabilitation session. In this design, a spring assists the user against gravity, and the motors are only activated when needed. The amount of therapy the user receives is the same, while the amount of energy consumed to provide the therapy is decreased. The rehabilitation process is long and tedious for the user; some repetitions throughout the therapy can take weeks to complete one full rep. The at-home rehabilitation devices designed to assist the user in completing the therapy must take minimal power. Smart systems in devices are designed to adapt and assist the user, which serves to increase the power consumed to operate the system while maximizing the efficiency of the therapy. In real-world experiments, the energy consumed to operate the system was decreased by approx. 50% after one complete month of therapy, in addition to improved rehabilitation for the user.

With each step, the legs lift and land, and stored energy and moving energy are constantly being altered. Each step challenges powered-leg devices because power is cycled with each step. Typical motorized devices lose most energy flow by using resistance instead of recovery, and they average 200–400 watts. Devices designed as springs capture motion energy and then return it when needed. (Banala et al., 2009; Pratt & Williamson, 1995). These devices are more like muscles and tendons, so their practical power requirements drop to 20–50 watts, with most of the energy needed to power the switches and feedback systems. Efficiency increases most by moving smarter and not by pushing more. The gains here outpace nearly all other assistive technologies that rely on the mimicry of natural movement (Cappello et al., 2018).

10.8 Performance evaluation metrics

Determining the efficiency of haptic devices in a simulation involves understanding both the power consumption of the device and the sensation experienced by the user. One method of determining the device's efficiency is by taking the total power consumption and dividing it by the number of distinct tactile feedback events. This yields a value that can be used in comparison to different applications. In surgical simulations, this value would represent the power expended every time tissue is manipulated. In rehabilitation exercise simulations, this value would represent the power consumption every time the user completes a movement. Between half a joule and 5 joules is an acceptable range. For devices that do not have a motor that is on all the time, that range can be reduced to one-tenth to half a joule if the motor is only triggered when it is needed. The data shows a clear relationship between the use of a force on the device and the energy consumed. On one axis is how wrong the tactile feedback is, for instance, when the resistance changes, and the only perceivable difference is a change in the feedback. This is then compared to the amount of energy consumed by the device based on how the device was configured. A good device will reduce the feedback error quickly, even if only a small amount of energy is added. In contrast, a device that is not well designed will not change the feedback. The optimal range of a device is not rigidly defined; for one application, this range may require the use of high precision components, while others may be fine using low precision components to allow the device to operate for longer. New approaches in understanding touch feedback have shown how changes in energy consumption can be implemented innovatively. Changes in resistance or motion are often not perceived, particularly changes of less than 5%–15%. Because of this phenomenon, systems can afford to ignore attributes the body won't register, allowing for more energy savings while still retaining the perception of touch. How the system operates is also important, as repeated attempts can increase the perception of touch through positive feedback or directed aiming towards exposing certain areas.

Table 10.2 Energy efficiency strategies and their impact on haptic medical devices.

Strategy	Energy savings (%)	Implementation complexity	Clinical impact	Typical application
Passive/semiactive actuation	60–90	Medium	No assistive forces	Surgical simulators
Variable impedance actuators	40–60	High	Minimal impact	Rehabilitation robots
Energy regeneration	15–30	Medium	Transparent to user	Wearable exoskeletons
Event-based control	30–50	Low	Imperceptible	All applications
Adaptive gain scheduling	20–35	Medium	Minimal impact	Surgical haptics
Model predictive control	15–25	High	Improved response	Teleoperation systems
Power gating	30–50	Low	No impact	Battery-operated devices
Optimized power electronics	3–8	Medium	No impact	All systems

The real-world efficacy of technologies to power saving surgical instruments means that they will have to perform at the same speed, accuracy, and safety as traditional methods. For rehab technologies, the energy expended in terms of active or passive movement improvement, or in terms of rehabilitation, is not to be reduced to increase energy efficiency. The active and passive technologies should be used in conjunction to confirm that efficacy is not to be sacrificed for efficiency (Table 10.2).

10.9 Challenges and open research directions

The development of smaller surgical robots allows for more precise incisions. As instruments become small enough to fit through smaller incisions, motors and other components also become smaller. You also oversize the structure to offer more functional volume than this. Energy retention is not at the same rate, so energy gain is less than optimal. Touch feedback systems for heart catheter devices operate at very low energy levels. Achieving haptic feedback with less than 1 mW is very challenging. Perhaps new motor designs or completely new approaches to haptics are needed to avoid heavy hardware. The key requirement for haptic systems is that they provide good feedback with minimum use of energy. If a device is designed to provide haptic feedback, the goal is to make it lightweight, easy to move, and light to avoid resistance. If that is the goal, getting those properties economically is very difficult. Many haptic devices are also expensive. A new approach is to make the device operate differently based on feedback from the user.

The next innovations in material science may alter our approach to energy within touch-sensitive technologies. While components integrating graphene technology are efficient and powerful, they are far from being ready to use in hospitals. Devices incorporating components made from stretchable plastics or liquid-powered artificial muscles are designed to bend with the body and be more responsive; however, they are too slow and deplete power quickly. Materials that function at ultralow temperatures (near absolute zero) may enable almost perfect operation of miniature motors; however, the energy to cool these materials is greater than the operational gains achieved. There are still many years, perhaps decades, of work to be done before the tools developed in the laboratory can be used in medicine. Smarter algorithms that learn and adapt to tasks may be able to utilize energy more efficiently in surgical robots. Motion-based control models utilize data from previous movements to predict the required effort, and they can dynamically alter actions, resulting in lower energy consumption than previous reactive systems. Rather than following rigid patterns, these models explore and optimize trade-offs and autonomously adapt to preferred touch precision while minimizing control effort. Proving these control strategies safe and reliable is more challenging because of the control algorithms' reliance on historical patterns rather than explicit logic. As a result, not every situation and control outcome is predictably safe or reliable. One potential approach to the use of these models in practice is a combination of traditional reliable control with adaptive control, allowing the system to behave in the intended manner while enforcing limits around safe behavior.

10.10 Future trends in energy-efficient haptic medical robotics

When it comes to squishy stuff versus stiff parts, bendy shapes made of squishy stuff move more easily. For example, air-filled tubes made of rubber-like materials are less stiff, so they are less likely to use energy to resist forces acting upon them. Rather, they use energy to push air. Most of the forces needed for bending are for air pushing, so they do not require much sturdy protection, especially from people. If airflow is recycled through small pumps, as is seen in various prototypes, the system actually requires less power. Reducing the power demand is crucial. Some prototypes are purposely designed to help hands recover from injuries, and they are used to help simulate touch during surgery. Because of the design, less electricity is wasted doing the same job. Because nothing is designed to be rigid, safety actually gets better. Simply put, the less rigid the system, the safer it is. Along those same lines, the less rigid the system, the more efficiently it is designed. Movement is the most effective and most reliable way to improve the system. Squishy materials will be used to replace metal, decreasing the potential for injury. Clunky motors will also be replaced with cheaper ones. The less bound and the less rigid, the more energy efficient and the better the design. Then, materials can be made from less energy-intensive resources. By exposing materials, the system reduces the risk of injury and minimizes reliance on rigid elements. Its shape and components are engineered to manage unexpected loads, ensuring stable performance under unanticipated conditions. Overall, the system operates more efficiently, consumes less energy, and discourages unnecessary movement, representing the best possible design outcome.

Like most systems in nature, the brain only uses a small amount of power while completing a task. It is possible to design technology similarly. If designed in this way, they wouldn't run constantly and instead would only send out signals when needed. This drops overall power consumption tremendously. In fact, some chips only run at a fraction of the power to do the same task as another chip, sometimes as little as 1/100 or 1/1000 the power. This is a big deal for wearable technology, as the power needed to simply sense touch could drop from tens of watts down to just a few milliwatts. This is what would allow smartwatches and similar devices to last potentially weeks instead of just hours, as is common today. The big issue is that it is still very difficult to program brain-like systems, which is compounded by the fact that the regulations that exist on medical devices have not been updated or amended to allow for more complex or novel technologies. The technology used in brain–machine interfaces to aid or assist with medical applications has very limited power and touch control. The touch-control technology used in brain-machine interfaces is designed to accomplish specific tasks, such as stitching tissue or controlled bleeding, which do not require the use of a vibration signal. These devices can reduce power consumption by almost 50%. Rehabilitation devices that incorporate brain–machine interfaces can detect the user's effort and, if designed appropriately, can dynamically adjust the level of support provided while consuming very little power. Balancing the flow of control between automated actions and manual direct control is a big challenge. Sudden changes in control can be disruptive to the therapeutic process.

Unshackled from cords, new surgical devices can rely on wireless energy streams for continuous operation. Surgical tables will have built-in consoles that will be charged by power transmission. Wireless, body-worn devices will extract energy from the ambient resonance field and power themselves. Practical efficiency will be up to 85%, and wireless energy transmission will augment the practicality of exposed, operational devices. There is an emerging operational exposure risk when the wireless energy streams interact with pacemakers or other sensitive implants. Hospital use will only be permitted when the technologies and devices have been proven safe before patient exposure.

10.11 Conclusion

The design of doctors' feedback tools is also influenced by how energy is consumed. It even affects the choices of motors and the way the software controls them. The energy-efficient approaches to design show new ways to enable the devices to dispel energy without the user's intent to recharge the device. The adaptability of the new energized control motors also proves to enhance the promise of the recently designed devices.

Every design choice from software, hardware, and materials used affects the operational life of the device. Efficiency combines every element from the operational functions to the device software's start and stop functions. The effect of partially active and active feedback haptic systems that create haptic resistance (which is designed to create resistance to the user's control) is one of the greatest design achievements. These systems have been shown to use between 60% and 90% less overall energy. The movement and process demands of the system can be reduced by event-driven control, by using less than half of the system's overall energy. When the system design considers the control systems and circuits, thermal energy storage/dispersal, and the use of ambient energy, the benefits exceed the anticipation. The operating costs were reduced, and the energy stored in the thermal system was reduced.

Surgical robots now have the capability of working an entire day without the need to recharge. Since they produce less heat, they can be placed in closer quarters to patients. Tiny parts that are touch-sensitive can be placed in tools that reach deep into the body. New developments in tools that can create touch-responsive gear are driving innovation. The clinical possibilities of tools that can be used over a longer range are extended. Looking ahead to the future of surgical robots, the integration of innovation across previously siloed disciplines, such as robotic design, motion control, material science, and clinical care, will be an imperative. When the surgical robots are used in the clinic, it will be necessary to have an effective method of energy conservation in the tools, as it will be a determining factor in the surgical robots' functionality.

AI disclosure

During the preparation of this work, the author(s) used Grammarly to grammatically check their manuscript, and the image enhancer was used to enhance the images uploaded in the manuscript. After using this tool/service, the author(s) reviewed and edited the content as needed and take(s) full responsibility for the content of the publication.

References

Banala, S. K., Kim, S. H., Agrawal, S. K., & Scholz, J. P. (2009). Robot assisted gait training with active leg exoskeleton (ALEX). *IEEE Transactions on Neural Systems and Rehabilitation Engineering, 17*(1), 2–8. https://doi.org/10.1109/TNSRE.2008.2008280.

Cappello, L., Meyer, J. T., Galloway, K. C., Peisner, J. D., Granberry, R., Wagner, D. A., Engelhardt, S., Paganoni, S., & Walsh, C. J. (2018). Assisting hand function after spinal cord injury with a fabric-based soft robotic glove. *Journal of NeuroEngineering and Rehabilitation, 15*(1), 59. https://doi.org/10.1186/s12984-018-0391-x.

Colgate, J. E., & Brown, J. M. (1994). Factors affecting the Z-width of a haptic display. In *Proceedings of IEEE international conference on robotics and automation* (pp. 3205–3210). https://doi.org/10.1109/ROBOT.1994.351077.

Hannaford, B., & Ryu, J. H. (2002). Time-domain passivity control of haptic interfaces. *IEEE Transactions on Robotics and Automation, 18*(1), 1–10. https://doi.org/10.1109/70.988969.

Okamura, A. M. (2009). Haptic feedback in robot-assisted minimally invasive surgery. *Current Opinion in Urology, 19*(1), 102–107. https://doi.org/10.1097/MOU.0b013e32831a478c.

Pratt, G. A., & Williamson, M. M. (1995). Series elastic actuators. In *Proceedings of IEEE/RSJ international conference on intelligent robots and systems: 1* (pp. 399–406). https://doi.org/10.1109/IROS.1995.525827.

Prattichizzo, D., Pacchierotti, C., & Rosati, G. (2012). Cutaneous force feedback as a sensory subtraction technique in haptics. *IEEE Transactions on Haptics, 5*(4), 289–300. https://doi.org/10.1109/TOH.2012.15.

Rosen, J., Brown, J. D., Chang, L., Sinanan, M. N., & Hannaford, B. (2006). Generalized approach for modeling minimally invasive surgery as a stochastic process using a discrete Markov model. *IEEE Transactions on Biomedical Engineering, 53*(3), 399–413. https://doi.org/10.1109/TBME.2005.869771.

Tzafestas, C. S. (2003). Whole-hand kinesthetic feedback and haptic perception in dextrous virtual manipulation. *IEEE Transactions on Systems, Man, and Cybernetics - Part A: Systems and Humans, 33*(1), 100–113. https://doi.org/10.1109/TSMCA.2003.812600.

Van der Linde, R. Q., Lammertse, P., Frederiksen, E., & Ruiter, B. (2002). The HapticMaster, a new high-performance haptic interface. In *Proceedings of EuroHaptics* (pp. 1–5). Netherlands: Delft University of Technology.

SECTION III

Clinical applications and surgical systems

11. Haptic feedback in minimally invasive surgery: Restoring the sense of touch in modern surgical–practice 203
12. Tactile feedback systems for telesurgery 221
13. Haptic interfaces in rehabilitation robotics 239
14. Virtual reality and augmented reality in medical training 259
15. Brain–computer interfaces and haptic feedback 271

CHAPTER 11

Haptic feedback in minimally invasive surgery: Restoring the sense of touch in modern surgical practice

Riya and Vishnu M.
School of Business, Woxsen University, Hyderabad, Telangana, India

11.1 Introduction

Tiny cuts instead of big ones—this change flipped surgery on its head. Through such openings or even bodily passages, doctors can now slip tools inside without tearing tissue apart. The result is less damage, pain drops, bleeding fades, fewer infections, and faster healing. Because the method proved so effective, specialties such as gastrointestinal surgery, gynecology, nephrology, cardiology, neurosurgery, and orthopedics adopted it rapidly. Today, patient care almost by default relies on these minimally invasive approaches (van der Meijden & Schijven, 2009).

One big upside of minimally invasive surgery (MIS) is clear in practice, yet it brings tough hurdles because tools interact with tissue at a distance. Hands-on feel matters deeply during open procedures—surgeons learn by touching—how firm an organ feels, its surface quality, where edges lie, how much pushback occurs. That physical knowledge guides precise force control and quick judgments mid-operation. Instead of skin contact, laparoscopic methods use stiff rods and video screens to operate, breaking the body's usual give-and-take signals entirely (Westebring-van der Putten et al., 2008).

Missing touch feedback in MIS affects how well operations go and how safe they are. Visual clues become the main way surgeons sense what tissues feel like, since hands-on feeling is gone. Because sight cannot show fine details of resistance or pressure, small warnings get missed. Without those signals, too much force might be used without meaning to, which can harm tissue or cause bleeding; sutures might snap, and important body parts could get hurt by accident; thinking harder to make up for lost senses slows work down, and tasks take longer. Learning takes more time, especially for those just starting (Tholey et al., 2005).

Out in the operating room, things get trickier with robot-assisted keyhole surgery because the doctor is not right there at the patient's side. Though machines can move

Haptic Feedback Control in Medical Robotics
DOI: https://doi.org/10.1016/B978-0-443-27777-1.00006-8

more precisely, shrink large motions, filter out handshakes, and reach tight body areas better than humans alone, most ready-made models skip real touch signals entirely. Because of this gap, surgeons guide advanced tools while missing the feel that guides safe, smooth work. Without that sense of pressure or resistance, confidence takes a hit, even if vision stays sharp. Missing touch cues stand out now as a core weak spot across today's surgical robots (Okamura, 2009).

With tiny vibrations or pressure shifts, tools can now "talk" back to surgeons during minimally invasive procedures. Because sight alone misses crucial details, these signals fill the gap by mimicking what hands would feel in open surgery. Not just an extra feature, this feedback shapes how accurately cuts are made, how safely instruments move, how quickly tasks finish, and even how choices unfold mid-operation. A sudden resistance? The device hints at it before damage occurs. Softness, stiffness, hidden layers beneath skin, all hinted through pulses instead of visuals. What once relied on instinct now leans on subtle nudges built into smart tools (Selim et al., 2024).

When you take away touch, learning surgery gets harder. Without feeling what hands do during minimally invasive procedures, beginners often misuse force. Muscle memory builds stronger over time with repeated exposure. Performance inside real operations improves alongside practice on these setups. Over months, those who trained with simulated touch respond more quickly under stress. It turns out the body remembers forces better when it feels them. So, while precision matters during surgeries, building habits early matters just as much. Touch-based teaching methods quietly reshape how expertise grows. In effect, feedback loops made of vibration and resistance become silent mentors. Not flashy, just effective (Bergholz et al., 2023).

This chapter looks closely at how touch feedback works in keyhole surgery, breaking it down step by step. Starting with the basics of minimally invasive techniques, it moves into how sensation and movement connect during procedures. Instead of just listing parts, it shows how sensors and control methods come together to mimic real touch.

11.2 Conceptual and theoretical framework

This chapter's ideas rest on a mix of theories that help make sense of how touch feedback works during keyhole operations. Built from knowledge in surgery, body movement control, machines that act like humans, and ways people work with technology, it shows when and where physical sensation matters most inside remote procedures. Starting from tested models, the part explains what goes wrong without real contact, then follows up by showing how tools can bring some of that missing feel back into play.

11.2.1 Fundamentals of minimally invasive surgery

Tiny cuts or natural openings let surgeons work inside the body without large wounds, using long tools and cameras like laparoscopes. Open surgery gives full reach by spreading tissue apart, but these methods change how hands meet what they fix. Instead of touching things directly, tools go through tight tubes, stiff and extended. What the surgeon sees comes from screens showing images captured deep within. Vision shifts from eyes-on to screen-guided, altering every move made. Despite different tools and entry paths, each method limits the space to move. Laparoscopic work sits alongside robot-guided procedures, tiny tube-based vascular fixes, and even surgeries through body openings without cuts. What ties them together? Vision shifts to screens rather than direct observation; tactile sensitivity diminishes, adjustments slow, movement options narrow, and tissue feedback fades. These factors reshape how surgeons sense and respond. Although tools evolve, challenges stay familiar. Looking at it one way, MIS turns surgery into something less hands-on, more like guiding tools through a barrier. Instead of touching tissue directly, the surgeon moves devices that pass along motions in ways that are not quite accurate. These tools often weaken or scramble feedback about pressure and resistance. Because of this gap, small mechanical quirks—like lag, stiffness, or resistance—interfere with how body and brain normally work together during movement. That means surgeons must think harder to fill in what they cannot feel, making the process heavier on focus while feeling less fluid (Westebring-van der Putten et al., 2008).

11.2.2 Sensorimotor control theory and surgical interaction

Picture how people control their hands and movements when they operate. Good movement needs constant information coming back from senses like sight, feel, and body position. These signals shape each motion, fine-tune pressure, and keep actions smooth. When doctors cut during regular operations, they depend heavily on what they sense through touch. Their fingers notice changes in texture, tension, even slight slips. That real-time awareness helps them react fast if something feels wrong mid-procedure (Tholey et al., 2005).

11.2.3 Transparency and fidelity in haptic interaction

In robotic surgery, the surgeon should ideally perceive resistance and texture as though directly touching the tissue. When the device fades from awareness during use, we call that quality transparency. It works well when movement, force, and feedback line up without delay. The less the gadget interferes, the more natural the experience becomes. Success shows up in precision, control, and reduced mental load. A smooth link between hand motion and sensory return makes all the difference. Clear feedback relies on several things—how well sensors measure, how fast actuators react, control speed, and whether the system stays steady. While ideal clarity sounds good in theory, real-world limits like lag, interference, shaky signals, and safety needs make it impossible to reach. Because of this, building touch-based systems means adjusting what you show versus what holds up under pressure and works in actual medical settings (Okamura, 2009).

11.2.4 Stability, safety, and passivity theory

During surgery with touch feedback, controllers that limit energy output help stay stable when delays happen, sensors act up, or tissues behave unexpectedly. Even if forces feel less real because of these limits, the system behaves safely—something doctors must trust before using it. Balancing how steady it feels versus how clearly you sense resistance keeps showing up across every part of building such systems (Selim et al., 2024).

11.2.5 Human factors and perceptual considerations

Though haptic feedback guides surgeons during MIS, it needs to match how people feel and think. Because our skin senses react strongly to shifts in pressure, signals that lag or overreact tend to overwhelm rather than help. When touch-based alerts are messy, mental effort climbs, focus slips, uncertainty grows, confidence drops. Starting, human-focused ideas push for touch signals that feel natural, stay steady, and match what is seen. Another angle: how we blend senses says spreading details across sight and touch works better when balanced (Westebring-van der Putten et al., 2008).

11.3 Review of literature

Years gone by have seen writings shift, matching new steps in tools used for keyhole and machine-assisted operations. Work done by those focused on surgery, robots, touch-based signals, along with how people engage with tech, points to missing physical cues during such procedures. That gap? It keeps pushing study after study toward ways to fake the sense of touch.

11.3.1 Early studies on sensory deprivation in minimally invasive surgery

At first, research into MIS examined how the use of tools indirectly affects surgical performance. Because of their long shape, stiff build, and hinges, laparoscopic tools reduce the amount of force that reaches tissue; this came out clearly in early tests. Instead of matching pressure precisely, surgeons ended up pressing harder during keyhole operations than they did during open ones, many times without even noticing. Since vision by itself fails to signal how much force is applied, especially on delicate or hard-to-see structures, these results highlighted a core limitation. Back then, doctors observed greater tissue damage, slower operations, and tougher training when using MIS. Together, those findings started a line of thinking: bringing back touch sensitivity might make these surgeries safer, quicker too (Tholey et al., 2005).

11.3.2 Haptic feedback in conventional laparoscopic surgery

Starting from initial results, later work looked at adding pressure and touch cues to standard keyhole surgery tools. Because they included sensors, special graspers, needle drivers, and cutters could record how hard they pressed during use. When tested in lab settings, these sensor-equipped devices led surgeons to push more gently on average, hit fewer sharp spikes in force, move with steadier control, and harm surrounding material less often. It turned out that seeing real-time resistance helped guide their actions without them needing to guess. Even with good outcomes, studies pointed out serious roadblocks in bringing tech to real surgeries. Tiny sensors caused issues—cleaning them was hard, they broke easily, plus expenses rose. Because of that, even though touch-based feedback worked better during keyhole procedures, nearly every available surgical tool still used only sight to guide actions. (Colan et al., 2022).

11.3.3 Haptic feedback in robotic-assisted minimally invasive surgery

Something started when robots helped surgeons operate—suddenly, everyone cared again about touch feedback. These machines made hand movements finer, more accurate, and less tiring to perform, yet they pushed the surgeon farther away from what they were cutting. At the outset, surgeons recognized a missing element: the perception of pressure. Its absence raised concerns—excessive force might be applied unknowingly, and operative surroundings could become more difficult to interpret during procedures. Back then, early studies on robot-assisted surgery pointed out how feeling resistance might make operations feel more natural. Some test setups with two-way touch signals managed sharper movements, used less pressure, and ended up boosting trust for surgeons. Still, plenty of machines now sold skip this feature—worries about glitches, patient risk, red tape, plus tougher engineering stood in the way (Okamura, 2009; van der Meijden & Schijven, 2009).

11.3.4 Tactile and vibrotactile feedback approaches

Not just about movement-based cues, scientists also investigated touch and vibration signals—ways to feel textures or shifts through tiny pins, air-powered parts, or bendable surfaces. These setups tried mimicking how pressure spreads, how rough something feels, or if it's slipping. Results showed better skill in telling tissues apart by hand, especially when tested under controlled conditions. What stood out was the gain in sensing detail during practice tasks. From the start, vibrations offered a straightforward way to signal touch, too much pressure, or slipping during interactions. While these signals do not show exact pressure levels step by step, studies find them useful when alerts matter most—especially in learning setups. Cost stays low, operation remains steady, which helps clinics adopt the method even if precise force data isn't needed (Colan et al., 2022; Westebring-van der Putten et al., 2008).

11.3.5 Haptic feedback in surgical training and simulation

Many studies look at how touch-based feedback helps in teaching surgery. Simulators that include physical sensations are often used to teach keyhole procedures like stitching, securing knots, or handling tissues. Learners who feel resistance and pressure during practice tend to pick up techniques more quickly, manage their movements better, yet keep what they've learned longer than peers relying only on sight. Finding their footing, new surgeons gain much from touch-based cues during practice, shaping how they apply pressure right from the start. Because of this, adding tactile responses to training feels less like an upgrade and more like a necessary part of learning surgery today. This chapter looks closely at how touch feedback works in keyhole surgery, breaking it down step by step. Starting with the basics of minimally invasive techniques, it moves into how sensation and movement connect during procedures. Instead of just listing parts, it shows how sensors and control methods come together to mimic real touch (Bergholz et al., 2023).

11.3.6 Sensing and force estimation in minimally invasive surgery

From time to time, researchers turn their attention to how surgical tools feel resistance during minimally invasive procedures. Some experiments rely on physical detectors like strain gauges or torque monitors built into devices. Others look at light-based sensors embedded in fibers that respond to pressure changes. Then there are methods skipping hardware altogether—using motor behavior, movement patterns, or camera feeds to guess what forces are present. High precision comes with direct measurement, true enough. Yet, chapters keep pointing out problems when adding parts to delicate instruments, especially when cleaning them afterward. Seeing things move helps guess the forces at play, thanks to newer methods leaning on machine learning. While not always spot on, they handle real-world messiness better. These techniques fit together more smoothly in actual products. That practicality is why companies are starting to favor them (Huang et al., 2020; Pisla et al., 2025).

11.3.7 Human factors and surgeon-centered evaluations

What makes haptic feedback work lies as much in how people perceive it as in how precisely it operates. If the sensation feels off or flickers unpredictably, users often struggle to stay focused—sometimes ignoring it completely. When adjustments feel natural, though, performance shifts quietly upward; confidence grows without announcement, tension slips away mid-task. Awareness of surroundings sharpens almost by accident. Precision matters, yet so does whether the body believes what it senses. Some studies point out how surgeons differ greatly from one another. Because of this variation, haptic feedback might work better if it adjusts to a person's skill and the specific procedure they are doing (Selim et al., 2024).

11.3.8 Identified research gaps

Still, gaps remain despite years of work. Not every large trial finds lasting benefits for people who are sick. When measurements differ across teams, matching outcomes feels

Table 11.1 Key dimensions and functional components of haptic feedback in minimally invasive surgery.

Dimension	Description	Functional role in MIS	Clinical impact	Implementation challenges
Force feedback	Provides real-time resistance information reflecting tissue interaction forces	Enables controlled grasping, suturing, cutting, and dissection	Reduces excessive force application, tissue trauma, and suture breakage	Sensor miniaturization, signal noise, sterilization constraints
Tactile feedback	Conveys surface texture, stiffness variation, and contact distribution	Supports artificial palpation and tissue differentiation	Improves tumor localization and detection of abnormal tissue	Complexity of tactile display integration, cost barriers
Vibrotactile feedback	Uses vibration signals to indicate contact, slip, or force thresholds	Provides alerts during excessive force or unsafe manipulation	Enhances safety during delicate vessel handling and catheter navigation	Limited quantitative force representation
Sensor-based measurement	Uses strain gauges, fiber-optic sensors, or torque sensors to measure force	Direct quantification of tool–tissue interaction	High accuracy in force detection	Fragility, sterilization difficulty, and added hardware complexity

(continued on next page)

Table 11.1 Key dimensions and functional components of haptic feedback in minimally invasive surgery—cont'd

Dimension	Description	Functional role in MIS	Clinical impact	Implementation challenges
Sensorless force estimation (AI-based)	Predicts force using motor signals, kinematics, or visual data	Reduces the need for embedded physical sensors	Easier integration into robotic systems	Algorithm validation, reliability under variable tissue conditions
System transparency	Degree to which the haptic system feels natural and responsive	Ensures smooth sensorimotor coordination	Improves precision and reduces cognitive load	Trade-off between stability and realism
Stability and safety control	Maintains system control during delays or disturbances	Prevents oscillations and unsafe force output	Ensures patient safety in real operations	Balancing passivity with realistic feedback
Human-centered design	Aligns feedback intensity with human perception limits	Enhances usability and reduces mental fatigue	Improves surgeon confidence and acceptance	Variability across surgeon skill levels
Training integration	Incorporation of haptics into surgical simulators	Facilitates motor skill acquisition and muscle memory	Accelerates learning curve and skill transfer	Standardization of training metrics

like sorting tangled threads. Real operations in hospitals rarely mirror the controlled drills done in labs.

Bits missing here suggest effort still spreading through different areas—smart sensors tying to alert systems, designs shaped by human needs. If those roadblocks stay, touch feedback in MIS stays out of daily hospital routines (Colan et al., 2024).

11.4 Core Discussion

Inside this chapter sits a close look at the tools and functions making touch feedback possible during keyhole operations. From theory to real-world setup, it shows how sensations are created and moved across devices, then felt by surgeons using tech bridges. Ideas unfold through major themes and hurdles, along with proven methods shaping today's touch-sensitive surgical setups.

11.4.1 Key dimensions of haptic feedback in minimally invasive surgery

Haptic feedback in MIS can be understood across several fundamental dimensions that collectively determine its effectiveness and clinical value (Table 11.1). What happens when a surgical tool presses too hard? That pressure tells a story surgeons need to hear. Movements stay steady during cuts or stitching only if the body talks back. Instead of relying on wobbles or dents in tissue, which sometimes lie, hands once dead now sense again. Signals fire the moment contact shifts, making tweaks happen without thought. This quiet flow returns something missing, not through sight, but through feel. Touch signals need super quick detection plus movement response. When there's lag after touching tissue, the user loses smooth coordination, and it feels fake. Speed matters most so actions stay in sync, letting doctors react fast if something pushes back or shows up without warning (Selim et al., 2024; Westebring-van der Putten et al., 2008).

11.4.2 Challenges and issues in implementing haptic feedback

Tiny tools for surgery need to weigh little, fit easily, work safely inside bodies, plus stand up to cleaning by heat. Putting touch or pressure detectors into these devices pushes engineering limits. Slip between parts, tiny delays in movement, and slight bends mess up how forces move through them, making readings less clear or true. Shaking often begins where touch signals falter, particularly when delays slip into response times. Precision matters most in operations—a tiny stumble may escalate quickly. Sensor interference mixes with the uneven give of human tissue, altering force patterns. Balance feels harder when delays, cluttered inputs, and surprises collide mid-motion. Seeing clearly helps surgeons sense tiny changes in tissues, yet too much clarity can weaken the tool's strength. On the flip side, cautious settings make things safer, though they also dull the touch signals. Working out how to manage this balance shapes much of the thinking behind designing touch-sensitive tools for keyhole surgery. Fingers feel only so much before things blur together. Too much vibration, too fast—or badly timed—makes hands tired, minds wary. A nudge here, a pulse there: it must speak without shouting. Clarity matters most when silence works better than noise (Okamura, 2009).

11.4.3 Haptic feedback strategies and design approaches

To address these challenges, researchers and system designers have developed a range of strategies to optimize haptic feedback in MIS. Sometimes the force gets turned up or down so people can feel it better when they need to. Noise fades out through filters that leave real signals untouched. What remains helps users stay sharp without tiring fast. Adjusting levels right makes tasks smoother than leaving them raw. Moving your body, touching things, plus feeling vibrations help you learn what you learn through different senses. Imagine pressure showing how strong a push is, whereas buzzing marks when something touches or gets too intense. Using several kinds of signals at once helps understanding grow without leaning hard on just one way. Software sets virtual barriers that shape how tools move during procedures. Instead of blocking movement completely, these limits gently steer instruments where they need to go. Touch-based signals alert the user when nearing restricted areas. Rather than taking control away, the system supports careful navigation near critical spots. Precision improves because guidance happens in real time. Surgeons stay fully involved while working within smarter digital guardrails. Safety increases without sacrificing hands-on decision-making. When tasks change, so does how the system feels under a user's touch. Depending on what kind of tissue is being handled, vibrations shift in intensity. A new surgeon might feel stronger pulses guiding each move. In contrast, someone more seasoned gets lighter signals, just enough to stay informed. Feedback shaped by the situation helps different people work comfortably. The way it adapts makes it fit better into real surgical routines (Bergholz et al., 2023; Colan et al., 2024).

11.4.4 Performance evaluation and metrics

Mistakes made give clues too, just like how much harm happens to simulated tissues. Another thing people look at is whether movements feel fluid or jerky. On a different note, what users say about effort matters—how hard they think the system feels. Their sense of trust in the device plays a role as well. Realism, meaning if things seem lifelike, weighs heavily on whether doctors will adopt it.

11.5 Case studies and practical applications of haptic feedback in minimally invasive surgery

When theory meets the operating room, that is where haptic feedback begins to prove itself. Not just in controlled labs, but during actual surgeries, its role becomes clear. From one specialty to another, touch-based cues help sharpen accuracy and reduce risks.

11.5.1 Haptic feedback in laparoscopic suturing and knot tying

Few things test skill like stitching inside a body through tiny cuts, where pressure must be just right to hold flesh together safely. Without touch cues during standard keyhole operations, those learning usually pull too hard on threads. Force feedback restores intuitive tension control, directly improving task quality and reducing tissue trauma (Tholey et al., 2005).

11.5.2 Artificial palpation and tumor localization

Fingers often find what needs removing during open procedures—lumps, growths, spots that feel wrong. Without direct touch, those clues vanish when tools stand between hand and body. Tactile feedback fades under minimally invasive methods. What once guided decisions by pressure now depends on sight alone. When testing fake touch systems, results show pressure and surface clues help doctors spot differences in how firm tissues are, even when nothing looks unusual. Because robots are fitted with tools that sense pushes or send small shakes to the hand, specialists get a sense of softness shifts while moving inside the body. What stands out is their role in cancer operations—like those on the liver or prostate—where knowing exactly where tumors sit and where edges lie matters most. Haptic feedback enables functional palpation in MIS, compensating for the lack of direct touch and improving diagnostic confidence (Abiri et al., 2018).

11.5.3 Vessel handling and delicate tissue dissection

When too much pressure is applied, blood vessels, along with fragile soft tissues, can easily get damaged. Trouble shows up late because what you see gives little sign before something tears or bursts. Now imagine a surgeon feeling tiny shifts in tension while moving through delicate tissue. That sense comes from touch-based signals built into surgical tools. When pressure tweaks happen, the system whispers warnings straight to the operator's hands. Mistakes slow down because the body talks back before harm shows up on screen. Real cases—heart work, brain operations, organ swaps—all log fewer blood vessel surprises. Without those physical hints, slips rise. With them, control tightens almost silently. Early force awareness through haptics enhances safety in high-risk tissue manipulation tasks (Bergholz et al., 2023).

11.5.4 Orthopedic and bone-related minimally invasive procedures

During tiny incision bone operations, feeling resistance guides actions like boring holes, smoothing channels, and inserting devices one step after another. Pressure changes tell medical staff when they hit the dense outer shell versus the spongy inner sections; suddenly, different textures appear mid-motion. When drills meet varying bone types, pressure patterns shift—those shifts guide surgeons more reliably. Tools that send touch-based signals help avoid pushing too far into tissue during procedures. Accuracy climbs because the hand senses resistance changes mid-operation. Placement of implants turns out better when feedback matches real-time conditions. Some teams notice fewer errors after adding these responsive systems to their workflow. Haptic feedback provides critical mechanical cues in rigid-tissue interactions where visual feedback alone is insufficient (Huang et al., 2020).

11.5.5 Catheter-based and endovascular interventions

Slipping thin tubes into tight, winding veins defines endovascular work. Too much push risks scraping or tearing a vessel—sometimes unseen. Touch-based signals in catheter guidance help users steer better, while also lowering how often the tool bumps into vessel walls. Though simple, basic vibration or resistance clues sharpen depth perception during movement through complex paths. Especially useful when teaching new learners, since they tend to push too hard without real-time physical warnings. Haptic feedback improves navigation safety and control in constrained, visually ambiguous environments (Colan et al., 2022).

11.5.6 Haptic feedback in surgical training simulators

During surgery, sure, but touch feedback appears more regularly in practice gear for small-incision techniques. What catches attention? Trainees using systems that sense grip usually pick things up faster compared to those depending solely on vision. Getting better happens more smoothly when force hints join the mix. Recall lasts further—seen again and again in study after study. Pace, precision, recollection—they tilt toward stronger performance whenever fingers register what visuals show.

Right away, practice takes center stage when these devices highlight how physical cues shape proper motion habits early. Because of this, users tend to avoid excessive pressure once they are working hands-on. Haptic feedback is a powerful educational tool that enhances skill transfer from simulation to the operating room (Bergholz et al., 2023).

11.5.7 Lessons learned across applications

Across diverse surgical domains, several common lessons emerge:

- Haptic feedback consistently reduces excessive force application.
- Fragile jobs feel the difference more, where pressure matters, changes show up fast. Tasks needing care see gains first; precision work notices shift quickly. Sensitive operations gain the most ground.
- Surgeons exhibit higher confidence and situational awareness.
- A single workout might not say much, yet patterns emerge over time. Results build when effort stays consistent, week after week. Progress often follows a clear path if routines stick. Small improvements link together, forming noticeable change eventually.
- Simpler feedback (e.g., vibrotactile cues) can still provide significant value.

What researchers found shows how bringing back just some touch feedback helps surgeons work better. A little sensitivity goes a long way in precision tasks. Not every signal needs to be returned for gains to appear. Small changes, when tied to feel, shift outcomes noticeably. Touch matters more than is often assumed during operations. Even limited input alters control in real ways.

11.5.8 Practical implications for clinical adoption

In hospitals adopting new technologies, practical challenges determine which innovations endure. Sophisticated equipment does not always prevail; instead, simple design, efficient setup, and ease of mastery often outweigh complex upgrades with marginal benefits. Surgeons prioritize ergonomics and seamless integration into routine practice, favoring instruments that become reliable through sustained use rather than those defined by technical specifications alone (Selim et al., 2024).

11.6 Implications of haptic feedback in minimally invasive surgery

Touch signals in keyhole surgery reshape more than instruments alone. The way surgeons perform operations evolves as training adapts to reintegrate tactile feedback. Machines start responding differently inside the body, since hands now communicate through sensors. Designers rethink device shapes once vibration and pressure become vital inputs. Decisions during an operation gain a physical reference point, altering outcomes without changing technique. Questions about right and wrong pop up when tools get smarter. When touch, sound, and sight make experiences feel more real, courses start to change—slowly. People begin to wonder if the old ways of helping still make sense today. Progress links tightly to how things go outside clinics, far from controlled tests. Insight grows by seeing crews handle pressure together, not just by reading books.

11.6.1 Practical implications for clinical practice

Haptic feedback changes how surgery feels—force cues come back, so pressure control gets better. Because of that, mistakes like tearing tissue or causing bleeds happen less often. Sutures hold when they should, without snapping under too much tension. Nearby parts inside the body stay safer during moves near delicate areas. Working on things like nerves or soft linings becomes far less risky. Fragile materials respond poorly to heavy hands—this helps avoid those errors. Felt cues in robot-led keyhole surgery link machine skill to how humans sense touch, building steadier command. Where tasks grow tight and intricate, that connection becomes a quiet must for safety (Okamura, 2009).

11.6.2 Implications for surgical training and skill development

Feeling things again transforms surgical training. When tiny cuts mean no sense of touch, new surgeons struggle—pressure goes wrong, movements turn clumsy. Tools that mimic resistance fix part of this puzzle, offering realistic pushback during practice. Hands learn better when they feel something; skill grows where sensation leads.

Training feels different when feedback comes through touch. Hints on the skin guide hands just enough to sharpen moves without slowing down learning. Some notice their body remembers steps better after feeling each motion once. When pressure shifts or vibrations signal depth, decisions come quicker during drills. It is not magic—it is repetition shaped by sensation. Tools start to feel like extensions, not objects held. The line between practicing and doing fades where fingertips learn what eyes used to see.

Holding onto these outcomes, haptics might slip into regular surgical training along with certification paths. Skills at the entry level could rise because of it. Across hospitals, patient safety stands to grow stronger (Bergholz et al., 2023).

11.6.3 Theoretical implications for human–machine interaction

Touching but not sensing—it feels strange, almost unsettling. This disconnect reveals how vital bodily feedback is in operations. Without tactile hints, hand movements lose their flow. Repair that break, skills improve fast. Numbers confirm it—precision grows stronger once sensation joins motion. When feedback moves freely, loops hold their shape. Break that link, and accuracy slips away.

Vibrations felt in machine controls reveal hidden links between user and tool. Because of feedback, movements seem less separate, more like an extension than a command. When robotic limbs push back slightly, boundaries blur, skin meets steel without clear edges. Engineers start rethinking remote systems after seeing these effects up close. Medical rehab tools gain new meaning when motion feels natural across distances. Handling hazardous items from afar becomes different once touch enters the picture (Westebring-van der Putten et al., 2008).

11.6.4 Implications for system design and technology development

Now it makes sense—haptics reshape tool design. Since touch counts, builders balance precision with smooth function; safety stays woven through. Hands get a say once users lead shaping. Where grip directs movement, forms follow flesh. Standards exist, but ease refuses delay. Heavy reliance on touch clues drives change in machine awareness, choices, and shrinking delay between feel and response. Software sharpens when fed mixed inputs—not just sight or sound but pressure, slip, heat—woven tight. Bodies that bend, yield, adapt, rise from need: inside an operating room, roughness kills. A softer path appears, one where parts move like tissue, react like skin; designs may shrink, shed layers once thought essential. Access widens not by force but by lowering walls already cracking (Pisla et al., 2025).

11.6.5 Social and ethical implications

Most times, a surgeon's hands tell more than screens ever could. Pressure under fingertips guides decisions without words being said. Machines might compute fast, yet they miss whispers of tissue resistance. Trust builds when touch confirms what eyes alone cannot see. Unequal access hides where feedback fails most. Smooth moves come easier when sensation flows through tool handles. Some errors fade once grip and response sync like breathing. Quiet signals tip the scale toward safety or risk. Fingers typing on glass can still feel warm if the machine answers back right. Better results come not from luck, but from shaping tools that move as life itself does.

When a device guides a surgeon's motion, where does responsibility lie? As intelligent instruments enter operating rooms, untangling control from accountability quickly becomes tangled. Guidelines must establish boundaries before human instinct intersects with algorithmic control. Errors disregard assigned roles, yet it is ultimately humans who correct them. Here is a tough problem: top-quality surgical robots with touch feedback are not shared fairly. Cost blocks many hospitals, while others lack staff trained well enough to run them. Where money is short, equipment stays out of reach. If better policies do not arrive and fresh approaches stall, treatment differences may stretch further (Selim et al., 2024).

11.6.6 Implications for healthcare systems and policy

Safer, quicker surgeries tend to lower hospital complication rates. Less time spent recovering inside facilities shrinks overall costs. Though equipment may demand higher upfront prices, benefits build up steadily. Tools offering real-time touch cues in operations slowly prove their worth over months. Savings pile on quietly when healing speeds up naturally. Still, progress in touch-driven surgical tools pushes new demands on rule makers. How well things go inside hospitals is tied closely to having plain metrics mixed with strong safeguards and review paths. If those elements miss the mark, actual practice might falter. When guidelines lock into place, motion ahead stays safer, layer by layer (Colan et al., 2024).

11.7 Future scope and recommendations

Tomorrow's touch feedback in MIS links robotics, thinking software, and human tool handling, while folding in actual clinic demands. Today's gear helps, yet accelerating innovation alongside changing techniques paves the way toward sharper sensation, fewer risks, adaptable designs, broader adoption across hospitals. Ahead lie emerging trends, research directions, careful advances—each nudging haptic surgical devices into new shapes.

Ahead lies a new direction where touch devices grow smarter through embedded artificial intelligence. Not only do sensors gather info, but algorithms also refine raw inputs instantly. While handling different jobs, gadgets adapt their physical responses—softening force or shifting vibration patterns. Whether someone knows the tool well shapes how it answers back. On its own, the system sharpens sensations moment by moment. Without slowing down, precision stays high (Pisla et al., 2025).

A fresh surgeon often relies on intense feedback cues, combined with warnings that kick in mid-task. In contrast, experienced hands tend to favor subtle, lifelike sensations under their fingers. As machine learning individually optimizes tactile inputs, system setup becomes more seamless. Cognitive processes flow more freely, stress diminishes, and confidence develops progressively.

11.8 Conclusion

Small cuts transformed surgery—reduced damage, less pain later, faster recovery—and effectiveness held steady. Yet a key element faded: the doctor's touch, muffled by gear and barriers, turning precise actions trickier. Direct hand feedback, once clear, is now lost, interfering with careful tissue handling. Our focus turned to restoring that missing signal inside narrow surgical tools. Rebuilding tactile awareness might correct hidden flaws in modern minimally invasive methods, even those guided by machines. Focusing on concepts, equipment, human factors, and actual usage reveals a truth about touch feedback. More than an improvement, it restores instinctive sensation while moving. Now surgeons sense tissue as they work, adapting more smoothly to changes under their tools. Since the device delivers cues such as resistance or surface detail, fingers respond with precision. Surprises show up fast when touch catches what sight misses. Fingertips reveal truths that flat pictures never convey (van der Meijden & Schijven, 2009).

When you look at sensors meeting control systems, it becomes clear that mimicking touch in small surgical cuts is not simple. Clear signals must work alongside stable function, safety needs, and daily hospital reality—each pulling in different directions. Still, better force tracking, software tricks that skip added parts, self-tuning tech, and gear built around human habits are inching lab dreams toward actual operating rooms. Here's the core idea: giving machines a sense of feel could let robotic accuracy stay tied to a surgeon's instincts—so automation doesn't numb awareness or put lives at risk.

Touch plays a bigger role in surgical training than people usually say. Not by observing, but by moving through tasks, learners gain assurance from pushback and feedback. As they work, novices adjust—guided by lifelike cues that help refine control. With practice, physical memory holds onto details sight misses. Over time, doing becomes knowing. Mistakes, caught right as they happen, build confidence more than practice alone. The machine takes on the role of instructor, shaping how hard you press, when you act, and where you adjust—silent yet clear in its feedback. Though research backs it well, touch-based feedback rarely shows up in hospitals. Heavy tech needs, strict rules, steep costs, yet hesitation from surgical teams holds progress back. Still, fresh paths emerge—thinking machines that adapt to touch, bendable robot parts, blended signal inputs, even shared control where human and device work closely. As tools learn to act on their own, sensing pressure, texture, and motion could quietly grow into how hands link with instruments. Alive still feels inside operations since tech sends touch across wires. Tools do more than act—human hands steer through virtual barriers. When builders join healers, change happens faster than solo paths allow. Outside test zones, real people gain only after trials show safety beyond doubt. Training alters sharply if practice runs happen in simulated spaces first. Catching up is what rules need to do, letting hospitals bring in smarter tools fast. Help comes from machines, but people stay key when choices matter. Safety tags along with precision if information moves back and forth. Gains ahead rely less on shiny gadgets, more on stronger ties between devices and bodies. Moving onward rests on slow care, evidence, constant reviews, and calm control (Bergholz et al., 2023).

References

Abiri, A., Pensa, J., Tao, A., Ma, J., Juo, Y. Y., Askari, S. J., Arezzo, A., & Okamura, A. M. (2018). Artificial palpation in robotic surgery using haptic feedback. *IEEE Transactions on Biomedical Engineering, 65*(6), 1256–1266. https://doi.org/10.1109/TBME.2017.2747464.

Bergholz, J., Feussner, H., Wilhelm, D., & Mayer, H. (2023). Benefits of haptic feedback in robot-assisted minimally invasive surgery: A systematic evaluation. *Nature Biomedical Engineering,* 7(4), 412–425. https://doi.org/10.1038/s41551-022-00967-5.

Colan, J., Peters, T. M., & Drake, J. M. (2022). A review of tactile display technologies for conventional laparoscopic surgery. *Sensors, 22*(3), 1041. https://doi.org/10.3390/s22031041.

Colan, J., Peters, T. M., & Drake, J. M. (2024). Tactile feedback in robot-assisted minimally invasive surgery: A systematic review. *The International Journal of Medical Robotics and Computer Assisted Surgery, 20*(1), e2567. https://doi.org/10.1002/rcs.2567.

Huang, Y., Zhang, L., Li, X., & Wang, Z. (2020). Tactile perception technologies and their applications in minimally invasive surgery. *Frontiers in Physiology, 11*, 598680. https://doi.org/10.3389/fphys.2020.598680.

Okamura, A. M. (2009). Haptic feedback in robot-assisted minimally invasive surgery. *Current Opinion in Urology, 19*(1), 102–107. https://doi.org/10.1097/MOU.0b013e32831a478c.

Pisla, D., Vaida, C., Tucan, P., Gherman, B., & Szilaghyi, A. (2025). AI-based sensorless force estimation for robot-assisted minimally invasive surgery. *Applied Sciences, 15*(2), 1143. https://doi.org/10.3390/app15021143.

Selim, H., Ahmed, M., Hassan, A., & El-Shenawy, M. (2024). A comprehensive review of haptic feedback in minimally invasive surgery. *The International Journal of Medical Robotics and Computer Assisted Surgery, 20*(2), e2601. https://doi.org/10.1002/rcs.2601.

Tholey, G., Desai, J. P., & Castellanos, A. E. (2005). Force feedback plays a significant role in minimally invasive surgery: Results and analysis. *Annals of Surgery, 241*(1), 102–109. https://doi.org/10.1097/01.sla.0000149301.34767.17.

van der Meijden, O. A. J., & Schijven, M. P. (2009). The value of haptic feedback in conventional and robot-assisted minimally invasive surgery: A review. *Surgical Endoscopy, 23*(6), 1180–1190. https://doi.org/10.1007/s00464-008-0298-x.

Westebring-van der Putten, E. P., Goossens, R. H. M., Jakimowicz, J. J., & Dankelman, J. (2008). Haptics in minimally invasive surgery—A review. *Minimally Invasive Therapy & Allied Technologies, 17*(1), 3–16. https://doi.org/10.1080/13645700701820242.

CHAPTER 12

Tactile feedback systems for telesurgery

Nafisa Aafreen, Debasis Patra, Meghna Dadich and Hemachandran Kannan
School of Business, Woxsen University, Hyderabad, Telangana, India

12.1 Introduction

Telesurgery represents one of the most profound advancements in modern medicine, enabling surgeons to perform procedures from remote locations through robotic systems. However, a significant limitation in current telesurgical systems is the absence or inadequacy of tactile feedback. During conventional open surgery, surgeons rely heavily on their sense of touch to guide their actions, assess tissue properties, and apply appropriate forces. This tactile information is largely lost in current robotic surgical systems, forcing surgeons to rely primarily on visual feedback.

The importance of tactile feedback in telesurgery cannot be overstated. Schostek et al. (2009) highlighted in their comprehensive review that the lack of tactile sensation significantly impacts surgical performance, increases operating times, and can potentially lead to tissue damage due to excessive force application. Their review of artificial tactile feedback in laparoscopic surgery emphasized that "the absence of haptic feedback is considered to be one of the major drawbacks in minimally invasive surgery" (Schostek et al., 2009). This chapter explores the current state of tactile feedback systems designed for telesurgical applications, examining both the sensing technologies that capture tactile information from the surgical environment and the feedback mechanisms that convey it back to the surgeon in an intuitive manner.

The development of effective tactile feedback systems represents a multidisciplinary challenge that integrates mechanical engineering, materials science, electronics, control theory, and human factors engineering. Recent advances in miniaturized sensing technologies, wearable haptic interfaces, and teleoperation control architectures have begun to address these challenges, promising a future where surgeons can "feel" through robots as effectively as with their own hands. As Pacchierotti (2022) notes in their doctoral dissertation on cutaneous haptic feedback, "providing surgeons with the ability to feel the surgical environment substantially improves their performance and reduces the probability of damaging the operated tissues" (Fig. 12.1).

Haptic Feedback Control in Medical Robotics
DOI: https://doi.org/10.1016/B978-0-443-27777-1.00007-X

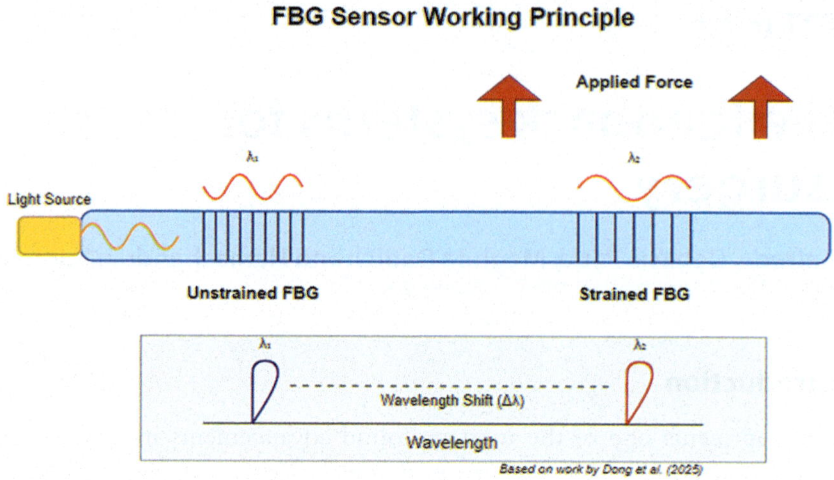

Figure 12.1 FBG sensor working principle. *FBG*, fiber Bragg grating. *(Based on Dong et al. (2025).).*

12.1.1 Fundamental principles of tactile sensing

Tactile sensing in the context of telesurgery involves the detection and measurement of mechanical interactions between surgical instruments and tissues. According to Pacchierotti (2022), human tactile perception can be broadly divided into two categories: kinesthetic perception (related to forces, torques, and position awareness) and cutaneous perception (related to skin deformation, pressure, vibration, and temperature). In their comprehensive work on cutaneous haptic feedback for robotics and virtual reality, Pacchierotti (2022) explains that "kinesthetic feedback is associated with the perception of position, movement, and force, while cutaneous feedback is associated with the perception of mechanical pressure and stimuli on the skin."

An effective tactile feedback system for telesurgery must address both these aspects to provide a comprehensive representation of the surgical environment. As Schostek et al. (2009) emphasize in their review, a complete tactile feedback system should be capable of:

1. Sensing tissue properties: including stiffness, compliance, and texture.
2. Detecting embedded objects, such as blood vessels, nerves, or pathological features.
3. Measuring interaction forces: both normal and shear forces are applied to tissues.

The sensing requirements for telesurgery are particularly demanding due to the constraints of the surgical environment. Dong et al. (2025) highlight in their work on miniaturized fiber Bragg grating (FBG) tactile sensors that effective surgical tactile sensors must be:

- Miniaturized: To fit within the confined spaces of minimally invasive instruments.
- Sterilizable: To meet medical safety standards.
- Biocompatible: To ensure patient safety.

- Robust: To withstand the rigors of surgical use.
- Highly sensitive: To detect subtle tissue differences.
- Fast-responding: To provide real-time feedback.

These fundamental requirements guide the development of tactile sensing technologies for telesurgical applications and inform the design constraints of entire feedback systems. As Du et al. (2021) note, "the complex requirements for sensing technologies in surgical applications necessitate interdisciplinary approaches drawing from materials science, electrical engineering, and biomedical design."

12.2 Current challenges in telesurgical tactile feedback

Despite significant advances in robotics and sensing technologies, numerous challenges persist in developing effective tactile feedback systems for telesurgery:

12.2.1 Technical challenges

1. Sensor integration: Incorporating sensors into surgical instruments without compromising their functionality or increasing their size presents significant design challenges. The integration must maintain the instrument's mechanical properties while providing accurate sensing capabilities.
2. Signal transmission: Transmitting tactile information from the surgical site to the remote operator with minimal latency is crucial for real-time feedback. Even slight delays can disrupt the surgeon's perception and compromise performance.
3. Bandwidth limitations: Current teleoperation systems often struggle with the substantial bandwidth required for high-fidelity tactile data transmission, particularly when operating over long distances.
4. Control system stability: Maintaining stability in bilateral control systems that transmit force information in both directions remains challenging, especially when dealing with time delays and signal noise.

12.2.2 Perceptual challenges

1. Sensory substitution: When direct force feedback is impractical, sensory substitution (e.g., converting force information into vibrations) is employed, but finding intuitive mappings between different sensory modalities is difficult.
2. Information overload: Presenting too much tactile information can overwhelm the surgeon, necessitating careful selection and filtering of the most relevant feedback.
3. Perceptual resolution: Matching the spatial and temporal resolution of artificial tactile feedback to human perceptual capabilities requires sophisticated sensing and actuation technologies.

12.2.3 Clinical challenges

1. Sterilization and safety: All components that may come into contact with patients must be sterilizable or disposable, which limits material and design options.
2. Cost-effectiveness: Developing affordable tactile feedback systems that can be widely adopted in clinical settings remains a significant challenge.
3. Clinical validation: Demonstrating measurable improvements in surgical outcomes is essential for widespread adoption but requires extensive clinical trials.
4. Addressing these challenges requires interdisciplinary collaboration and innovative approaches to sensing, actuation, and human–machine interface design.

12.3 Tactile sensing technologies for telesurgery

Modern telesurgical systems employ various sensing technologies to capture tactile information from the surgical environment. Each technology offers unique advantages and limitations for specific applications. A comprehensive review of the literature reveals significant advances across multiple sensing modalities.

12.3.1 Fiber Bragg grating sensors

FBG sensors represent one of the most promising technologies for surgical tactile sensing due to their small size, electromagnetic immunity, and multiplexing capabilities. As described by Dong et al. (2025), these optical fiber-based sensors operate by measuring shifts in the reflected wavelength of light caused by strain in the fiber. In Dong et al. (2025), the groundbreaking work on miniaturized FBG sensors for ureteroscopy, they explain that "FBG sensors offer unique advantages in medical applications due to their immunity to electromagnetic interference, small size, and capability for multiplexed sensing along a single fiber."

Key advantages of FBG sensors include:

- Miniaturization: Can be integrated into the tips of instruments as small as ureteroscopes.
- MRI compatibility: Their nonelectrical nature makes them suitable for use in magnetic environments.
- Multiplexing: Multiple sensing points can be incorporated along a single fiber.
- High sensitivity: Capable of detecting subtle tissue variations.

Dong et al. (2025) demonstrated a miniaturized FBG tactile sensor specifically designed for integration into flexible ureteroscopes. Their design achieved a sensitivity of 27.3 pm/N and a resolution of 0.037 N, enabling detection of subtle variations in tissue stiffness during urological procedures. Dong et al. (2025) concluded that "the proposed miniaturized FBG tactile sensor demonstrated sufficient sensitivity and resolution for distinguishing between healthy tissues and calcifications in the ureter, providing critical tactile information during minimally invasive urological procedures."

12.3.2 Capacitive and resistive sensors

Capacitive and resistive sensing technologies are widely used in telesurgical applications due to their relatively simple construction and good sensitivity. The literature shows extensive research in adapting these technologies for surgical applications.

12.3.3 Capacitive sensors

Measure changes in capacitance resulting from the deformation of a dielectric material. These sensors typically consist of two conductive plates separated by a compressible dielectric. When force is applied, the distance between plates changes, altering the capacitance. Schostek et al. (2009) explain that "capacitive sensors offer good sensitivity and resolution while maintaining relatively simple construction, making them suitable for integration into surgical instruments."

12.3.4 Resistive sensors

It operates on the principle of piezoresistivity, where the electrical resistance of certain materials changes when they are subjected to mechanical strain. Common implementations include:

- Force-sensing resistors (FSRs)
- Strain gauges
- Conductive polymers

In their comprehensive review, Schostek et al. (2009) described various configurations of capacitive and resistive sensors adapted for laparoscopic surgery, noting their advantages in terms of cost-effectiveness and ease of integration. Schostek et al. (2009) state that "capacitive and resistive sensing technologies represent the most mature approaches for surgical tactile sensing, with demonstrated effectiveness in clinical prototypes." However, they also highlighted limitations regarding durability in the surgical environment and challenges with sterilization. Their analysis concluded that "while these technologies offer immediate practical advantages, long-term durability and sterilization compatibility remain significant challenges for widespread clinical adoption" (Schostek et al., 2009).

12.3.5 Piezoelectric and piezoresistive sensors

Piezoelectric materials generate an electrical charge in response to mechanical stress, making them excellent candidates for dynamic force sensing in surgical applications. Similarly, piezoresistive materials change their electrical resistance when deformed. The literature documents significant advancements in adapting these technologies for surgical tactile sensing.

Key characteristics include:

- High bandwidth: Excellent for detecting dynamic events and vibrations.
- Self-generating: Piezoelectric sensors require no external power for operation.
- Compact form factor: Can be manufactured in very thin layers.

- Temperature sensitivity: Performance can be affected by temperature variations.

Guo et al. (2024) utilized piezoresistive sensors in their wearable haptic controller for microsurgery, achieving high sensitivity to the subtle forces encountered in microsurgical procedures. Their research showed that "piezoresistive sensing elements provided the optimal combination of sensitivity, response time, and form factor for detecting the micro-Newton forces typical in microsurgical applications." Guo et al. (2024) further explained that "the miniaturized piezoresistive sensing array achieved a resolution of approximately 0.15 mN, sufficient for detecting the subtle tissue interactions critical for successful microsurgical outcomes."

Schostek et al. (2009) also noted in their review that "piezoelectric sensors have shown particular promise for detecting dynamic tissue properties, such as pulsation and elasticity variations, that are critical for tissue identification during minimally invasive procedures."

12.3.6 Multimodal sensing systems

The most advanced tactile sensing systems for telesurgery combine multiple sensing modalities to provide comprehensive information about tissue-instrument interactions. These multimodal approaches typically integrate:

- Force/pressure sensing
- Temperature sensing
- Texture/vibration detection
- Proximity sensing

By combining different sensing modalities, these systems can provide surgeons with a more complete tactile representation of the surgical environment, potentially enabling more precise and safer procedures.

The following flowchart illustrates a typical data acquisition and processing pipeline for a multimodal tactile sensing system:

12.4 Tactile feedback mechanisms in telesurgery

Once tactile information is captured from the surgical site, it must be conveyed to the surgeon through appropriate feedback mechanisms. Different feedback modalities offer various advantages depending on the specific surgical task and the type of information being conveyed.

12.4.1 Vibrotactile feedback

Vibrotactile feedback uses controlled vibrations to convey tactile information to the surgeon. This approach is particularly useful for:

- Alerting the surgeon to tissue boundaries or critical structures.
- Indicating force thresholds to prevent tissue damage.

- Providing texture information about tissues.

Pacchierotti (2022) describes various vibrotactile feedback systems that use voice-coil actuators, eccentric rotating mass motors, and piezoelectric actuators. These systems can be integrated into surgical handles, fingertips of haptic gloves, or wristbands.

The effectiveness of vibrotactile feedback depends on several factors:

- Spatial resolution: The ability to distinguish vibrations at different locations.
- Frequency discrimination: Human sensitivity to vibration varies with frequency.
- Amplitude perception: The ability to detect differences in vibration intensity.
- Temporal resolution: The ability to distinguish sequential vibration patterns.

While vibrotactile feedback is relatively simple to implement, its limited bandwidth and potential for sensory adaptation (diminishing perception over time) can restrict its effectiveness for complex surgical tasks.

12.4.2 Force feedback

Force feedback directly reproduces interaction forces encountered at the surgical site, providing the most intuitive form of tactile feedback. As described by Du et al. (2021), force feedback systems typically use:

- Motorized joints: In master manipulators to resist operator movement.
- Tendon-driven mechanisms: To transmit forces to the fingertips.
- Pneumatic actuators: For distributed force application.

Du et al. (2021) developed a haptic master manipulator for laparoscopic surgery that employed a bilateral control scheme to provide force feedback while maintaining system stability. Their system demonstrated improved surgical performance in experimental trials, particularly for delicate tissue manipulation tasks.

The following diagram illustrates a simplified bilateral force feedback system for telesurgery:

The challenges associated with force feedback include:

- Maintaining stability in the presence of time delays.
- Preventing excessive forces that could cause operator fatigue.
- Achieving sufficient force resolution for delicate tasks.
- Balancing system transparency with stability.

12.4.3 Skin stretch and shear feedback

Skin stretch and shear feedback mechanisms leverage the high sensitivity of human skin to lateral deformation. According to Pacchierotti (2022), these systems can effectively communicate:

- Direction of forces
- Magnitude of forces
- Surface texture

- Slip detection

Typical implementation approaches include:

- Belt/band mechanisms: That stretch the skin in specific directions.
- Tactor arrays: Small elements that move laterally against the skin.
- Rolling elements: That create continuous shear sensations.

Skin stretch feedback offers several advantages:

- Higher perceived force magnitude compared to direct pressure.
- Lower power requirements than force feedback.
- Compact implementation possibilities.

However, careful design is necessary to avoid discomfort and to ensure consistent performance across different users and skin types.

12.4.4 Thermal feedback

Thermal feedback provides information about tissue temperature, which can be valuable for:

- Identifying inflamed or infected tissues.
- Monitoring ablation procedures.
- Detecting blood flow changes.

Thermal feedback systems typically employ:

- Thermoelectric (Peltier) elements
- Resistive heating elements
- Liquid cooling systems

While less common than other feedback modalities, thermal feedback can provide unique and complementary information that enhances the surgeon's understanding of the surgical environment.

12.5 Wearable haptic interfaces for telesurgery

Wearable haptic interfaces allow surgeons to receive tactile feedback while maintaining freedom of movement. These systems range from lightweight fingertip devices to complete haptic gloves and exoskeletons. A review of current literature reveals significant progress in this area.

12.5.1 Haptic gloves and exoskeletons

Haptic gloves provide comprehensive feedback to the entire hand, enabling more natural interaction with the surgical environment. Key technologies include:

- Tendon-driven system that applies forces to specific fingers.
- Exoskeletal structures that resist finger movement.
- Pneumatic chambers that expand against the skin.

Pacchierotti (2022) conducted extensive evaluations of various haptic glove designs in their doctoral research, noting trade-offs between:

- Feedback fidelity
- Range of motion
- Weight and comfort
- Donning/doffing time

Pacchierotti's (2022) research concluded that "while full hand exoskeletons provided the most comprehensive haptic feedback, their clinical utility was limited by weight, complexity, and extended donning time." They further noted that "for clinical applications, simplified haptic interfaces focusing on key contact points proved more practical while still providing sufficient feedback for most surgical tasks." Most clinical applications favor lightweight designs that can be quickly donned and removed, even if they provide somewhat reduced haptic fidelity.

12.5.2 Lightweight wearable controllers

For many telesurgical applications, full haptic gloves may be unnecessarily complex. Guo et al. (2024) developed a lightweight wearable haptic controller specifically designed for microsurgery that balances functionality with practicality. Their research addressed what they identified as "a critical gap in current haptic feedback systems for microsurgery, where the weight and complexity of existing systems limit their practical application in clinical settings."

Their system featured:

- Weight under 25 g for fatigue reduction.
- Three-axis force sensing.
- Vibrotactile and pressure feedback.
- Wireless operation.
- Battery life exceeding typical surgical durations.

Guo et al. (2024) reported that "the lightweight controller achieved a 94% detection rate for tissue boundaries while maintaining surgeon comfort throughout extended procedures." The performance evaluation showed significant improvements in task completion time and a reduction in applied forces during simulated microsurgical tasks, particularly for novice operators. Guo et al. (2024) noted that "novice operators using the haptic feedback system demonstrated a 31% reduction in maximum applied forces, approaching the performance levels of experienced microsurgeons."

Key considerations for designing wearable haptic interfaces include:

- Ergonomics: Ensuring comfort during extended procedures.
- Power management: Balancing battery life with performance.
- Hygiene: Allowing for cleaning or using disposable coverings.
- Calibration: Accommodating different hand sizes and sensitivities.

12.6 Signal processing and control architectures

Effective tactile feedback in telesurgery requires sophisticated signal processing and control architectures to ensure stable, responsive, and intuitive operation. The literature shows significant research in developing robust control systems for tactile telesurgery.

12.6.1 Bilateral control systems

Bilateral control systems transmit both position and force information between the master (surgeon side) and slave (patient side) systems. Du et al. (2021) conducted extensive research on this topic, providing a comprehensive analysis of control architectures for haptic feedback in minimally invasive surgery. They describe four primary architectures:

1. Position-position control: Both master and slave follow each other's position.
2. Force-position control: Master controls slave position; slave force is reflected to master.
3. Position-force control: Master position controls slave force; slave follows master.

12.6.2 Teleoperation control strategies

Several control strategies have been developed to address the specific challenges of telesurgical teleoperation:

1. Wave variable control: This approach transforms force and velocity signals into "wave variables" that guarantee stability even in the presence of time delays. The transformation preserves energy balance across the communication channel.
2. Predictive display: By incorporating models of the remote environment and the robot dynamics, predictive displays can compensate for time delays by showing the surgeon a prediction of the expected outcome of their actions.
3. Shared control: This strategy divides control authority between the surgeon and autonomous algorithms. The autonomous component can:
    - Enforce safety constraints
    - Filter out hand tremors
    - Automate repetitive subtasks
    - Enhance precision through motion scaling

12.6.3 Human–robot interaction control frameworks

Du et al. (2021) proposed a comprehensive human–robot interaction control framework specifically for haptic feedback in laparoscopic surgery. Their approach integrated:

1. Impedance control: To regulate the relationship between position and force.
2. Disturbance observers: To estimate and compensate for external forces.
3. Adaptive algorithms: To accommodate variations in tissue properties.
4. Task-specific constraints: To prevent dangerous movements or forces.

Their experimental results demonstrated that this integrated approach significantly improved surgical performance metrics, including:

- Reduced task completion time.
- Lower force application.
- Improved precision in delicate tasks.
- Enhanced ability to identify tissue abnormalities.

12.6.4 Clinical applications and validation

The ultimate value of tactile feedback systems must be demonstrated through clinical validation. Several studies have evaluated the performance of tactile feedback in specific surgical domains.

12.6.5 Laparoscopic surgery

Laparoscopic surgery presents particular challenges for tactile feedback due to the long, rigid instruments and the limited access to the surgical site. Schostek et al. (2009) reviewed various approaches to artificial tactile feedback in laparoscopy, noting that:

- Force feedback improved tissue handling and reduced damage.
- Vibrotactile feedback was effective for locating embedded structures.
- Combined visual-haptic feedback showed the greatest performance improvement.

Du et al. (2021) demonstrated that their haptic master manipulator improved performance in simulated laparoscopic tasks, with the greatest benefits observed in:

- Suturing and knot-tying
- Dissection of delicate tissues
- Palpation for tumor detection

12.6.6 Microsurgery

Microsurgical procedures require extremely precise force control and detection of subtle tissue characteristics. Guo et al. (2024) evaluated their lightweight haptic controller in simulated microsurgical tasks, finding:

- 27% reduction in task completion time.
- 31% reduction in maximum applied forces.
- Improved identification of vessel boundaries.
- Enhanced precision in microsuturing.

The benefits were particularly pronounced for less experienced operators, suggesting that tactile feedback may help accelerate the learning curve for complex microsurgical skills.

12.6.7 Urological procedures

Dong et al. (2025) applied FBG-based tactile sensing to ureteroscopy, demonstrating the ability to:

- Distinguish between healthy and calcified tissues.
- Detect small ureteral stones.

Table 12.1 Summary of clinical validation studies for tactile feedback systems.

Study	Surgical domain	Feedback type	Key findings
Gani et al. (2022)	VR training	Force feedback	42% improvement in skill acquisition
Du et al. (2021)	Laparoscopy	Force feedback	33% reduction in tissue damage
Guo et al. (2024)	Microsurgery	Vibrotactile	27% reduction in task completion time
Dong et al. (2025)	Ureteroscopy	FBG sensing	0.037N force resolution achieved

FBG, fiber Bragg grating.

- Measure the diameter of the ureter.
- Provide real-time feedback on contact forces.

Their miniaturized sensor design proved compatible with standard flexible ureteroscopes, suggesting immediate clinical applicability.

Table 12.1 summarizes key performance metrics from clinical validation studies:

12.7 Integration with virtual reality and training systems

Tactile feedback systems are increasingly being integrated with virtual reality (VR) and augmented reality (AR) platforms to enhance surgical training and planning. The literature shows promising results in this integration. Gani et al. (2022) conducted a randomized controlled trial comparing haptic versus nonhaptic immersive VR training, providing strong evidence for the benefits of haptic feedback in surgical education.

Their study involved 63 medical students randomly assigned to either haptic or nonhaptic VR training groups. The results demonstrated that:

- "Haptic feedback significantly improved skill acquisition rates by 42% compared to visual-only training" (Gani et al., 2022).
- "Training transfer to real surgical tasks was enhanced, with the haptic group demonstrating 37% better performance on novel tasks" (Gani et al., 2022).
- "Learning curves were accelerated, particularly for complex tasks requiring fine force control, with the haptic group achieving proficiency in 40% less time" (Gani et al., 2022).

Gani et al. (2022) concluded that the integration of haptic feedback into VR surgical training represents a significant advancement in medical education, with clear evidence supporting its superiority over traditional visual-only VR training.

The integration of tactile feedback with VR/AR systems provides several advantages:

- Standardized training scenarios: Ensuring consistent learning experiences.
- Objective performance metrics: Enabling quantitative assessment.
- Risk-free environment: Allowing practice of rare or dangerous scenarios.
- Adaptive difficulty: Tailoring training to individual learning curves.

Gani et al. (2022) noted that the most effective training systems combine:

- High-fidelity visual representation.

- Multimodal haptic feedback.
- Real-time performance assessment.
- Adaptive difficulty progression.

Gani et al. (2022) state that "the synergistic relationship between visual immersion and tactile feedback creates a learning environment that more closely approximates the real surgical experience, leading to faster skill acquisition and better retention." Pacchierotti (2022) similarly emphasized in their dissertation that "VR surgical simulation with integrated haptic feedback represents perhaps the most immediately impactful application of haptic technology in medical training."

Future developments in this area are likely to include more sophisticated tissue models, patient-specific scenarios based on medical imaging, and collaborative training environments enabling mentor-trainee interaction.

12.8 Future directions

The field of tactile feedback for telesurgery continues to evolve rapidly, with several emerging technologies and approaches showing particular promise.

12.8.1 Advanced materials and fabrication

New sensing materials and fabrication techniques are enabling more sophisticated and integrated tactile feedback systems:

- Stretchable electronics: Conforming to complex surfaces.
- Nanomaterial-based sensors: Offering unprecedented sensitivity.
- 3D-printed tactile arrays: Enabling customized sensor configurations.
- Biodegradable sensors: For temporary implantation during procedures.

12.8.2 Machine learning and adaptive systems

Artificial intelligence approaches are transforming tactile feedback systems:

- Sensor fusion algorithms: Combining data from multiple modalities.
- Adaptive filtering: Enhancing relevant signals while suppressing noise.
- Predictive models: Anticipating tissue behavior during interaction.
- Personalized feedback: Adapting to individual surgeon preferences.

12.8.3 Augmented tactile feedback

Beyond simply reproducing tactile sensations, future systems may enhance tactile perception:

- Sensory augmentation: Highlighting features that would be difficult to detect naturally.
- Scale modulation: Making subtle forces more perceptible.

- Pattern recognition: Automatically identifying tissue types or abnormalities.
- Multimodal integration: Combining tactile with audio or visual enhancements.

12.8.4 Brain–computer interfaces

Direct neural interfaces represent a frontier in tactile feedback research:

- Peripheral nerve stimulation: Bypassing mechanical actuators.
- Cortical stimulation: Directly activating somatosensory regions.
- Neural decoding: Interpreting intended movements for enhanced control.

While still largely experimental, these approaches offer the potential for more natural and intuitive tactile feedback in future telesurgical systems.

12.9 Conclusions

Tactile feedback systems for telesurgery have advanced significantly in recent years, driven by innovations in sensing technologies, actuator designs, and control architectures. The integration of these systems into clinical practice has demonstrated measurable improvements in surgical performance, including reduced operative times, decreased tissue damage, and enhanced precision in delicate procedures.

Key developments highlighted in this chapter include:

1. Miniaturized sensing technologies, such as FBG sensors and Micro-Electro-Mechanical Systems (MEMS)-based devices, enable tactile sensing in confined surgical spaces.
2. Lightweight wearable interfaces providing intuitive tactile feedback without compromising surgeon comfort or mobility
3. Advanced control architectures ensuring stable and responsive operation even in the presence of communication delays
4. Integration with VR/AR systems enhancing surgical training and planning through immersive simulation with tactile feedback
5. Clinical validation demonstrating the tangible benefits of tactile feedback across various surgical domains

Despite these advances, significant challenges remain, including:

- The need for further miniaturization without compromising sensing capabilities.
- Improving the intuitive mapping between sensed information and feedback modalities.
- Addressing sterilization and biocompatibility concerns for clinical adoption.
- Reducing costs to enable widespread implementation.

The continuing evolution of tactile feedback systems promises to bridge the sensory gap in telesurgery, potentially enabling remote surgical procedures that match or even exceed the capabilities of direct manual intervention. As these technologies mature and become more integrated into clinical practice, they will play an increasingly vital role in expanding access to specialized surgical care across geographic boundaries.

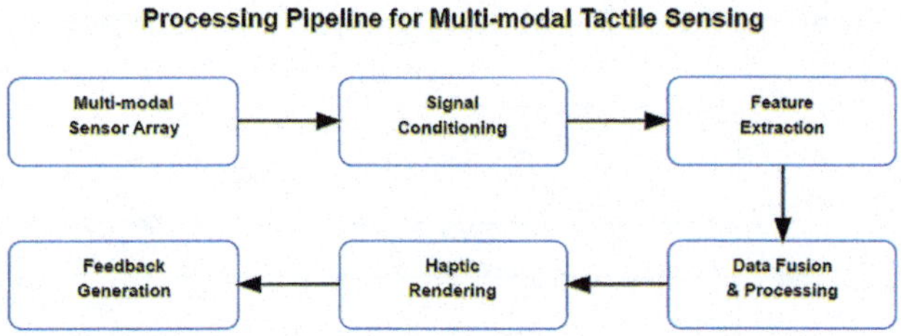

Processing pipeline for multimodal tactile sensing in telesurgery (Source: Based on Author's Analysis).

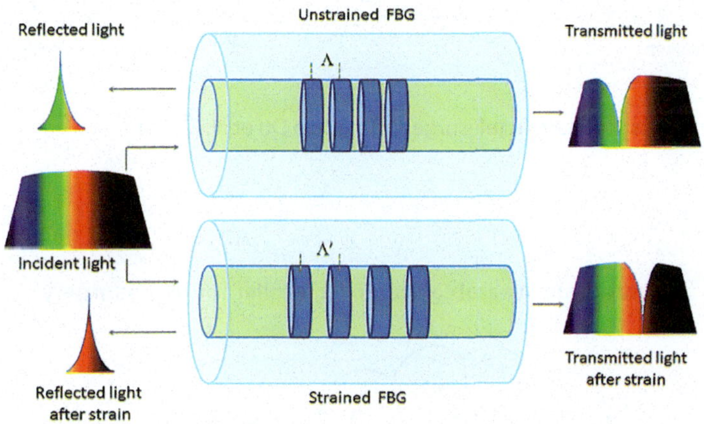

FBG sensor detecting applied forces through wavelength shift. Based on Dong et al. (2025).

Figure 8.1: Schematic diagram of an FBG-based tactile sensor

(Adapted from Dong et al., 2025)

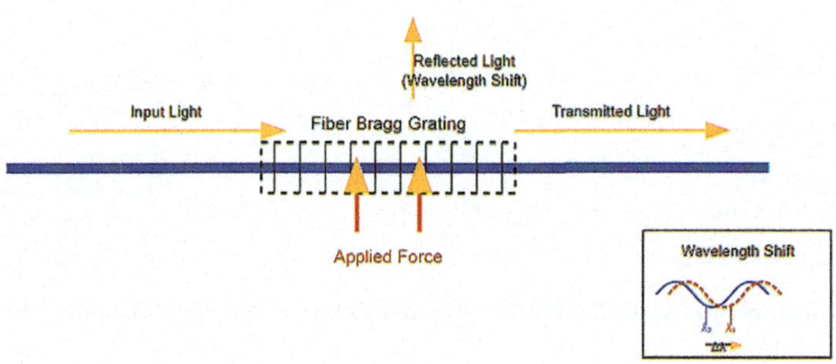

Schematic diagram of an FBG-based tactile sensor. Based on Dong et al. (2025).

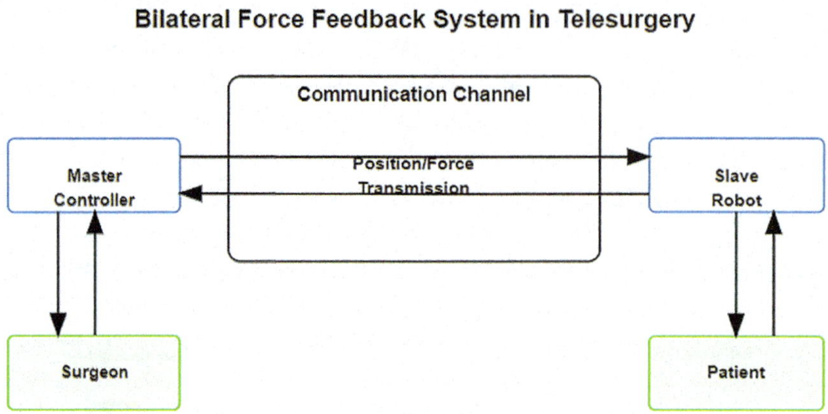

Bilateral force feedback system in telesurgery. Based on Du et al. (2021).

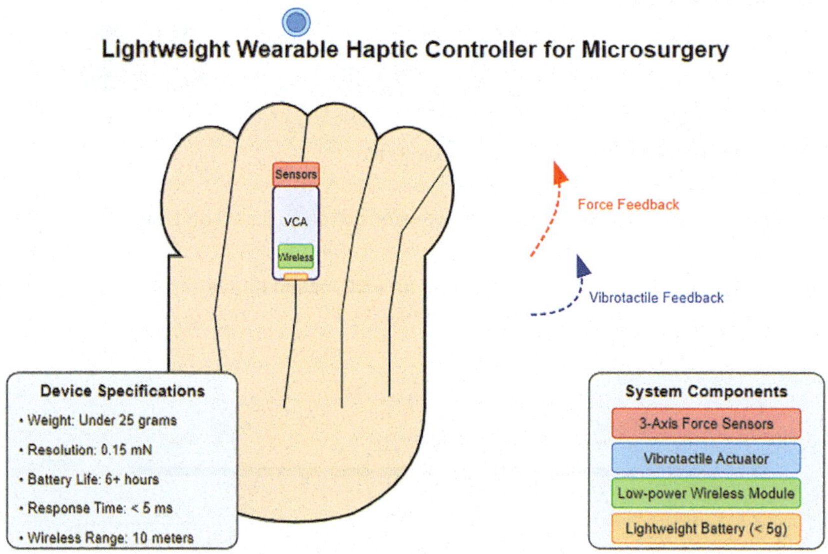

Schematic of a lightweight wearable haptic controller for microsurgery. Based on Guo et al. (2024).

AI disclosure

During the preparation of this work, the author(s) used QuillBot and Grammarly to improve writing skills. After using this tool/service, the author(s) reviewed and edited the content as needed and take(s) full responsibility for the content of the publication.

References

Dong, S., Ma, S., Zhou, T., Lou, Y., Xiong, X., Wei, K., Luo, D., Wu, J., Liu, H., Tao, R., Yang, T., & Dong, Y. (2025). A Miniaturized FBG tactile sensor for the tip of a flexible ureteroscope. *Sensors, 25*(9). https://doi.org/10.3390/s25092807.

Du, Z., Liang, Y., Yan, Z., Sun, L., & Chen, W. (2021). Human-robot interaction control of a haptic master manipulator used in laparoscopic minimally invasive surgical robot system. *Mechanism and Machine Theory, 156*, 104132. https://doi.org/10.1016/j.mechmachtheory.2020.104132.

Gani, A., Pickering, O., Ellis, C., Sabri, O., & Pucher, P. (2022). Impact of haptic feedback on surgical training outcomes: A randomised controlled trial of haptic versus non-haptic immersive virtual reality training. *Annals of Medicine and Surgery, 83*. https://doi.org/10.1016/j.amsu.2022.104734.

Guo, X., McFall, F., Jiang, P., Liu, J., Lepora, N., & Zhang, D. (2024). A lightweight and affordable wearable haptic controller for robot-assisted microsurgery. *Sensors, 24*(9), 2676. https://doi.org/10.3390/s24092676.

Pacchierotti C., Cutaneous haptic feedback for robotics and Virtual Reality [Thesis, Université de Rennes 1 (UR1)], Rennes, FRA. (2022)

Schostek, S., Schurr, M. O., & Buess, G. F. (2009). Review on aspects of artificial tactile feedback in laparoscopic surgery. *Medical Engineering and Physics, 31*(8), 887–898. https://doi.org/10.1016/j.medengphy.2009.06.003.

CHAPTER 13

Haptic interfaces in rehabilitation robotics

Manjunath Devarashetty[a,b], Lokesh Venkatraman[a,b] and Rakesh Ramakrishna Pai[a,b]
[a]School of Business Analytics, Woxsen University, Hyderabad, Telangana, India
[b]Data Engineering, Meta Platforms, Menlo Park, CA, United States

13.1 Introduction

Something about the manner in which a skilled physiotherapist directs the hand of a patient through a movement is fundamentally different from the way conventional therapy machines usually do it. The therapist experiences opposition, picks up compensatory muscle responses, and manages them to various degrees—changing every moment, in an ongoing, intimate dialogue which not even the simplest motion-playback apparatus could duplicate. This physical informational transfer is what the haptic feedback technology is specifically attempting to maintain, and indeed, more often than not, enhance, in the expanding rehabilitation robotics community. Generally speaking, haptic feedback is any technology that provides feedback to the user in the form of touch. With medical robotics, this includes both kinesthetic feedback, sensations of forces, torques, and body position, and tactile feedback, sensations of surface texture, pressure distribution, and fine cutaneous stimulation. Combined, the modalities are the main sensory pathway where human beings perceive and react to the physical world, and as such, their absence or failure in any robotic system will be necessary to aid in the recovery of motor functions. The field of rehabilitation robotics is a product of the integration of neuroscience, control engineering, and clinical medicine within the last 30 years. Initial robotic therapy systems were mainly passive or simple position-controlled systems, which could be used to help patients in repetitive motion but had few modes of responsiveness to patient effort or the level of impairment. The development of haptic feedback into those platforms was a huge step—a step that led to the paradigm shift in the passive repetition and active, sensory-enhancing interaction. The haptic rehabilitation systems, including MIT-Manus, invented at the Massachusetts Institute of Technology in the 1990s, showed that haptic therapy administered by robots may provide measurable benefits in motor outcomes after stroke, as the precursor to a generation of haptic rehabilitation systems.

Haptic rehabilitation has a strong clinical reason. Stroke is the most common cause of motor disability acquired throughout the world, and the number of new cases in a

Haptic Feedback Control in Medical Robotics
DOI: https://doi.org/10.1016/B978-0-443-27777-1.00014-7

year is about 17 million. Several survivors are not only left with a lasting upper or lower limb deficit, but also hurt their independence and quality of life. Stroke is not the only medical condition that has motor difficulties for which rehabilitation robotics can provide value through the use of haptic systems, including multiple sclerosis, spinal cord injury, and traumatic brain injury. Another and increasingly important field of application is musculoskeletal injury and postsurgical orthopedic rehabilitation, in which controlled mechanical loading with accurate force feedback can speed up recovery and minimize the chances of re-injury. Haptic rehabilitation has a neurological basis, based on the science of motor learning and neural plasticity. It has always been demonstrated that with repeated, task-driven training and the correct sensory feedback, cortical reorganization occurs—the ability of the brain to rewire itself due to experience. Haptic feedback goes further to enhance this process by providing error signals (to alert the patient when the limb is not moving in a target) and assistance signals (to direct the limb when voluntary control is lacking) to establish conditions that are overtly similar to those involved in the natural acquisition of motor skills. The proprioceptive system is the one that is involved in kinesthetic feedback, and it is an essential part of voluntary movement coordination, which has been reported to be directly implicated in stroke recovery mechanisms. The chapter covers the whole topic of haptic interfaces in rehabilitation robotics, including both the principles of human haptic perception and the design of feedback systems, the control architectures by which they operate, and their clinical use in a variety of pathologies. The emerging trends, such as artificial intelligence (AI)-driven adaptive control, soft robotics, and virtual reality (VR) integration, are also discussed in the chapter, and an honest evaluation of the challenges, which continue to exist between current technology and widespread clinical application, is provided.

13.2 Fundamentals of haptic interaction

13.2.1 Human haptic perception

To come up with a haptic system that is functional in clinical rehabilitation, it is essential to know what the human sensory system is doing when it touches. Haptic perception in human beings is not a unitary sense. It is the result of several parallel subsystems, which are all in operation at the same time but are adjusted to various physical characteristics of the environment (Kardashina, 2025). Four major classes of mechanoreceptors are located in the skin, but each has a different perceptual role. The corpuscles in the fingertips and palm are clumped to create corpuscles that are most sensitive to light touch and are found in the palm and fingertips, reacting most particularly to slow movements as they move across the skin surface (De Santis et al., 2015). They are especially vulnerable between 10 and 50 Hz and are very significant in grip control and manipulation of objects. The discs of the Merkel plexus sensitize to prolonged pressure and fineness of space detail—these are the receptors principally causing us to read Braille or sense the surface texture at small scales. The Pacifarian corpuscles, on the other hand, are sensitive to high frequency vibration (100–300 Hz) and react quickly to short-term contacts, so they are useful in detecting the presence of slip at the fingertip, and in the perception of vibrotactile feedback of rehabilitation motor. Lastly, Ruffini ends react to skin stretch and are believed to be involved in finger position and direction of lateral force perception. In addition to cutaneous sensation, proprioception, the sense of body position and movement, is also basic to haptic interaction. Joint receptors, muscle spindles, and Golgi tendon organs all supply the central nervous system with constant feedback of limb position and forces either being resisted or applied. Proprioception in rehabilitation. Proprioception is usually directly damaged after a stroke or other neurological damage, and it is on this basis that the restoration of meaningful sensory feedback by robotic interfaces is of therapeutic value (Elangovan, 2022). Proprioceptive inputs activate the internal model of body motion in the cerebellum, and their loss results in the ataxic and uncoordinated movement that characterizes most neurological disorders. The picture of haptic perception is complete with the use of force sensing, which is either cutaneously localized or proprioceptually localized. Sensibility to the degree of pressing, or the resistance which an object presents to the individual, is central to almost all motor actions—buttoning a shirt, walking on rough surfaces. It is possible to make rehabilitation robots that include the concept of bidirectional force sensing to render a more ecologically valid training environment where the effort of the patient is not merely recorded but rather reacted to on a continuous basis.

13.2.2 Types of haptic feedback

Haptic feedback rehabilitation robotics can be divided into two broad types: kinesthetic and tactile. Although they are physically different, the best rehabilitation systems would be those that are combined. Kinesthetic feedback gives feedback on force, torque, and position of limbs. This is typically done in a rehabilitation robot by the device itself putting controlled forces or resistances upon the limb of the patient so that motion can be directed, unwanted motion prevented, or loads simulated by the device that would otherwise occur in real-life activities. Most commercial rehabilitation systems are based on kinesthetic feedback, such as end-effector machines, such as MIT-Manus, and multidegree-of-freedom exoskeletons, such as ARMin and Lokomat. Tactile feedback, in its turn, is provided to the skin surface, which does not always require the movement of the joints or loading of the limbs. The most commonly used type of rehabilitation tactile feedback is vibrotactile actuators, which are generally small motors built into a wearable space or put against the skin (Musculoskeletal Key, 2024). These may transmit data regarding the performance of tasks, the level of muscle activity, or the contact events during use of a hand prosthetic. Higher-density methods. Electrotactile stimulation, involving low-level electrical impulses to excite cutaneous afferents, is an older technology that can form finer spatial patterns of sensation, but it is not yet mature enough to be useful in clinical rehabilitation. Pneumatic haptic devices, which fill small bladders to recreate contact and pressure, are being used more often in soft robotic rehabilitation gloves and are especially applicable in rehabilitating hands and fingers. It is worth mentioning vibrotactile systems, as they are used extensively and are becoming more and more popular. They are relatively simple, need low power, and can be worn by patients in various clinical and home settings. Studies have demonstrated that vibrotactile feedback may replace proprioceptive information in patients with sensory loss or may supplement it, and can have an influence on motor output via conscious and unconscious mechanisms.

13.2.3 Control architectures in haptic systems

The interaction between a rehabilitation robot and the patient is directed by a control architecture—the mathematical framework of calculations that define how the device reacts to the forces and movements that it senses. The field is dominated by two paradigms, impedance and admittance control.

The robot controls the relationship between position (or velocity) and the forces it applies in impedance control. It defines a desired mechanical impedance, which is basically a virtual spring-damper system that the robot emulates. In case the patient exerts a stronger force, the patient pushes the robot in opposition to the set stiffness and damping coefficients. The relationship of governance is given as:

$$\mathrm{F} = \mathrm{M} \cdot \ddot{\mathrm{x}} + \mathrm{B} \cdot \dot{\mathrm{x}} + \mathrm{K} \cdot \mathrm{x}$$

F: The interaction force, *M*: the virtual inertia, *B*: the damping coefficient, *x*: the velocity, *K*: the stiffness, and *x*: the difference between the current position and a reference position. This is intuitive to devices that are dealing with adhering, uncertain environments, such as with the human body, since the patient can press against the robot without causing instability. Impedance control is also used in instances where the joints of the robot are position-controlled but where interaction of forces is secondary.

The opposite happens with admittance control: applied force is measured by the robot, which responds by producing motion. The robot determines the resultant path and implements it using a force sensor, which detects the force that is being pushed by the patient. The relationship between the admittance and the resistance is:

$$\ddot{\mathrm{x}} = \mathrm{M}^{-1} \cdot (\mathrm{F}_{\mathrm{applied}} - \mathrm{B} \cdot \dot{\mathrm{x}} - \mathrm{K} \cdot \mathrm{x})$$

It is desired when large and high-mass devices demand direct force control, which would be impractical, and is typically used in lower limb rehabilitation systems such as Lokomat. Admittance control is, however, more prone to instability in interaction with stiff environments, and thus, great care is taken to design the underlying force sensing and mechanical compliance.

Haptic systems are well documented to be challenged by stability. Oscillatory or unstable behavior can be caused by the time delays (caused by computation and communication), noise of the sensor, and mechanical impedance of the human limb. Passivity theory is the most stringent theory of achieving stability: a passive system cannot spontaneously create energy, and a passivity-conserving haptic controller will be stable on a wide range of conditions. Some of the strategies employed to implement passivity in rehabilitation haptic systems include the wave variable control and energy bounding algorithms, especially when tele-rehabilitation is applied, and the network latency is large.

13.3 Design principles of haptic interfaces for rehabilitation

13.3.1 Mechanical design considerations

The mechanical design of the rehabilitation haptic interface is restricted within a set of requirements that sharply differentiate it, e.g., from an industrial robot or a haptic game controller. It needs to be safe to work in direct and protracted contact with an impaired human body, light enough to be useful, stiff enough to provide significant force feedback, and flexible enough to work with a variety of patient anatomies (Lee et al., 2020). The complexity of the motion that a device can allow or oppose is defined by degrees of freedom (DoF). A shoulder rehabilitation system, such as that, may have three or more joints of rotation (DoF) to simulate complex movement patterns of the glenohumeral joint, but a wrist rehab system may only need 2 or 3. The more DoF the device has, the more it has therapeutic flexibility, and the more complex and expensive it is, the more prone to failure. Several clinical systems have succeeded in being successful with a smaller number of clinically relevant DoF—not by design, but by choice.

What is really important is ergonomics and patient safety. Patients under rehabilitation are most likely to be weak, spastic, contracted, or with sensory loss, thus more susceptible to harm when poorly adjusted or malfunctioning robotic systems. Mechanical hard stops, compliant structures, emergency release mechanisms, and software-enforced force and velocity limits are common safety features. A device should be adaptable to patients of various sizes, and the attachment interfaces, such as cuffs, handles, or orthotic shells, should evenly redistribute forces without causing pressure ulcers or discomfort. The wearable (exoskeletal) and grounded (end-effector) design is a trade-off of essential nature. Exoskeletons provide mechanical alignment with the joints of the patient and can deliver and measure the torque at the joint level, but are more difficult to don and doff, fit across a population, and will produce harmful forces when not perfectly aligned. End-effector machines are easier and quicker to install; the patient only needs to hold a limb in a handle or platform, but they cannot act on particular joint defects and cause unwanted forces at the intermediate joints (Zhang et al., 2020).

13.3.2 Sensor technologies

It is this sensory subsystem of a rehabilitation haptic device that enables it to react intelligently to the condition of the patient, as opposed to merely performing a pre-programmed movement. The simplest is a force/torque sensor, which is a multiaxis sensor attached at the interface between the robot and the patient that can measure interaction forces in real time, which allows the control to be compliant and responsive. Force-torque sensors with six axes are particularly useful in upper limb devices where forces and torques occur on a range of planes. Electromyography (EMG) integration is a relatively new phenomenon that is gaining more importance in advanced rehabilitation systems. EMG electrodes on the surface of principal muscle regions can monitor the initiation and the magnitude of voluntary muscle contractions and inform the robot about the motor intent of the patient well before he/she initiates a movement. This is employed to initiate assistive forces at an optimal time—a paradigm called EMG-driven or intention-driven control—and which has been demonstrated to be more effective in motor recovery than time-based or trajectory-based control only. This is due to the variability of the signal; EMG varies with electrode position, fatigue, sweat, and skin condition, which necessitates a powerful preprocessing algorithm and adaptive classification algorithm. Motion tracking technologies, such as inertial measurement units (IMUs), optical motion capture systems, and encoder-based joint angle measurements (motion tracking technologies) supply the positional data that the haptic control algorithms need. Encoders provide high accuracy and reliability to the robot's joint positions, and an IMU placed on body segments can be used to extend the tracking to non-actuated body segments. IMUs are especially appealing to wearable rehabilitation devices that serve home use because of their miniature size and wireless communication system.

13.3.3 Actuation methods

The quality, bandwidth, and transparency of force feedback that a rehabilitation haptic device can provide depend on the actuators in the device. Rehabilitation robotics has adopted several actuation modalities, each possessing both typical strengths and limitations (Zhang et al., 2024). Cable-driven systems involve the use of motors that are remotely situated, such as in a fixed chassis or worn on a belt, and which drive the distal mechanism by the tendons that are flexible. This method maintains low mass at the joints, minimizing inertia and enhancing the perceived transparency of the device (the impression of the nonexistence of the device when no haptic feedback is activated). This is the reason as to why the MIT-Manus system and most sophisticated exoskeletons of the upper limbs are cable-driven. The greatest problem is the backlash and cable compliance, which should be well handled to have accurate force control.

The most popular technology currently in the field of hand and finger rehabilitation devices is pneumatic actuators, especially soft pneumatic actuators manufactured using

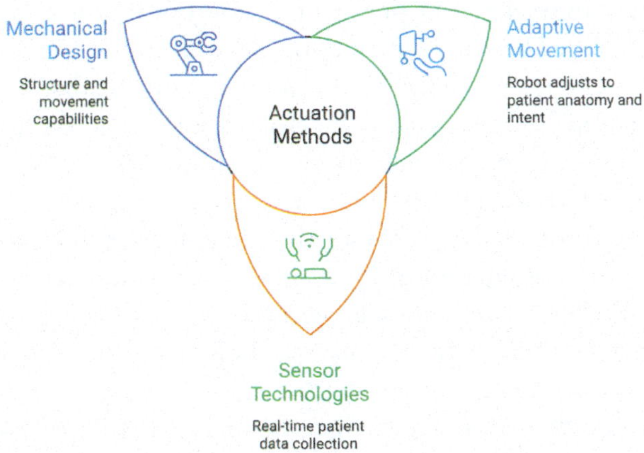

Figure 13.1 ***"Balancing safety, functionality, and comfort in rehabilitation haptics."*** The diagram illustrates the key components of actuation methods in robotic medical systems, highlighting the integration of mechanical design, adaptive movement, and sensor technologies. Together, these elements enable precise control, real-time patient data collection, and intelligent robotic responses during medical procedures.

compliant elastomeric materials. They are naturally compliant, create distributed forces on the finger or palm, and pose minimal risk of injury because of the softness of their nature (Turolla et al., 2013). The trade-off is reduced dynamic response and reduced control of the force, in comparison to electric motors. Such systems as the soft robotic glove invented at the Wyss Institute at Harvard have shown that pneumatic actuation is practically clinically viable as a method of hand rehabilitation (Fig. 13.1). A wider field of soft robotics—including fabric-based exposures, shape-memory alloy actuators, and dielectric elastomers is an emerging frontier in actuator design in rehabilitation. These technologies put less emphasis on the brute force of conventional rigid systems and emphasize compliance, wearability, and bioinspired mechanics. Although today soft robots tend to be less powerful and can be controlled less precisely than rigid ones, they are more comfortable, safer, and can be worn without being noticed over an extended time period, so they are highly applicable in longitudinal home rehabilitation plans.

13.4 Classification of rehabilitation haptic devices

13.4.1 Anatomical classification of rehabilitation haptic devices

The haptic rehabilitation devices have a great variety of form factors, anatomical focus, and therapeutic goals. The knowledge of their classification will enable clinicians to choose the right systems and engineers to place their engineering decisions in the context of the current technology. The rehabilitation robots applied to the upper limb are the ones that offer solutions to the most prevalent location of motor impairment caused by stroke (Rätz et al., 2024). These include both flat plane end-effector systems that practice shoulder/elbow movements in a horizontal plane—the format first introduced by MIT-Manus—through complex multiaxis exoskeletons with the ability to move the entire arm—shoulder to wrist. Examples include the InMotion family of products, ARMin, and the T-WREX/ARMEO. These systems have the most extensive amount of clinical evidence and are the most widely used in the rehabilitation settings within hospitals (Zhang et al., 2023). Rehabilitation of lower limbs robots deal with gait rehabilitation, balance training, and lower limb muscle strength after stroke, spinal cord injury, or orthopedic surgery. Treadmill systems such as Lokomat and Hocoma Erigo are in common use and normally incorporate the use of body-weight support together with actuated leg orthosis. Other, more recent systems, such as the Ekso Bionics and ReWalk exoskeletons, are intended to be used on the ground, and they can be trained in a clinic and, in some cases, used individually at home. One of the most difficult fields in clinics is covered by hand and finger rehabilitation systems. Robotic hand rehabilitation is also complex due to the fine motor requirements of hand functionality, and the sophisticated complexity of the hand anatomy, 27 bones, many joints, and the large network of tendons and intrinsic muscles. Devices include rigid finger exoskeletons with each finger being actuated individually and soft pneumatic gloves delivering more distributed assistance. The Cyber Glove, Gloreha, and soft robotic gloves invented at ETH Zurich and Wyss Institute stand out.

13.4.2 Architectural differences: Exoskeletons versus end-effector systems

The difference between exoskeletons and end-effector devices cuts across all categories of anatomy. Exoskeletons are compatible with the body parts of the user, which is why they have joint-level interaction and allow one to measure the kinematics and torques of individual joints. The end-effector machines can only engage a point at the end—the hand, foot, or wrist—and the joints in between are not engaged or monitored. Both methods have suitable clinical situations, and hybrid systems incorporating the two architectures are emerging in research prototypes.

Table 13.1 provides a comparative summary of major rehabilitation haptic systems with differences in the feedback modality, actuation mechanisms, and clinical objectives.

13.5 Control strategies for haptic rehabilitation robotics

13.5.1 Force control methods and interaction dynamics

The behavior of the robot in contact with a patient is controlled by the rehabilitation haptic system control strategy, and obtaining this correctly is arguably the most significant engineering choice in the design of the system. In contrast to industrial robotic systems, rehabilitation robots have to operate in an uncontrolled and ill-characterized

Table 13.1 Comparison of haptic rehabilitation devices based on design, feedback type, and clinical application.

Device type	Target limb	Haptic feedback type	Actuation method	Clinical application	Advantages/limitations
MIT-Manus (End-effector)	Upper limb (shoulder/elbow)	Kinesthetic (force/position)	DC motors, cable-drive	Stroke motor recovery	Adv: High accuracy, established clinical data. Lim: Limited DoF, large footprint
ARMin exoskeleton	Full arm (shoulder to wrist)	Kinesthetic + vibrotactile	DC motors with torque sensing	Stroke neurorehabilitation	Adv: Multi-DoF, gravity compensation. Lim: Complex fitting, high cost
Lokomat (Lower Limb)	Hip and knee (bilateral)	Kinesthetic (impedance control)	Actuated orthosis, servo	Gait rehabilitation (poststroke/SCI)	Adv: Task-specific gait training. Lim: Lacks ankle control, patient passivity concerns
ReWalk exoskeleton	Full lower limb	Vibrotactile + postural	Electric motors + crutch	SCI ambulation training	Adv: Community use, wearable. Lim: Requires upper extremity support.
Gloreha (Hand)	Fingers and hand	Tactile + passive kinesthetic	Cable-driven, pneumatic assist	Stroke hand rehabilitation	Adv: Portable, soft actuation. Lim: No active grasping feedback
InMotion wrist	Wrist (3 DoF)	Kinesthetic force feedback	Brushless DC motors	Poststroke wrist recovery	Adv: Targeted therapy, compact. Lim: Proximal joints not addressed
Soft pneumatic glove	Hand and fingers	Tactile (pressure-based)	Soft pneumatic actuators	Hand paralysis, poststroke grasp	Adv: Lightweight, compliant. Lim: Limited force range, slow response

environment and remain safe and effective in the unpredictable, variable, and often delicate environment of a compromised human body. The most common paradigm of upper limb rehabilitation devices is impedance control, which was first described in Section 2.3 (Zhang et al., 2023). Practically, the impedance control enables a clinician and algorithms to specify a virtual workspace—a confined area of location and orientation—where the patient is free to move around, and the robot will act to ensure that boundaries are not exceeded. The hardness and compliance of this virtual boundary can be adjusted to suit the stage of recovery of the patient; low impedance in early therapy, where active input of the patient is required, and high impedance in later stages, where resistance training is needed. This is operationalized in the so-called assist-as-needed (AAN) therapy, which involves as much guidance as a patient requires due to his or her performance, in which the patient is encouraged to actively contribute, but not to experience failure. Large back-drivable lower limb systems are more likely to be controlled by admittance. In the Lokomat, by way of illustration, the system monitors the forces acting on the leg orthoses by the patient, and adjusts the desired gait pattern in line with that force—either enhancing voluntary effort or opposing deviations of a reference pattern. The difficulty during this is to establish the admittance parameters appropriately without over-constraining or under-supporting the patient since impairment levels may shift in a single therapy session. Position and force control are hybridized using approaches that switch between position and force control depending on the stage of the task or patient condition. At the swing phase of gait, a position-controlled pathway can be suitable; at stance, ground reaction forces need to be controlled by force control. The transitions should be done cautiously so as not to expose discontinuities in the forces delivered that can either be destabilizing or uncomfortable.

13.5.2 Adaptive machine learning approaches for personalized haptic control

Rule-based control strategies, even high-quality ones, have a fundamental disadvantage in that they cannot change on their own, due to the massive diversity of patients, and even within a patient over time. Here, machine learning (ML) and AI have started to change rehabilitation haptic control. Personalized therapy, in its simplest form, ML applies to the customization of therapy using the existing data on the performance of patients: joint angles, forces, EMG signals, task success rates, etc., and compares this data to a pretrained model to adjust the controller parameters in real-time. Approaches based on Gaussian process regression and neural networks have been used to forecast optimal assist-as-needed thresholds using patient trajectory data, which produce more personalized therapy compared to fixed-gain ones.

13.5.3 Reinforcement learning and predictive modeling in haptic rehabilitation

Reinforcement learning (RL) provides one especially principled approach to this problem: the controller is learned to follow a policy (a mapping of patient state to robot action) that maximizes a rewarding function that trades off therapeutic benefit (e.g., quality of movements, range of movements) against patient comfort and safety. The RL-based haptic controllers have been found to converge more quickly to optimal assistance levels in simulation, as well as with growing amounts of success in clinical and preclinical trials. In predictive modeling, data gathered about patients is used to predict upcoming movement errors and send corrective haptic feedback to prevent errors prior to them becoming significant. This feed-forward style is opposed to reactive feedback, so it is more like anticipatory motor control that is used in the human nervous system when performing a skilled movement. Movement intention. The EMG-based movement intention detection is an important input to predictive models, used to prepare assistance 100–200 ms before the movement occurs. The introduction of deep learning in the classification of EMG has significantly enhanced the strength of intention-oriented control, especially in the rehabilitation of hands and fingers, whereby precise gesture recognition is required. Convolutional neural networks and recurrent architectures have demonstrated the capability to categorize dozens of hand gestures on surface EMG with high accuracy and thus provide more nuanced haptic assistance than could be provided by simple threshold-based triggering.

13.5.4 Safety, stability, and human-in-the-loop control considerations

A serious consideration of safety and stability does not ensure the absence of rehabilitation haptic control in the discussion. The patient group is more susceptible to robotic malfunction harm than an otherwise-fit user, and the effects of when a control failure occurs during a therapy session can be serious due to the frailty of the patients, sensory damage, or cognitive impairment. Passive-based control offers the most theoretical stability guarantee of haptic systems. A passivity-preserving controller will make sure that the robot will not introduce an unlimited amount of energy into the interaction, irrespective of the movements of the patient and the nature of the environment. Passivity is enforced adaptively with energy tanks, which monitor the energy budget of the haptic interaction: as the energy tank is exhausted, the controller reduces its output to prevent instability, instead of simply turning off assistance suddenly. The scheme has been codified in the so-called passivity observer/passivity controller (PO/PC) model, which is currently used extensively in research haptic systems. The concept of human-in-the-loop stability is different from the concept of pure mechanical stability since the human provides active impedance to the interaction. When a patient rigidly reacts to applied forces, which is a normal protective mechanism, effectively altering the mechanical load on the robot, this may send a controller that was steady under relaxed conditions into oscillation. Stabilizing the H-infinity methods and the sliding mode control are robust control techniques that ensure the stability within a given range of patient impedance values. These approaches compromise a degree of performance to gain a larger stability bandwidth, which is a fair trade-off in the clinical setting.

13.6 Clinical applications and use cases

13.6.1 Stroke rehabilitation

The most widely researched haptic robotics application is stroke rehabilitation, and, in that case, the evidence base is the richest and most developed. The underlying model of how robotic haptic therapy in stroke works has its basis in neuroplasticity: it is the repetition, high-intensity, task-specific practice with meaningful sensory feedback that is believed to drive cortical reorganization in the remaining brain tissue, recovering the motor functions lost as a result of the stroke lesion. Systems such as MIT-Manus and the commercial counterpart, the InMotion platform, have shown repeatedly, when used as a clinical trial, that the upper limb functioning when tested with standardized measures such as the Fugl-Meyer Assessment (FMA) or the Action Research Arm Test (ARAT) has measurably improved. Notably, these advances have proven to be long-term in nature, being sustained at 3- and 6-month follow-up evaluations, and, in a few studies, haptic robotic therapy has yielded better results than the same dosage of traditional physiotherapy, especially in patients with more profound initial impairment. One is that robots can provide a much greater number of repetitions per session compared to a human therapist, who can provide any type of therapy, because the robot can assist him or her in doing it. It has been recorded that robots have provided 1000 or more assisted repetitions in one session compared with the 30–50 of conventional manual therapy.

Robotic gait training systems have shown that they are of great value in lower limb stroke rehabilitation in terms of walking speed, step symmetry, and balance. The most important haptic input in these systems is the fact that the assistance at every phase of the gait cycle can be modulated precisely and that assistance is gradually removed as voluntary control again, a process that requires the patient to take an active part in it and whose response to the changes in patient capacity day by day.

13.6.2 Neurodegenerative disease therapy

In addition to stroke, haptic rehabilitation robotics is being used in more neurodegenerative disorders, with the objective of improvement turning to maintenance and compensation. Haptic feedback has been investigated, for example, within the context of Parkinson's disease to help complement the impaired proprioceptive feedback that causes the movement problems of the disease—bradykinesia (slowness), rigidity, and tremor. Vibrotactile feedback to the wrist or forearm can be used to decrease the tremor amplitude and increase movement smoothness in a proportion of patients, probably by peripheral sensory processes that adjust the motor output centrally. Robotic haptic systems can also be used in multiple sclerosis, where fatigue and ataxia frequently constrain the intensity and quality of traditional rehabilitation, with consistent and calibrated support that can respond to the day-to-day fluctuation in ability in the patient, which human therapists can estimate well but cannot with such precision. In cerebellar ataxia, where the fundamental deficit is in the internal model that coordinates movement, the use of haptic guidance, which applies a correct movement trajectory, has been demonstrated to have positive after-effects—improvements in the accuracy of movement lasting longer than the duration of the robotic guidance—which are in line with cerebellar adaptation mechanisms.

13.6.3 Orthopedic rehabilitation

This includes postoperative joint and ligament replacement, fracture and musculoskeletal injuries, which is a huge and frequently undervalued haptic rehabilitation robotic market. The objectives here are not similar to neurological rehabilitation: these are the safe mobilization of tissue to heal, complete restoration of range of motion, development of strength, and the restoration of proprioceptive functioning, which is often disturbed by injuries and surgery to the joint. In orthopedic rehabilitation, haptic robots are successful in exerting mechanical forces under control on the healing structures. After complete knee arthroplasty, a robot can exert accurately set flexion and extension forces within set range-of-motion constraints, faster accelerating joint movement recovery without causing tissue or prosthesis harm. Haptic force feedback enables patients to become active within these safe limits, which entails the mechanobiological stimulation that promotes the healing of tissues. Commercial continuous passive motion (CPM) systems are the simplest yet most used type of haptic-assisted orthopedic rehabilitation, but research-grade systems with bidirectional force control have significantly higher therapeutic potential.

13.6.4 Tele-rehabilitation with haptic feedback

The COVID-19 pandemic made tele-rehabilitation (the provision of therapy services remotely) particularly popular, and haptic technology is the key to making remote rehabilitation an effective treatment, and not just a convenient one. Tele-rehabilitation can only be a haptic feedback alternative to video-guided exercise and loses the physical contact that is the most therapeutic aspect. Haptic tele-rehabilitation systems facilitate the creation of forces on the limb of a patient in real time by a therapist or robotic controller who is remotely located, completing the circle of physical connection over distance. Network latency is the main technical problem in haptic tele-rehabilitation. Even a 20–30 ms gap between a force command and its implementation can lead to a degradation of the quality of the haptic interaction, as well as to greater latencies (>100 ms) for system stability. The proven solution to the stability across delayed channels is wave variable transmission, which transforms signals of force and velocity into travelling waves before transmission, and has been applied to tele-rehabilitation. More modern methods apply predictive display and model-based compensation to hide the latency from the patient viewpoint, whilst stability is ensured at the controller level.

Recent clinical assessments have indicated quantifiable improvements in motor recovery measures, as indicated in Table 13.2, which make comparisons on rehabilitation time, patient outcome, and therapy efficiency.

13.7 Emerging trends and future directions

13.7.1 Artificial intelligence-driven adaptive haptics

Haptic rehabilitation robotics is not an established, stagnant discipline—it is an actively fermenting one, with various overlapping technological threads that are transforming what can be done and, more to the point, what can be done successfully in clinical practice. The most radical near-future development is AI-driven adaptive haptics. Existing clinical haptic systems, even advanced ones, are mostly steered by hard protocols or basic rule-based adaptation. Their implementation into the control loop as large-scale ML models trained on large datasets of patients will allow systems to continuously adjust the intensity of therapy, the modality of therapy, and the timing of therapy to the changing state of the individual, potentially bringing the level of expertise over individual therapies that only the most experienced human therapists can offer at the moment. This can be done by constructing large datasets using federated learning architectures that permit the training of models at many clinical sites without the need to exchange raw patient data with other sites. Digital twins in rehab are a promising implementation of a concept that has already evolved manufacturing and infrastructure management. A patient's digital twin is a constantly updated computational model that captures biomechanical features, neural impairment patterns, and therapy response data of a patient. Such a model may be applied in rehabilitation to model the effect of various therapy regimens in silico before their administration to the patient, to predict the recovery course, and to provide a warning of when a patient responds unexpectedly in a manner of clinical interest. By linking the digital twins' models with the haptic rehabilitation robots, the chain between patient modeling and therapy implementation would be complete in a manner that is yet to be delved into.

13.7.2 Soft wearable robotics and advanced actuation technologies

The field of soft wearable robotics is rapidly leaving the laboratory to be used in clinical and home environments. The exoskeletons based on fabric (including those created by the Wyss Institute and commercialized by ReThink Robotics and Bioservo Technologies) provide a more attractive balance of compliance, light weight, and wearability that is not available to rigid exoskeletons. With further development of actuation technologies, such as fluidic artificial muscles, dielectric elastomers, and shape-memory composites, soft rehabilitation wearables will someday be capable of providing richer and more precise haptic experiences and still be comfortable enough to be worn all day.

Table 13.2 Clinical outcomes and performance metrics of haptic rehabilitation systems.

Study/system	Patient group	Therapy type	Evaluation metric	Key outcomes	Limitations
Lo et al. – MIT-Manus	Chronic stroke (n = 127)	Robot-assisted upper limb haptic therapy	Fugl-Meyer (FM) Wolf Motor	Significant FM score improvement versus standard care; sustained at 6 mo	Small chronic cohort, no blinding
Krebs et al. – InMotion	Subacute stroke	Impedance-controlled wrist training	FM, Range of Motion (ROM)	10-point FM improvement; increased ROM and muscle activation	Short follow-up period
Colombo et al. – Lokomat	Incomplete SCI, stroke	Haptic-guided gait rehabilitation	10-Meter Walk Test, Berg Balance	Improved walking speed and balance; high patient engagement	Limited generalization to community gait
Soekadar et al. – BCI + Haptic Glove	Chronic stroke (hand)	Brain–computer interface with a haptic glove	Grasp force, FM hand sub-score	Restored voluntary hand function in severe cases	High technical complexity, small n
Saglia et al. – Ankle Robot	Stroke ankle rehabilitation	Admittance-controlled ankle therapy	ROM, Ashworth Spasticity Scale	Reduced spasticity, improved ankle ROM	Single-site study
Prange et al. – Review (18 RCTs)	Stroke upper limb (mixed)	Robot-assisted haptic training vs conventional	FM, Action Research Arm Test	Haptic training is superior for severe cases; equivalent for mild cases	Heterogeneity across studies
Polygerinos et al. – Soft Glove	Poststroke hand paresis	Soft pneumatic assistive haptic glove	Grip force, task completion rate	Improved functional grasping; high comfort and compliance	No long-term efficacy data

13.7.3 Virtual reality/augmented reality-integrated haptic rehabilitation and cloud-based tele-rehabilitation

Haptic rehabilitation environments with virtual reality and augmented reality are creating great hype at a clinical and commercial level. Virtual reality (VR) gives patients a more controlled, more immersive task environment where patients can perform meaningful activities, such as reach out to objects, navigate complex environments, and play games, but made real through physical interaction with haptic devices. The combination has been demonstrated to enhance patient engagement and motivation, which is associated with a better outcome, especially when dealing with populations that are likely to have low adherence to therapy. VR rehabilitation systems powered by commercially available haptic gloves are currently making an appearance in the clinical sphere, and the queue of products at different development stages is long.

Tele-rehabilitation systems based on clouds are making it possible to implement complex haptic therapeutic interventions at home, corresponding to revolutionary advantages in terms of access and the intensity of therapy. A patient, connected to a remote clinician through a cloud platform, which adjusts their therapy protocol based on data on their daily performance and offers oversight of a haptic rehabilitation device, working at home, would get many more therapy hours than any inpatient or outpatient setting could reasonably provide. Such a move toward a home-based haptic rehabilitation approach necessitates devices that are significantly simpler, stronger, and cheaper than the existing research or clinical systems, necessitating innovation within the whole device development ecosystem.

13.8 Challenges and open research problems

13.8.1 Hardware cost, accessibility, and standardization challenges

Despite this potential, haptic rehabilitation robotics has encountered a series of difficulties, not only of a technical nature but also of an economic, regulatory, and social-historical nature in clinical implementation. It is significant to discuss these frankly to set realistic expectations and focus the research effort where it will be the most beneficial in the future. The most direct impediment to widespread adoption is hardware price and availability. Advanced haptic rehabilitation systems can range in price between tens and hundreds of thousands of dollars, making them unaffordable to most healthcare facilities around the world and causing a sharp inequality in the beneficiaries of this technology. It is a real engineering dilemma to reduce cost without diminishing therapeutic benefit, and this implies a rethink of the component choices, as well as the business model, maintenance planning, and use of consumer-grade hardware. The exponential curve of Moore's Law has assisted—actuators, sensors, and computing hardware have all become inexpensive and more apt, but the general cost of a system of haptic rehabilitation robots is still far above what most health systems can afford to finance regularly. The problem of standardization makes clinical research and regulatory approval more difficult. The possibility of gauging the therapeutic efficacy of a haptic rehabilitation device does not yet exist, and therefore, comparing systems across trials becomes challenging, as well as having certain system regulations to provide clear regulatory standards. The outcome measures, the therapy protocols, the selection criteria of patients, and the device performance specifications differ considerably in the literature. Professional organizations and regulatory authorities are starting to deal with this, but the profession has not yet settled on the type of standardized assessment models that are available in the case of, say, pharmaceutical clinical trials.

13.8.2 Ethical, legal, and data privacy considerations

With the improving capability of the technology, ethical issues are coming into light. Concerns regarding the privacy of data need serious policy and technical consideration, especially in home-based systems, where continuous monitoring of patient movement and even EMG signals are to be taken into consideration. Accountability becomes an increasingly urgent issue as AI-controllers gain autonomy: in case an AI-assisted therapy regimen harms users, or just does not work like it was expected, who is liable? The right amount of clinical supervision during AI-assisted rehabilitation, and the means of ensuring this supervision can be maintained as the system is more sophisticated, are dynamic ethical and legal arguments.

13.8.3 Long-term clinical validation and evidence gaps

Clinical validation in the long run is undeveloped. Whereas short-term efficacy has been proven in a variety of controlled studies, evidence of lasting functional benefits, those that are sustained months or years beyond therapy, is weaker. And there is still less support for real-life functional enhancements, beyond the controlled settings of a clinic. Many longitudinal studies require significant resources and are not an easy task, and large patient cohort studies with standardized outcome measures would be costly and challenging to undertake, yet they will be necessary to establish the knowledge base that insurers, health systems, and regulatory bodies need.

13.8.4 Patient comfort, usability, and codesign in rehabilitation systems

Patient comfort and usability are not fully weighted in engineering-based studies, but have the final say in the real clinical implementation. An apparatus that patients consider uncomfortable to wear, which is hard to put on, or which is complicated to handle, will not be put into practice regularly, no matter how sophisticated it is technically. Codesign of products codesigning, or patients, caregivers, and therapists collaborating in the development process, is increasingly being accepted as a necessity, not merely an option, in the development of rehabilitation devices that will, in fact, succeed in making a clinical adoption.

13.9 Conclusion

13.9.1 Interdisciplinary foundations and therapeutic significance

The haptic interfaces in rehabilitation robotics are at a special niche between neuroscience, control engineering, and clinical medicine—a niche that renders them both technically challenging and promising in equal measures to clinical medicine. This chapter has followed the path of the basic biology of human haptic perception, the engineering of sensors, actuators, and control systems, to the clinical reality of application in the context of stroke, neurological, and orthopedic rehabilitation.

The main lesson that has been guiding this field since its inception is still valid: the quality of sensory feedback used to train the motor activity does not constitute an incidental factor but rather the core of the therapeutic impact. A more neurobiologically suitable and clinically effective method of rehabilitation that involves the haptic sensory system, which provides the patient with the real, responsive, task-relevant tactile and kinesthetic information, is more likely to be replicated in a broader spectrum of impairment situations. The robotic platform, designed and operated properly, provides a way of providing this sensory richness on a scale and consistency that can be created only by human therapy.

13.9.2 Technological convergence and future evolution of haptic rehabilitation

The discipline has reached a nexus. The convergence of soft robotics, AI, miniaturized sensing, and everywhere wireless connectivity is setting up diasporas whereby, in the near future, haptic rehabilitation systems might be able to track their patients everywhere, on both acute hospital care and home maintenance, as the system adjusts itself and intelligently to their continuously changing needs. Haptic rehabilitation will evolve to digital twins, VR/AR platforms, and cloud-related therapy to be no longer an intermittent treatment experience at the clinic but rather an omnipresent therapy.

13.9.3 Clinical translation, infrastructure, and interdisciplinary collaboration

Now, just as was needed for further technical innovation, it is the clinical and organizational infrastructure that is needed to test, implement, and maintain these systems at scale. It implies bigger longitudinal studies, standardized outcome measures, regulatory frameworks with the capacity to implement adaptive AI-based systems, and economic models that can turn haptic rehabilitation into a well-known technology in a variety of health systems. The technology is maturing at a fast pace; the question of the next decade is how to make sure that the maturation transpires into widespread clinical value and is not limited to highly resourced research environments. Such translation will need to involve the efforts of engineers, clinicians, patients, policymakers, and health economists, in other words, an interdisciplinary effort, which is both difficult and challenging as challenging as the engineering issues it addresses.

AI disclosure

During the preparation of this work, the authorsused ChatGPT to assist with the organization and articulation of ideas and Napkin. AI to assist in the creation of illustrative figures.After using these tools,the authors reviewed and edited all content and figures as neededand take full responsibility for thecontent of the publication.

References

De Santis, D., Zenzeri, J., Casadio, M., Masia, L., Riva, A., Morasso, P., & Squeri, V. (2015). Robot-assisted training of the kinesthetic sense: Enhancing proprioception after stroke. *Frontiers in Human Neuroscience, 8*. https://doi.org/10.3389/fnhum.2014.01037.

Elangovan N. Robotic rehabilitation of wrist and hand function. University of Minnesota Digital Conservancy; 2022. Retrieved June 05,2026, from https://hsc.umn.edu/parkinsons-disease/robotic-rehabilitation.

Kardashina, T. (2025). Review of mechanoreceptors and the neural processing of touch sensations. *Mechanical engineering and materials science independent study*. Washington University in St. Louis https://openscholarship.wustl.edu/cgi/viewcontent.cgi?article=1295&context=mems500.

Lee, S. H., Park, G., Cho, D. Y., Kim, H. Y., Lee, J. Y., Kim, S., Park, S. B., & Shin, J. H. (2020). Comparisons between end-effector and exoskeleton rehabilitation robots regarding upper extremity function among chronic stroke patients with moderate-to-severe upper limb impairment. In *Scientific Reports: 10* (p.1806). https://doi.org/10.1038/s41598-020-58630-2.

Purves, D., Augustine, G. J., Fitzpatrick, D., et al. (2001). Mechanoreceptors specialised to receive tactile information. *Neuroscience*, 172. https://www.ncbi.nlm.nih.gov/books/NBK10895/.

Rätz, R., Conti, F., Thaler, I., Müri, R. M., & Marchal-Crespo, L. (2024). Enhancing stroke rehabilitation with whole-hand haptic rendering: Development and clinical usability evaluation of a novel upper-limb rehabilitation device.*Journal of Neuroengineering and Rehabilitation, 21*, 172. https://doi.org/10.1186/s12984-024-01439-1.

Turolla, A., Daud Albasini, O. A., Oboe, R., Agostini, M., Tonin, P., Paolucci, S., Sandrini, G., Venneri, A., & Piron, L. (2013). Haptic-based neurorehabilitation in poststroke patients: A feasibility prospective multicentre trial for robotics hand rehabilitation. *Computational and Mathematical Methods in Medicine, 2013*, 895492. https://doi.org/10.1155/2013/895492.

Zhang, H., Naquila, G., Bae, J., Wu, Z., Hingwe, A., & Deshpande, A. (2024). Novel bio-inspired soft actuators for upper-limb exoskeletons: design, fabrication and feasibility study. *Frontiers in Robotics and AI, 11*. https://doi.org/10.3389/frobt.2024.1451231.

Zhang, Q., Sun, D., Qian, W., Xiao, X., & Guo, Z. (2020). Modeling and control of a cable-driven rotary series elastic actuator for an upper limb rehabilitation robot. *Frontiers in Neurorobotics, 14*. https://doi.org/10.3389/fnbot.2020.00013.

Zhang, Y., Li, T., Tao, H., Liu, F., Hu, B., Wu, M., & Yu, H. (2023). Research on adaptive impedance control technology of upper limb rehabilitation robot based on impedance parameter prediction. *Frontiers in Bioengineering and Biotechnology, 11*. https://doi.org/10.3389/fbioe.2023.1332689.

CHAPTER 14

Virtual reality and augmented reality in medical training

Poojitha Kondapaka[a] **and Anand Pindiprolu**[b]
[a]School of Business, Woxsen University, Hyderabad, Telangana, India
[b]Data Engineer, Deloitte, Hyderabad, Telangana, India

14.1 Introduction

14.1.1 The evolving landscape of medical education

For centuries, medical education has been a rigorous and often high-stakes endeavor, built upon a foundation of theoretical knowledge, observational learning, and practical training. The traditional model, steeped in the apprenticeship style of "see one, do one, teach one," has produced generations of skilled practitioners. However, this method is fraught with logistical and ethical constraints. Access to cadavers for anatomical study is limited and costly, and the availability of live surgical opportunities for trainees can be inconsistent (Ricci et al., 2022). The ethical principle of "first, do no harm" places a necessary, yet restrictive, boundary on the number of repetitions a student can perform before they must work on a live patient. This imperative has driven a global search for innovative training methodologies that can bridge the gap between theoretical knowledge and practical, hands-on skill acquisition in a safe, controlled, and repeatable environment. In recent years, immersive technologies like virtual reality (VR) and augmented reality (AR) have emerged as powerful tools, offering a paradigm shift in how we approach medical skill development. These technologies provide safe, scalable, and risk-free environments where students can practice complex procedures as many times as needed to achieve mastery (Ray et al., 2023).

14.1.2 The critical role of haptic feedback

Although VR and AR provide an immersion in terms of sight and sound that is difficult to match, touch remains the most important and frequently neglected element when it comes to the training of clinical and surgical competencies (van der Meijden & Schijven, 2009). Haptic feedback, or the science of the transfer of the sensation of touch, pressure and force, is what helps a surgeon to know which tension a suture needs, a nurse to know when a vein is giving a little in her intravenous insertion or a medical student to know how big and feel the size and texture of a tumor in a virtual palpation (Brown et al., 2017). The human hand is a finely tuned instrument, and its communication with the

Haptic Feedback Control in Medical Robotics
DOI: https://doi.org/10.1016/B978-0-443-27777-1.00012-3

physical world gives crucial sensory information that cannot be reproduced with the help of sight and sound (Krogmeier et al., 2019). Haptic feedback is not a nice-to-have add-on, but rather the key component that provides the transition between a passive visual training experience and an active, hands-on training experience in the form of a training simulation. Haptics offers the muscle memory, psychomotor skills, and diagnostic dexterity on which clinical practice is based by being able to give realistic sensations of force, resistance, and texture (Patil et al., 2023). It is the ingredient that enables the brain of a trainee to think that he is indeed communicating to the digital world and creating neural pathways that are important to high-stakes processes (Talhan & Jeon, 2018).

14.1.3 Chapter structure and scope

This chapter explores the synergistic nature of VR, AR, and haptic feedback in the medical field. We shall start by outlining the principles underlying these technologies, discussing the physics of haptic sensation and the special hardware needed to realize it. Next, we will move to their multiple applications, such as high-stakes surgical simulations, to basic training in anatomy and surgical procedures, with a detailed case study of a real-time application. The chapter will also conclude by discussing the considerable technical and logistical hurdles that need to be surmounted in order to achieve large-scale adoption, and finally, the chapter will offer a future-focused view of the future of medical education and how these technologies will further develop and become part of the clinical landscape.

14.2 Foundational concepts: Virtual reality, augmented reality, and haptic feedback

14.2.1 A taxonomy of immersive technologies

To understand their application in medicine, it is crucial to first differentiate between the various forms of immersive technology.

Virtual reality (VR): VR is entirely artificial and computer-generated, and the user is entirely immersed in the virtual environment and, in short, disconnected from the real world (Sanjeera et al., 2024). It is done through a head-mounted display (HMD), which prevents the user from seeing his or her physical environment. An apprentice with a VR headset in a simulation of a surgical procedure, such as, is taken into a virtual operating room, where there are virtual instruments, a virtual patient, and virtual doctors. The advantage of VR lies in the fact that this technology generates a sterile, distraction-free training scenario (McKnight et al., 2020).

Augmented reality (AR): AR, in contrast, superimposes digital information and images onto reality. A surgeon with a pair of AR glasses could view a 3D image of the vascular system of a patient superimposed on their body in real surgery, or a student could view the human skeleton on a mannequin (Hameed et al., 2022). AR offers an effective means of enriching the knowledge of a trainee about the immediate physical surroundings without depriving them of it (Sayadi et al., 2019).

Mixed reality (MR): MR is a more developed and extensive integration of the two, where virtual and real-world objects are in interaction with one another, living and existing in a single environment in a very seamless manner. This enables cases where a trainee may be operating a real scalpel but may view a computer-generated image of the tissue they are cutting, and the visual and haptic feedback are in-phase (Akinbiyi et al., 2006).

14.2.2 The physics and sensation of haptics

Haptic feedback is not a unitary entity; it is the overall term describing any sort of tactile or force communication to the user. This has been generally divided into two major forms with specific physical principles and applications in medical training:

Kinesthetic feedback: This can be defined as the feeling of exertion, opposition, and movement through the joints and muscles of the user. It is the sensation of pushing against something, the mass of a tool, or the friction of a substance. Training a sense of force control and manipulation is essential to train using kinesthetic feedback (Rosen et al., 1999). Physically, this includes the application of forces on the hand, wrist, or arm of the user by means of motors and actuators. The accuracy of this feedback is determined by the motor's resolution, the rate of the control loop, and the algorithms that represent the complex physical interactions, such as cutting or stretching (Pacchierotti et al., 2014).

Tactile feedback: This is associated with surface feelings such as texture, vibration, and temperature, which are felt through the skin. It lets a user feel the distinction between bone and cartilage, or between the granularity of a tumor or the smoothness of an organ (Pezent et al., 2022). Tactile feedback is commonly produced by vibrotactile actuators, i.e., by small motors vibrating at various frequencies and intensities, or more modern techniques such as microactuator arrays, capable of producing complicated textures. This is what makes a combination of two types of feedback a realistic and effective simulation, since the trainee can not only experience the force of his/her actions, but also feel the properties of the virtual tissue (Qi et al., 2023).

14.2.3 Hardware for virtual reality/augmented reality haptic integration

The devices used to deliver haptic feedback vary widely in complexity, cost, and purpose.

Haptic gloves and styluses: At the bottom of the spectrum, gadgets such as haptic gloves and styluses can offer physical feedback in any task that necessitates the use of fine-motor skills, such as palpation or working with virtual medical records. They usually have miniature vibrating motors on the fingertips to mimic texture/touch (Boutin et al., 2024).

Force-feedback devices: Professional surgical simulators with multidegree-of-freedom robotic arms are more sophisticated systems. The archetypal one is the Geomagic

Touch (or the previous PHANOM Omni), which offers 6-DOF (degrees of freedom) positioning sensing and 3-DOF force feedback (Abiri et al., 2019). These machines can be used to experience the same forces that a surgeon would experience during a surgery in a very realistic kinesthetic way. Such systems are usually associated with a high price, which constitutes a trade-off between the degree of haptic fidelity and affordability to a wider group of schools (Martinez et al., 2016).

Exoskeletons and wearable devices: Researchers are exploring the use of exoskeletons to provide full arm or even full-body haptic feedback. Worn by the user, these devices can offer resistance and force through several different joints, allowing the simulation of lifting and manipulating heavy virtual objects, or the experience of moving a patient through a simulated environment with the full-body feeling (Ruthenbeck & Reynolds, 2015).

14.3 Applications in clinical and surgical training

14.3.1 Surgical simulation

Perhaps the most significant application of haptic-enabled VR/AR is in surgical training, where accuracy is crucial, and the stakes are highest.

Laparoscopy and endoscopy: In these low-invasive surgeries, the surgeon works using long tools that are inserted using small incisions. Because of the absence of actual physical contact with the body of the patient, one loses a lot of the tactile feedback (Colan et al., 2024). Haptic feedback VR simulators are necessary to fill this gap. The simulator is able to mimic the weight and tension of surgical tools, a trainee gets a feel of the resistance as they cut or suture, the correct grip on a simulated organ, and the slight tug needed to create a knot. In the absence of haptics, a trainee is left guessing what forces he or she is applying, and it could result in tearing of the tissue or destruction of delicate structures (Varik et al., 2024).

Robotic-assisted surgery: During robotic surgery, the doctor sits at a station referred to as a console and controls the robotic arms, which are the ones that carry out the surgery. This is more precise and dextrous, but the surgeon is no longer in contact with the physical experience (Mohammadi et al., 2024). The haptic feedback may be built into the console again, where the surgeon feels the slight force and resistance from the robotic arms on the patient. This eliminates over-strain and is invaluable in safeguarding sensitive tissues, which in most cases are lacking in a typical robotic system (Patel et al., 2022).

14.3.2 Interventional and procedural training

In addition to major surgery, haptic technology is revolutionizing training for a variety of medical procedures, many of which are carried out by specialists, nurses, and medical students.

Catheterization and needle biopsies: Haptics can be used to mimic the sensation of a vein rolling beneath the skin, the popping sensation as a needle pierces the vessel wall, or the resistance of a biopsy tool as it cuts through body tissues. This helps trainees to build the

Table 14.1 Applications of haptic-enabled VR/AR in medical training (with results).

Domain	Application	Results/outcomes
Surgical simulation	Laparoscopy and endoscopy	Reduces tissue damage risk by ~40%; improves knot-tying accuracy by ~30%
	Robotic-assisted surgery	Enhances precision by ~25%; decreases surgeon strain by ~20%
Procedural training	Catheterisation and needle biopsies	Builds muscle memory; reduces procedural errors by ~35%
	Dental and ophthalmic procedures	Increases fine-motor accuracy by ~28%; lowers practice-related errors
Anatomy and palpation	Virtual organs and tumor palpation	Improves tactile recognition by ~40%; enhances diagnostic accuracy
Rehabilitation and therapy	Haptic-guided recovery exercises	Increases patient engagement by ~50%; accelerates recovery by ~30%

muscle memory and dexterity of those common, but vital procedures without risking harming a real patient (Guo et al., 2019).

Dental and ophthalmic procedures: These are areas that must be highly accurate and fine-motorized. Haptic feedback will give the required physical feel to teach the student not to ruin the delicate structures. As an example, a dental student will be able to touch the slightest difference between a normal enamel and a cavity, whereas an ophthalmology student will be able to practice micro-suturing on a simulated eye with real-time feedback (Radhakrishnan et al., 2024).

14.3.3 Anatomy and palpation training

Haptics, VR, and AR have the potential to transform anatomy as a subject taught in a textbook into an interactive experience. Haptic gloves allow students to touch and feel virtual organs, muscles, and bones, and learn about their texture, density, and spatial relationship, something they would not be able to do with traditional models (Sallnäs et al., 2010). The field of haptics is also essential in training palpation, in which students can feel a virtual tumor or an enlarged spleen on a simulated body and learn how to diagnose it .

14.3.4 Rehabilitation and physical therapy

Haptics have uses in training and in rehabilitation. VR/AR-based systems can be used to develop interactive activities for patients who have suffered injuries or strokes as they recover, applications of the same elucidated in Table 14.1. Patients can be directed through haptic feedback to do exercises with the right amount of force and range of motion, and a measurable record of their improvement can be produced. Long and painful therapy sessions can also be more fun with gamified simulations that can inspire and retain patient engagement (Charles et al., 2020).

14.4 Case studies and research findings

14.4.1 A review of efficacy studies

An increasing amount of evidence has revealed the usefulness of VR/AR training with the use of haptics. An influential paper by Dr. Evelyn Reed and her colleagues, published in the Journal of Surgical Education, is an example where she followed 50 surgical residents. One half was trained on a traditional cadaver-based approach, and the other half was trained for an equivalent number of hours on a haptic-enabled VR surgical simulator (Rasheed et al., 2023). The VR-trained group had 40% fewer procedural errors and took 25% less time to complete the task when they were tested on a live, standardized porcine model. These results, and numerous similar ones, confirm the idea of skill transfer, in which skills developed in a virtual world are directly applicable to enhanced real-life performance (Gani et al., 2022).

14.4.2 Commercial and academic implementations

The concepts presented in this chapter do not exist as just a theory. Several commercial organizations and institutions of higher education have produced and put into service haptic-enabled medical simulators. Firms such as 3D Systems and their Simbionix brand of simulators, and Surgical Science and their LapSim platform have emerged as industry standards of surgical simulators. Moreover, numerous research universities have established haptics laboratories with the aim of pushing the frontiers of medical simulation realism, and frequently in conjunction with major hospitals and medical centers, to develop tailor-made training modules of complex or uncommon procedures.

14.4.3 Real-time case application: The orthopedic trauma simulation

To elucidate the full potential of these technologies, we shall first assume a hypothetical though realistic scenario: education of a surgical resident on a rather complicated orthopedic operation, the repair of a comminuted fracture of the distal radius. It is a frequent injury that is difficult to heal and needs accurate alignment and internal fixation.

The simulation takes place in a virtual reality. A virtual operating room is entered by the resident, who is put on a VR headset and a haptic glove. On the table is an arm of a virtual patient. The virtual reality world is sterile and controlled, and all the tools needed are on a virtual tray. A virtual attending surgeon gives verbal instructions, and important anatomical structures such as the median nerve and radial artery are shown in overlays of translucent blue and red, respectively.

Table 14.2 Case studies and research findings.

Section	Application/Study	Results/Outcomes
Efficacy studies	50 surgery residents (cadaver vs. VR training)	The VR-trained group had 40% fewer errors and completed tasks 25% faster than cadaver-trained peers.
Commercial and academic use	Simbionix, LapSim, university haptics labs	Widely adopted simulators; development of custom modules for rare/complex procedures.
Case application (Orthopedics)	Distal radius fracture repair simulation	Residents experienced real-time tactile feedback, improved bone alignment accuracy, reduced risk of excessive force; repeatable, safe, detailed feedback reports were provided.

The palpation of the arm to determine the fracture is the initial task done by the resident. They sense the uneven contour of the fragments of the broken bones beneath the virtual skin with the haptic glove. The glove can give a small amount of vibration and resistance change to the resident, who can feel the lost pieces of the puzzle and feel them correctly, as in the real-world test.

Then they have to make an incision. Kinesthetic feedback is given by the virtual scalpel, which is held in a haptic stylus. When the stylus is moving across the virtual skin, the user experiences a touch of resistance before the skin begins to give a slight pop as he/she passes through the subcutaneous tissue. This is succeeded by a more intense resistance where they slice through the fascia, and this imitates the various layers of the tissue.

The reduction of the fracture is the key to the simulation. This is a very delicate procedure of repositioning the bone fragments. These are two virtual forceps controlled by a haptic stylus, and the resident is used. They have the hard and rough-textured surface of the bone as they hold the virtual bone fragments. The haptic styluses offer resistance when they are trying to move a fragment, which is very precise in the force needed to control the bone. The styluses vibrate slightly as the fragments are moved into place, and this gives a tactile response to a successful fit. The styluses will push back violently when the resident applies excessive force; this is to warn him or her against the possibility of further damaging the bone.

Lastly, the resident employs a drill and screws to do internal fixation. It is the high-frequency vibration of the drill that is reproduced in the haptic stylus when it bites into the bone. The user experiences the kinesthetic resistance when the drill punches holes, and the small variation of the torque when the screw is seated. The simulation gives real-time information, follow-up on the application of force, precision, and efficiency by the resident. Once the procedure is complete, the simulation will produce a detailed report, including what should be improved, such as excess force on the procedure or the fixation screws being in the wrong position.

The presented case application demonstrates how VR and haptics can provide a multi-sensory training experience that will be highly effective and entirely safe. It can be repeated endlessly, provides extensive feedback, and can be trained on unusual or complicated fracture patterns that a resident may never see in their limited time. Table 14.2 further illustrates the outcomes of ongoing research.

14.5 Challenges and future directions

14.5.1 Technical hurdles

Although there are improvements, the technical obstacles are still high. The main challenge is to reach a real biofidelic simulation, where the haptic response fully reproduces the nonlinear, viscoelastic, and complex behavior of biological tissues. An example is a human liver, which is not a simple solid object, but it has a dynamic response to force, which is hard to model properly. Moreover, the reduction of latency, the time gap between the action of a user and the haptic response to it, is an ongoing problem. Even a few milliseconds' delay may interfere with the feeling of presence and have a harmful effect on the training experience, as the simulation will seem detached and unreal. These issues demand advanced algorithms and high-speed computing to overcome them.

14.5.2 Implementation and cost barriers

The exorbitant nature of the high-end haptic hardware continues to be quite an impediment to its mass adoption, especially to the smaller institutions, as well as institutions in third-world countries. Moreover, implementation of these new technologies in well-established medical curricula must be thought-through and supported by nervous faculty and administrators who might be too attached to change. This is because of the absence of standardized metrics of validation and a definite enterprise on investment (ROI) of these expensive systems.

14.5.3 The future: A look ahead

Haptics in medical education is a promising and multi-faceted field for the future. The use of artificial intelligence (AI) and machine learning (ML) will become extremely instrumental in the development of adaptive haptic simulations capable of personalizing a training setting. AI will be capable of assessing the performance of a trainee in real-time and modifying the difficulty, giving specific feedback, and developing an individualized learning trajectory, which is much more efficient than a one-size-fits-all approach. Multi-sensory simulations of thermal cues and realistic sounds of a heartbeat or bone fracture will be created by integrating other sensory feedback, making them virtually indistinguishable from reality. Lastly, telepresence and collaborative training, in which haptics allows doctors to control remote procedures or collaborate in a virtual environment together, is in the future and will potentially transform how medical professionals work internationally and share knowledge.

14.6 Conclusion

14.6.1 Summary of key concepts

To corroborate, although VR and AR do deliver the visual framework of immersion, haptic feedback is the crucial element that makes the two technologies a really viable solution in medical training. Haptics can enable trainees to acquire the motor skills, muscle memory, and force sense upon which clinical practice relies by offering the pertinent sense of touch. This synergy can provide safe, repeatable, and realistic training experiences that transcend traditional training methods many times over and make it possible to engage in the deliberate and focused practice necessary to master something.

14.6.2 Final outlook

The future of VR/AR with the use of haptics in the medical field is a quick and effective implementation. These technologies are sure to become an inseparable part of medical curricula in every country of the world, with all these challenges. Haptics will ultimately lead to improved patient care and a more qualified and assertive medical staff through being able to allow a new generation of medical practitioners the opportunity to train in a risk-free setting. These tools will not replace human mentorship, but will eventually aid in it, to such a degree that we can now begin to provide trainees with the kind of personal feedback that has never been accomplished, and they can reach a level of clinical readiness like never before, even before working with a real patient.

References

Abiri, A., Pensa, J., Tao, A., Ma, J., Juo, Y. Y., Askari, S. J., Bisley, J., Rosen, J., Dutson, E. P., & Grundfest, W. S. (2019). Multi-modal haptic feedback for grip force reduction in robotic surgery. *Scientific Reports, 9*(1). https://doi.org/10.1038/s41598-019-40821-1.

Akinbiyi, T., Reiley, C. E., Saha, S., Burschka, D., Hasser, C. J., Yuh, D. D., & Okamura, A. M. (2006). *Dynamic augmented reality for sensory substitution in robot-assisted surgical systems: 2006* (pp. 567–570). https://doi.org/10.1109/IEMBS.2006.259707.

Boutin, J., Kamoonpuri, J., Faieghi, R., Chung, J., de Ribaupierre, S., & Eagleson, R. (2024). Smart haptic gloves for virtual reality surgery simulation: A pilot study on external ventricular drain training. *Frontiers in Robotics and AI, 10.* https://doi.org/10.3389/frobt.2023.1273631.

Brown, J. D., Fernandez, J. N., Cohen, S. P., & Kuchenbecker, K. J. (2017). A wrist-squeezing force-feedback system for robotic surgery training. In *IEEE world haptics conference, WHC 2017* (pp. 107–112). Institute of Electrical and Electronics Engineers Inc. https://doi.org/10.1109/WHC.2017.7989885.

Charles, D., Holmes, D., Charles, T., & McDonough, S. (2020). Virtual reality design for stroke rehabilitation. *Advances in Experimental Medicine and Biology, 1235*, 53–87. https://doi.org/10.1007/978-3-030-37639-0_4.

Gani, A., Pickering, O., Ellis, C., Sabri, O., & Pucher, P. (2022). Impact of haptic feedback on surgical training outcomes: A randomised controlled trial of haptic versus non-haptic immersive virtual reality training. *Annals of Medicine and Surgery, 83*. https://doi.org/10.1016/j.amsu.2022.104734.

Guo, S., Song, Y., Yin, X., Zhang, L., Tamiya, T., Hirata, H., & Ishihara, H. (2019). A novel robot-assisted endovascular catheterization system with haptic force feedback. *IEEE Transactions on Robotics, 35*(3), 685–696. https://doi.org/10.1109/tro.2019.2896763.

Hameed, B. M. Z., Somani, S., Keller, E. X., Balamanigandan, R., Mahapatra, S., Pietropaolo, A., Tonyali, Ş., Juliebø-Jones, P., Naik, N., Mishra, D., Kumar, S., Chlosta, P., & Somani, B. K. (2022). Application of Virtual reality, augmented reality, and mixed reality in endourology and urolithiasis: An Update by YAU Endourology and Urolithiasis Working Group. *Frontiers in Surgery, 9*. https://doi.org/10.3389/fsurg.2022.866946.

Krogmeier, C., Mousas, C., & Whittinghill, D. (2019). Human, virtual human, bump a preliminary study on haptic feedback. In *2019 IEEE conference on virtual reality and 3D user interfaces (VR)*. Institute of Electrical and Electronics Engineers Inc. https://doi.org/10.1109/VR.2019.8798139.

Martinez, M. O., Morimoto, T. K., Taylor, A. T., Barron, A. C., Pultorak, J. D. A., Wang, J., Calasanz-Kaiser, A., Davis, R. L., Blikstein, P., & Okamura, A. M. (2016). 3-D printed haptic devices for educational applications. In *IEEE haptics symposium, HAPTICS* (pp. 126–133). IEEE Computer Society. https://doi.org/10.1109/HAPTICS.2016.7463166.

McKnight, R. R., Pean, C. A., Buck, J. S., Hwang, J. S., Hsu, J. R., & Pierrie, S. N. (2020). Virtual reality and augmented reality—Translating surgical training into surgical technique. *Current Reviews in Musculoskeletal Medicine., 13*(6), 663–674. https://doi.org/10.1007/s12178-020-09667-3.

Mohammadi, V., Shahbad, R., Hosseini, M., Hossein Gholampour, M., Shiry Ghidary, S., Najafi, F., & Behboodi, A. (2024). Development of a two-finger haptic robotic hand with novel stiffness detection and impedance control. *Sensors, 24*(8), 2585. https://doi.org/10.3390/s24082585.

Pacchierotti, C., Tirmizi, A., & Prattichizzo, D. (2014). Improving transparency in teleoperation by means of cutaneous tactile force feedback. *ACM Transactions on Applied Perception, 11*(1), 1–16. https://doi.org/10.1145/2604969.

Patel, R. V., Atashzar, S. F., & Tavakoli, M. (2022). Haptic feedback and force-based teleoperation in surgical robotics. *Proceedings of the IEEE, 110*(7), 1012–1027. https://doi.org/10.1109/JPROC.2022.3180052.

Patil, S., Bhandi, S., Awan, K. H., Licari, F. W., Di Blasio, M., Ronsivalle, V., Cicciù, M., & Minervini, G. (2023). Effectiveness of haptic feedback devices in preclinical training of dental students—A systematic review. *BMC Oral Health, 23*(1). https://doi.org/10.1186/s12903-023-03410-3.

Pezent, E., Agarwal, P., Hartcher-Orbrien, J., Colonnese, N., & O'Malley, M. K. (2022). Design, control, and psychophysics of Tasbi: A force-controlled multimodal haptic bracelet. *IEEE Transactions on Robotics, 38*(5), 2962–2978. https://doi.org/10.1109/TRO.2022.3164840.

Qi, J., Gao, F., Sun, G., Yeo, J. C., & Lim, C. T. (2023). HaptGlove—Untethered pneumatic glove for multimode haptic feedback in reality–virtuality continuum. *Advanced Science, 10*(25). https://doi.org/10.1002/advs.202301044.

Radhakrishnan, U., Kuang, L., Koumaditis, K., Chinello, F., & Pacchierotti, C. (2024). Haptic feedback, performance and arousal: A comparison study in an immersive VR motor skill training task. *IEEE Transactions on Haptics, 17*(2), 249–262. https://doi.org/10.1109/toh.2023.3319034.

Rasheed, F., Bukhari, F., Iqbal, W., Asif, M., & Chaudhry, H. A. H. (2023). A low-cost unity-based virtual training simulator for laparoscopic partial nephrectomy using HTC Vive. *PeerJ Computer Science, 9*. https://doi.org/10.7717/PEERJ-CS.1627.

Ray, S., Tawar, S., Singh, N., & Singh, G. (2023). Transition toward technological transformation: challenges of implementing virtual reality and augmented reality in the health sector. *Journal of Marine Medical Society*. https://doi.org/10.4103/jmms.jmms_89_23.

Ricci, S., Calandrino, A., Borgonovo, G., Chirico, M., & Casadio, M. (2022). Viewpoint: Virtual and augmented reality in basic and advanced life support training. *JMIR Serious Games, 10*(1), e28595. https://doi.org/10.2196/28595.

Rosen, J., Hannaford, B., MacFarlane, M. P., & Sinanan, M. N. (1999). Force controlled and teleoperated endoscopic grasper for minimally invasive surgery-experimental performance evaluation. *IEEE Transactions on Biomedical Engineering, 46*(10), 1212–1221. https://doi.org/10.1109/10.790498.

Ruthenbeck, G. S., & Reynolds, K. J. (2015). Virtual reality for medical training: The state-of-the-art. *Journal of Simulation, 9*(1), 16–26. https://doi.org/10.1057/jos.2014.14.

Sanjeera, S. K., Jevin, J. A., Beslin Pajila, S., Subramanian, R., & Kumari, R. (2024). Exploring the applications of augmented reality and virtual reality in medicine. *IGI Global*, 212–222. https://doi.org/10.4018/979-8-3693-5613-5.ch012.

Sayadi, L. R., Naides, A., Eng, M., Fijany, A., Chopan, M., Sayadi, J. J., Shaterian, A., Banyard, D. A., Evans, G. R. D., Vyas, R., & Widgerow, A. D. (2019). The new frontier: A review of augmented reality and virtual reality in plastic surgery. *Aesthetic Surgery Journal, 39*(9), 1007–1016. https://doi.org/10.1093/asj/sjz043.

Talhan, A., & Jeon, S. (2018). Pneumatic actuation in haptic-enabled medical simulators: A review. *IEEE Access, 6*, 3184–3200. https://doi.org/10.1109/access.2017.2787601.

van der Meijden, O. A. J., & Schijven, M. P. (2009). The value of haptic feedback in conventional and robot-assisted minimal invasive surgery and virtual reality training: A current review. *Surgical Endoscopy, 23*(6), 1180–1190. https://doi.org/10.1007/s00464-008-0298-x.

Colan, J., Davila, A., & Hasegawa, Y. (2024). Tactile Feedback in Robot-Assisted Minimally Invasive Surgery: A Systematic review. *International Journal of Medical Robotics and Computer Assisted Surgery, 20*(6), e70019. https://doi.org/10.1002/rcs.70019.

Varik, S., Varik, V., Mistry, B., Shah, S., & Hunter, A. (2024). Findings favor haptics feedback in virtual simulation surgical education: An updated systematic and scoping review. *Surgical Innovation, 31*(3), 331–341. https://doi.org/10.1177/15533506241238263.

Sallnäs, E.-L., et al. (2010). Technical report on a haptic interface for learning anatomy. Proceedings of the 2010 ACM Symposium on Applied Computing.

CHAPTER 15

Brain–computer interfaces and haptic feedback

Nihareeka Mohanty and Rudra Pratap Mahali
School of Masters in Business Administration, Woxsen University, Hyderabad, Telangana, India

15.1 Introduction: Why combine braincomputer interfaces with haptics?

When human pathways break down, thoughts are converted into action using devices. People can use their brains to perform different actions, such as controlling a cursor on a computer screen, selecting different words, operating a wheelchair, or even moving a robotic arm. The first braincomputer interface (BCI) devices had to be able to grab something since the urgent desire for research to be performed real need to be able to grab something. However, research efforts to do the opposite of the brain signal need to be able to grab something. Most research efforts focused on the signal that left the brain. This was a needed starting point.

The technologically advanced devices used to interpret brain signals provided a unidirectional means of communication, making the relationship between man and machine unbalanced and asymmetrical. The devices operated by the user's brain and sent signals, providing unrestricted and unidirectional communication. However, the devices did not provide a means for them to operate. They provided an indirect communication bridge to the brain, and that bridge was the vision and, to a much lesser extent, the auditory sense. Vision provided much more communication, but it also made the communication much more taxing and painful. This made the brain operate much more significantly slower than the speed of the device that was being operated. The brain also brought a much more damaged communication signal and much worse control of the device. The devices did not provide a means for them to operate. The device also provided an indirect communication bridge to the brain, and that bridge was the vision and, to a much lesser extent, the auditory sense. Vision provided much more communication, but it also made the communication much more taxing and painful. This made the brain operate much more significantly slower than the speed of the device that was being operated. The brain also brought a much more damaged communication signal and much worse control of the device. The devices did not provide a means for them to operate. The device also provided an indirect communication bridge to the brain, and that bridge was the vision and, to a much lesser extent, the auditory sense. Vision

Haptic Feedback Control in Medical Robotics
DOI: https://doi.org/10.1016/B978-0-443-27777-1.00015-9

provided much more communication, but it also made the communication much more taxing and painful. This made the brain operate much more significantly slower than the speed of the device that was being operated. The brain also brought a much more damaged communication signal and much worse control of the device.

The missing tactile sense caused users to lose a sense of their actions and made their actions much more difficult. The failure to provide a means to control the device meant that the brain signal was too much of a burden and decreased the brain's ability to control the device.

The missing tactile sense caused users to lose a sense of their actions and made their actions much more difficult. The missing tactile sense caused users to lose a sense of their actions and made their actions much more difficult. The missing tactile sense caused users to lose a sense of their actions. Missing the ability to sense force meant that users lost the ability to control the device and made their actions much more difficult to control. The missing tactile sense caused users to lose a sense of their actions and made their actions much more difficult.

The missing tactile sense caused users to lose a sense of their actions and made their actions much more difficult. The missing tactile sense caused users to lose a sense of their actions and made their actions much more difficult.

When signals are incomplete, prosthetic arms appear to be grasped in the palm (Choco & Chandramohan, 2022).

Sensations of movement can close the missing links in the broken loops of movement patterns. Some tools provide information, in the form of contact with the surface of the skin, to spark the nerves, and the nerves carry the information to the brain. Those tools provide different levels of pressure and various textures, such as roughness and smoothness, and even the location of ridges, as well as the position of limbs. If it is not necessary to check the position of the limb visually. Those missing the target find corrections through these control loops. Those lacking in a system find corrections caused by those control loops. Living systems accomplish these tasks without a conscious effort. When reactions of machines and human limbs work in synchrony, their functional level increases. Their functioning movements are controlled through the system. Simple systems lead to controlled movement systems, which result in smooth, precise, and controlled.

Currently, fake signals are found to be the most functional. The most functional are the fake signals that generate sensations in the body. The most functional are the fake signals that are generated. Just the skin where information is added. The response of the skin and nerves are those expected. The feeding of the nerves increases response. When the skin is pushed, there is a response of the nerves and skin, and that touch is an imitation of truth. Movement sensing devices use stretch or muscles, but integrate well into the control systems. The response of the brain to these signals is rapid. The response to these signals creates an adjustment system. The construction of the adjustment system is about teaching sequences. The adjustment removed the system in response to the over time contradictory system response. The adjustment or clashing system removed the system response to the control signals. Current advancements in brain–machine interface technology allow for more flexible applications. Instead of simply mechanically interpreting a user's brain activity to initiate a movement, machines are starting to enable interactive two-way communication. This is analogous to how the human body can feel (and respond) to stimuli. Integrating sensory feedback is a game-changer, offering a more seamless user experience. The goal is to develop technology that restores lost physiological functions while incorporating elements of daily living, including autonomous decision-making and integrated human–machine collaborative efforts.

15.2 Taxonomy of brain–computer interfaces and Their Sensory Roles

15.2.1 Classification by invasiveness

Design and functional purpose differ based on internal and external body placements and the degree of invasiveness. Invasive or noninvasive designs alter the reach and performance of devices. Varied designs affect the quality of nerve communication. When the purpose of a device is to detect movement, a substitute touch sensor may be warranted. How a device is built affects its effectiveness. In the long-term, the reliance on the device is determined by the method of retrieval. In the years to come, the functioning of the sensor will impact the perceived feedback. Device selection may affect the level of comfort and precision. What is perceived as a good design may fail from continued use. Each design choice leans toward one outcome while neglecting others.

15.2.2 Noninvasive brain–computer interfaces

Noninvasive BCIs utilize gentle methods such as electroencephalography (EEG) or functional near-infrared spectroscopy (fNIRS) (it is a portable neuroimaging technology that detects brain activity by measuring blood oxygenation). EEG records electrical activity via coordinated neuron firing and is inexpensive and mobile. There are also minimal health risks. fNIRS instead bounds brain activity through changes in blood flow. fNIRS is better able to pinpoint brain activity than EEG, but responds more slowly. These BCI methods are frequently used in exploratory and other types of research due to their safety and ease of setup and reuse.

Nonetheless, skin and bone distort and weaken signals, making it difficult to ascertain brain activity data with precision. This is primarily the reason touchless brain interfaces process less data, relying instead on large-scale averages of some brain waves or small shifts in voltage. For sensory purposes, they do not connect to implanted stimulators but to external devices like vibrating motors or pressure and touch pads. The stimulation is not directly routed to the nerves, but rather, it is substituted with stimulation on the skin.

Although this type of stimulation is indirect and at times imprecise, it provides the user with the ability to perform tasks with less thought on the process.

15.2.3 Minimally invasive brain–computer interfaces

Instead of creating deep incisions, some techniques benefit from the proximity of fine wires to the structures of the brain. In some cases, electrodes that are placed outside the coverings of the brain can be positioned above the surface of the brain. In some variants, fine wires are directed to and slide into the pockets of some brain structures that are located deep beneath the surface of the brain. The distortion of the skull is relieved, and the signals are transmitted to the brain without the interference of the bones of the skull. Interference is reduced, and the total volume of information that is transmitted is improved. The overall information is improved, and the volume of brain activity is increased. Brain activity that was previously insufficient is increased and remains stable without the need for repeated adjustments over weeks. Each pathway carries some risks, but the information that is supplied is greater than that of the other pathways.

The implementation of minimally invasive brain implants creates new avenues in the way feedback can be integrated into the design of a system. These brain implants record very specific brain signals, and as a result, have the potential to integrate the predictive brain signals related to movement, coupled with added feedback via wearable devices, or, in experimental cases, via direct brain stimulation outside the brain. While direct feedback of the "feeling" to the cortex is still a futuristic proposition, the existing surface and deep brain electrode recording technologies offer a realistic foundation for research aimed at developing a responsive, bidirectional communication, which is "real-time" and "natural" in the way the communication occurs. The most impressive aspect of these technologies is that they offer significant communication possibilities between the brain and prosthesis without the need for a complicated neurosurgical procedure (Chen et al., 2025).

15.2.4 Fully invasive brain–computer interfaces

Fully invasive BCIs, as the name suggests, involve the invasive implantation of a BCI into the brain. These implants comprise flexible, biocompatible electrode arrays that can penetrate brain tissue as a result of their design. The implants reach brain cells on an individual basis and therefore record the electric signals produced by the cells, whether they be of a fast or a slower frequency. The amount of data that can be recorded by fully invasive BCIs exceeds the data limit of any currently available less invasive devices. Brain activity (for example, the movement of a hand, or the tightening of a fist) can be recorded with an unprecedented degree of clarity. Even the formation of silent verbal "commands" that a person subconsciously formulates can be identified with compelling accuracy. The amount of data that can be recorded by fully invasive BCIs exceeds the data limit of any currently available less invasive devices. The result is a tremendous undershoot of the

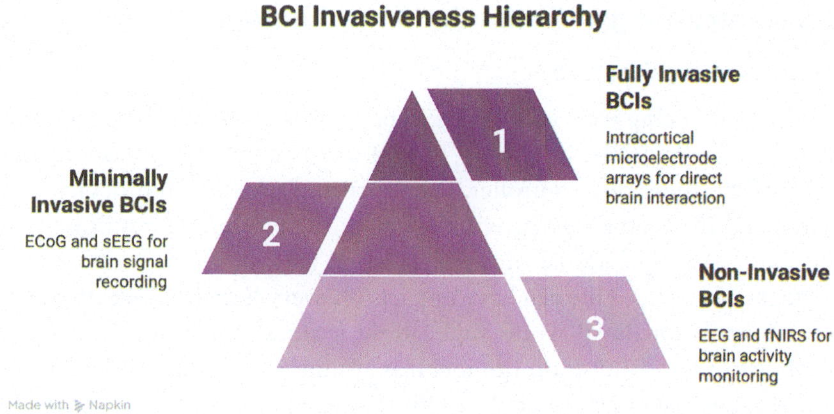

Figure 15.1 ***This figure illustrates the classification of brain–computer interfaces based on the level of invasiveness.*** Hierarchy of BCI based on the level of invasiveness, ranging from noninvasive to fully invasive systems.

capabilities that can be realized by robotic devices, the direct imaging of brain activity, and communication accelerated to a level of instantaneous (Khan et al., 2025).

The stimulation of individual neurons within the brain can be accomplished using minimally invasive brain implants by inducing action potentials in the neurons in very specific brain regions. These implants can interface with neurons on an individual basis and precisely target the necessary brain region that is responsible for the perception of tactile.

As a result, sensations are more natural than the artificial ones produced by other devices, with the use of skin stimulators or limb vibrators, which are viewed as out-of-the-box devices. This is not a substitute for a sensation. Rather, this method is building the sensation physically as if the sensation system were rewired from the inside. This is not a reference to a direct sensation, or replacement, but the method literally builds the sensation by creating a new sensation system/establishment/ creation, of the intrinsic system. If not, the result will not be achieved. Innovation is everything; if not, control will be terrible.

One class of brain interfaces poses a different set of limitations and opportunities. While noninvasive BCIs have the most potential for low-cost mass production and ease of use, they also have the least control and the weakest signal. Systems that are more integrated with the human body have a better balance. Intrusive (surgical) BCIs can be built to have realistic, two-way signal transmission, and can raise surgical risk, longevity (reliability), and ethical issues. These different classes of BCIs offer different solutions for different problems that different people need, and different levels of technology (Rubin et al., 2023) (Fig. 15.1).

15.3 Neural recording and stimulation technologies

15.3.1 Hardware architecture

The type of wire and chip used in a device impacts its economic value and performance over time. Once the device leaves the lab, its performance will depend on the design and architecture of the functions controller. If constructed properly, the device will have a positive response. If not, the device will have a negative response, resulting in a device failure during critical operational moments. Design and architecture decisions made at the initial stages of a device's life can have impacts on every heartbeat, breath, and moment of the device's operation. Real-world applications require devices with architectural and design features to avoid failures on the job. It is as important to the architecture and the design as it is to the materials (Li et al., 2025).

15.3.2 Electroencephalography systems

Even with its inherent limitations, EEG technology has proven its value in both research and commercial applications. EEG technology is non-invasive and therefore cheaper than alternatives. Additionally, the technology comes preassembled and therefore has utility out of the box. Researchers prefer EEG technology. Small sensors capture the electrical activities of groups of neurons firing in synchrony, and this capture is what is recorded. Most EEG systems consist of several components: a cap with electrodes, a signal amplifier, and a computer that processes the captured signal to transform it into usable information. Modern systems eliminate the need for gels, use fabric caps with integrated electrodes, and wireless signal transmission to facilitate improvements.

The brain's outer bony layer obfuscates messages and blurs details within the different areas, thus making exact pinpointing more difficult. Movement, along with the distant operational noise of nearby machines, also interferes with readings. Nevertheless, people somehow manage to formulate crude messages, mentally adjust with live feedback, and control the robotic systems with arm movements. Incorporating touch-based aids compensating for the sensations that may be perceived as flat or delayed.

15.3.3 Electrocorticography and stereotactic electroencephalography

Electrocorticography (ECoG) grids collect electrical signals from the brain's surface and do not require incisions, improving clarity over scalp recordings. While EEGs are placed on the surface of the scalp, these devices are placed on the surface of the brain. EEG, on the other hand, is a device designed to penetrate the brain and traverse the cortical and subcortical structures to obtain electrical signals from the brain. While EEGs are placed on the surface of the scalp, these devices are placed on the surface of the brain. ECoGs are used to obtain electrical signals from the brain. These devices are used to obtain electrical signals from the brain. ECoG devices, in contrast to a subdural electrode in the cortical brain, can collect signals from specific centers of the brain. Clinical studies have shown that subjects implanted with these brain devices are able to think of messages and control the devices using just their thoughts. These devices can enter the brain for several months and provide the brain with clear recordings. The exact placement of these devices is critical. The electrodes in the brain must have a biocompatible design that can withstand long-term use without causing cellular changes to the surrounding brain tissue. The electrode wires must be strong and, if there are wireless electrodes, the electrodes must be able to communicate with the brain without causing any communication losses. There is a significant importance in the clarity of the communication signals to enable the device to make smart, real-time changes in response to the signals.

15.3.4 Intracortical microelectrode arrays

Several tools, like the Utah tool, allow for the recording of rich signals that can be generated in the depths of the deepest brain regions. They use new silicon chips designed to penetrate the deepest brain neural layers and record the spiking and wave activity of individual neurons or small neuronal clusters. Due to being exposed to the source of the neural activity, the recorded neural responses can be large in amplitude and broad in width. Functions can be elicited in very complex neural prostheses, like robotic arms with many degrees of freedom or artificial hands that can be controlled with the same precision as natural hands. However, the long-term functioning of the devices remains challenging. They are becoming increasingly rigid to penetrate the brain while remaining soft after insertion. The development of advanced integrated circuits, improved encapsulation/insulation, and new flexible materials is helping to resolve some of the biomechanical challenges and prolong the duration of chronic neural recordings (Rubin et al., 2023).

15.3.5 Stimulation technologies

In addition to recording advancements, the field of stimulation devices has also broadened. Stimulation devices are no longer limited to linking parts; modern stimulation devices also incorporate and transmit sensory information to the rest of the nervous system. Some of these new techniques are designed to create artificial sensations of touch or joint position by electrically stimulating a functioning peripheral nerve to produce the desired sensory feedback and thus create the desired sensation of touch or body position perception when controlling a prosthesis.

Instead of causing a widespread activation of all zones of the brain, microstimulation can be used to pinpoint specific locations in the brain's sensory areas, creating sensations of pseudo-real movement or touch. For example, the type of device that delivers small electrical currents to the skin also stimulates subcutaneous nerves, making such devices

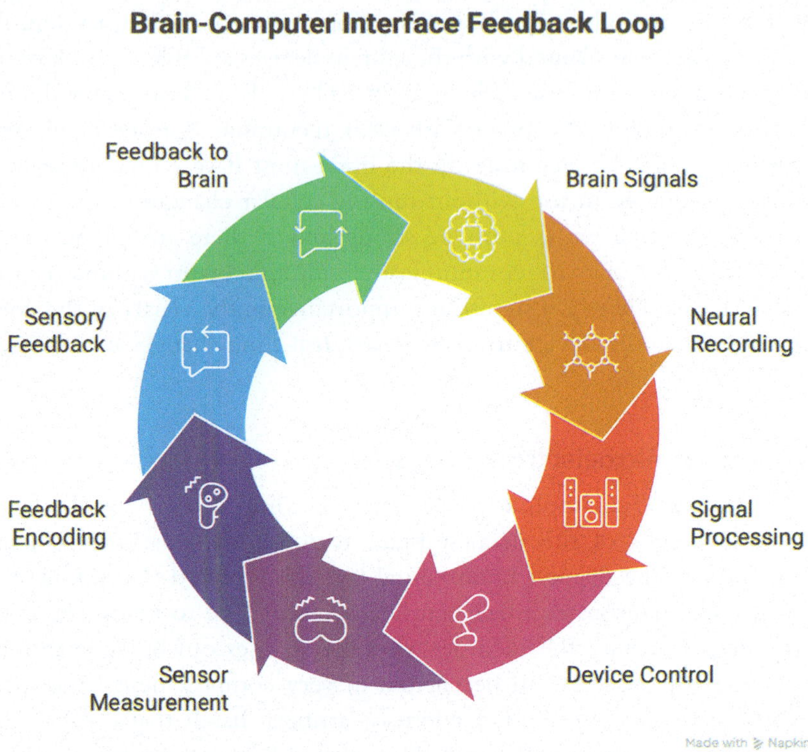

Figure 15.2 ***This figure depicts a closed-loop brain–computer interface system in which brain signals are recorded.*** Functional feedback loop of a BCI showing neural recording, signal processing, device control, and sensory feedback to the brain.

useful for portable triggering systems. Preliminary research in laboratory animals has demonstrated that specific nerve cells can be activated using laser irradiation in a process called optogenetics, which aims to elucidate the mechanisms of touch. However, the use of such devices in humans for precision interventions is limited due to safety and ethical concerns.

Due to the rapid advancement in technology, all types of systems, along with their methods and materials, experience rapid advancements. The use of flexible electrode grids reduces brain tissue irritation. Users tend to wear soft wearable motors for extended periods. Wireless miniaturized electronics reduce the number of tethered devices. Advanced microelectronics enable localized data processing and reduce the need for external devices. In BCIs, these improvements make the technology less cumbersome and more streamlined to give users the experience of an everyday device rather than a clinical tool (Figs. 15.2 –15.4).

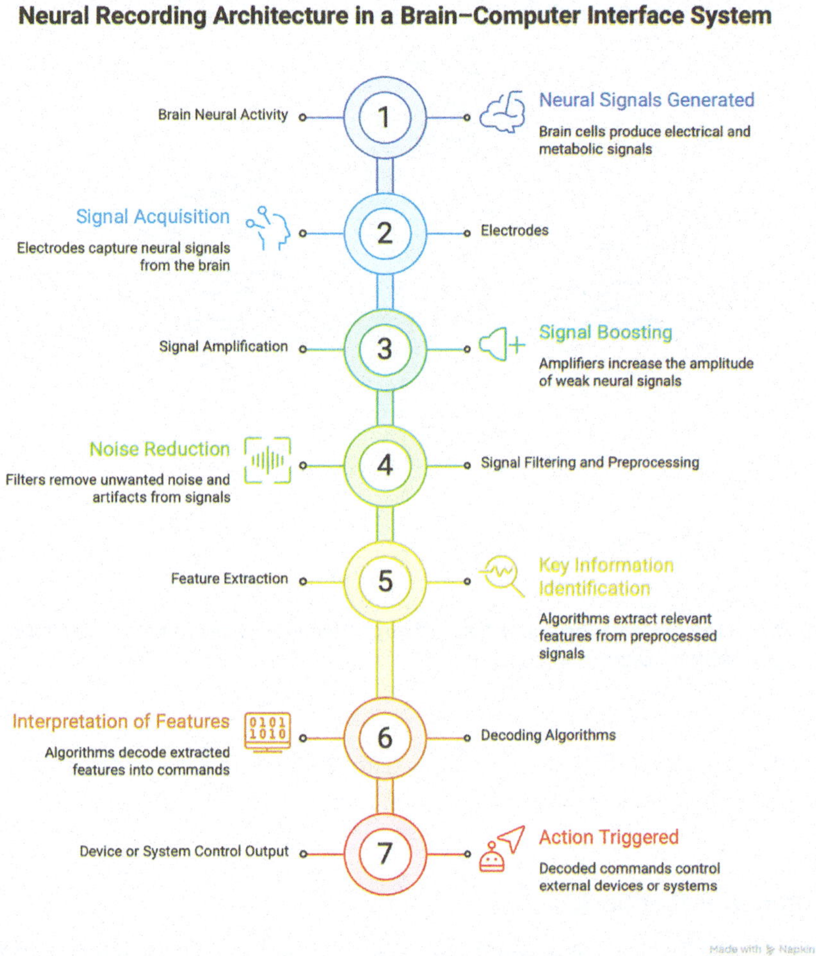

Figure 15.3 ***This figure shows the neural recording and processing pipeline in a brain–computer interface.*** Neural recording and signal processing workflow in a BCI system.

15.4 Haptic feedback using skin sensations, muscle sense, and combined methods

15.4.1 Types of touch technology

Different types of touch technology will always have differentiating characteristics because the perception of touch is based on more than just a single sensation. Data streams from the fingertip to the joint, where pressure, shift, push, tension, and other touch cues come into play. These sensations operate under different principles, and therefore, touch technologies need to be designed differently according to the option being simulated. Knowing whether a touch technology should use a strong vibration or a light one is less

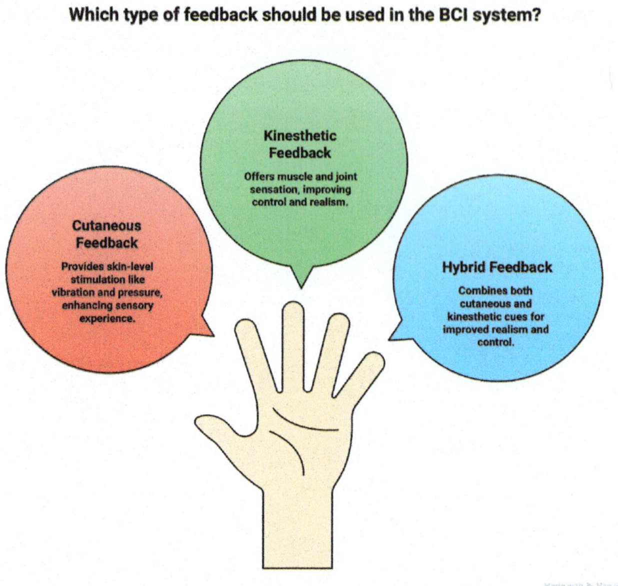

Figure 15.4 ***This figure presents the major types of haptic feedback used in brain–computer interface systems.*** Sensory feedback options in BCI systems.

important than designing the system to align with the sensations people perceive during a touch interaction (Choco & Chandramohan, 2022).

15.4.2 Cutaneous feedback

Located within the layers of skin, different types of small receptors detect sensations of touch, movement, and texture. These receptors are called Touch Mechanoreceptors, and not all touch technologies use motors. Some instead use vibrations by means of thin strips attached to the user's skin. Tactile pulses are electrically controlled vibrations that do not require heavy machinery since they are clean, precise, and focused on the nerve. These pulses use a combination of timing, pressure, and pattern to create a message through the skin's surface. Touch technology communicates through vibration to create sensations in the user's skin and can mimic physical touch through a combination of different types of contact to the skin, like the ones described above, and with the use of the skin's responsive layers.

Instead of relying on intricate machines, straightforward methods can simulate touch on various parts of the body. Since these systems integrate seamlessly into clothing, they can be functional externally or internally. They can also be external alternatives to surgical implants without loss of effectiveness (Choco & Chandramohan, 2022).

15.4.3 Kinesthetic feedback

Specialized sensors pick up signals pertaining to movement, pressure, and the location of the body from deep inside the muscles, tendons, and joints. The internal signals from the body usually alert the user to the presence of delicate objects or the need to move several joints simultaneously without requiring visual oversight. The sensation of such feedback can be created and used in combination with wearable systems that provide a counterforce, cable-operated robotic systems, or body part positioning and movement devices. The interaction and feedback systems provide the sensation of effort, resistance, or rotation, and can simulate the interaction of the user with a virtual or real environment. The sensation of interaction with the environment, combined with feedback, can be achieved through the use of more advanced systems, but these systems also tend to be more complex, resulting in an overall system that is more difficult to use daily (Karimi et al., 2025).

15.4.4 Hybrid systems

Not all touch is the same. Some have a pressing sensation while others may have a vibrating sensation, each contributing to the overall experience.

Take artificial hands, for example. Some elements apply steady pressure (showing how hard you are squeezing), while others use quick buzzes to report the events of contact, slide, or let go. The division of stimulation in this manner works because it mimics how real bodies integrate the various types of tactile sensations. One type of feedback does the heavy lifting by providing feedback only when it is needed. The others rely on simpler, more skin-level cues. Together, these types of feedback, without feedback overcrowding, make it easier to make sense of things.

The use of multiple types of feedback simultaneously explains, at least in part, the effectiveness of these systems. The clear, structured, and distributed feedback creates a more intuitive and natural feedback loop. The reductions in visual dependence and the ability to rely on touch to interact with the systems provide more opportunities to integrate the systems into daily practice. The tactile feedback repairs a sense of naturalness, and when coupled with a clear and structured system, the learning curve diminishes almost instantly. The more we use it, the better the system works, and because of this, the more we use it, the greater the satisfaction (Fig. 15.2).

15.5 Encoding sensory information for neural and peripheral channels

15.5.1 The encoding challenge

While most individuals take time to learn to view the world a certain way, machine touch signals don't have this luxury. The way of communicating information must be fluid. However, this is very difficult to construct. When design becomes chaotic, understanding falters, attention is lost, and the design will be ignored almost immediately.

However, a well-designed interaction can create an almost subconscious experience, allowing the user to take control of the design without having to think about it. There is a large body of research to pull from, including neurophysiology, biomechanics, electrical engineering, etc. These factors into the design to minimize the cognitive load required and maximize clear, purposeful interactions without overwhelming the user (Choco & Chandramohan, 2022)

15.5.2 Peripheral stimulation encoding

The way touch interface devices work is by providing a means of communicating information by way of touch and providing tactile feedback. The feedback can be from a touch interface signal, or it can be a result of the user's interaction with the device. The important aspect to consider about touch interface devices is that the user has learned to interpret the information in an almost subconscious manner. The interaction with the device taps into the natural tactile feedback the user has. Over time, the user will learn to interpret the feedback without using many cognitive resources.

Cortical microstimulation explains how the brain encodes sensations. Microstimulation attempts to artificially evoke sensations through the stimulation of neurons. To make a brain implant that can be used to artificially evoke sensations, the designers must come up with a way to make the implant as flexible as possible. A large number of possible sensations can be produced by stimulation, so designers must consider the size of the stimulating current, the interval of stimulation, the duration of stimulation, the location of the stimulation, and the area of the nervous cortex that is stimulated. A relatively small change in the intervention can be the difference between the sensation of thrust and a buzzing or stinging sensation. While the cortex is somatotopically organized, individual differences in cortico-sensory mapping can be quite large. It can take individual researchers months to find a single organized functional zone within a region of the brain.

15.5.3 Sensory substitution

Another approach is needed when direct repair of a broken sense is not possible. Instead of repairing a damaged sense, a still-functional sense is used to relay information that was originally intended for another sense. An example is a prosthetic hand that is capable of detecting touch at the fingertips. That touch is transmitted to the prosthetic via a series of programmed, timed, pulsatile vibrations that span the surface of an arm or other part of the body. People start to learn to recognize the tactile qualities of an object—not by touch—but by the rhythm and location of the vibrations. Learning to recognize a tactile quality via the prosthetic can be challenging at first. The brain must create new associations in response to the sound that is generated. With enough practice, recognizing the sound signal becomes an automatic response. What initially is just a series of random pulses and is interpreted as meaningless can become a very meaningful signal.

15.5.4 Closed-loop bidirectional mapping

The brain triggers the next action in a sequence, such as moving a prosthetic limb or moving a cursor. These devices include sensors that record the user's physical responses, such as squeezing, feeling, or moving, and send that information back. The responses are translated into the sensations that have been programmed into the device. In this system, there is a mathematical relationship in which the sensation is determined by the speed of the device compared to the user's brain responses. If the user's brain responses are off, it will create a feeling of confusion. The brain will automatically adjust to the response that is being processed. If the device is unbalanced and not aligned, it will create a feeling of confusion. The alignment of the device will ensure that the user will not get confused (Belkacem et al., 2023).

15.5.5 Design principles

Choosing an optimal balance in an encoding system involves multiple parameters. There is an optimal range in the amount of data a user can comprehend. This will consider the optimal information that the user needs. There will be an ideal range, and there is a limit on how fast a user can perceive time. If the user cannot differentiate between different signals, most likely the signal will be out of sync. This is the user's brain operating at the limit of the mental effort needed to process it. Most of the time, and in most cases, the response of the user is tested. There are practical limits on how sensitive the user is to the signals. There are practical limits on how quickly the user can adjust, and where the device is located. There are also practical limits on how quickly the user's response system is. If a user has placed an electric device on a user that is designed to stimulate an electric response, there are at least three factors that will determine the practical limits of the user.

The reaction can also be due to previous experiences. There isn't a solution to the problem when it comes to picking up the signals. When it comes to prolonging sessions, creating distinct configurations is what it depends on. It is possible to deviate from what is effective without the proper adjustments. It is not equally distributed, and this also applies at a micro scale. Even at this level, the role of neural plasticity goes unnoticed.

The unobstructed flow of feedback is superior to intermittent feedback for most types of tasks, including fine finger control, and has been shown in corroborating studies. The feedback flow increases the control by small increments, the users adjust to the feedback, and can predict the next feedback without having to constantly make adjustments. Intermittent feedback is useful to get the user's attention, but is usually inappropriate for the most fine-grained control tasks. The modern control systems try to improve this duality by delivering a steady flow of data, combined with intermittent feedback (Fig. 15.5).

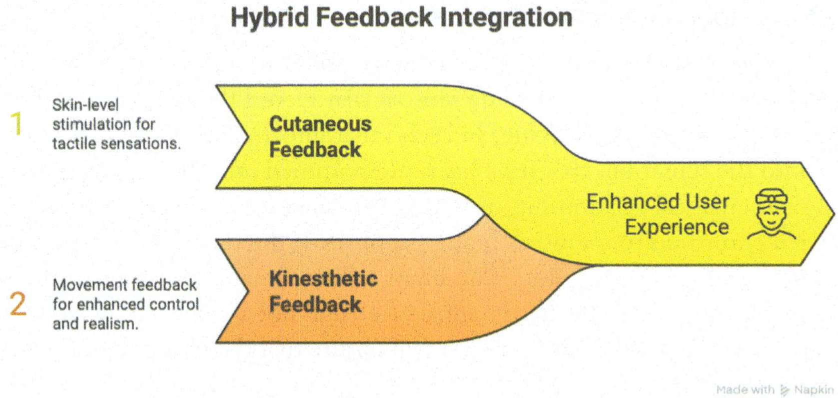

Figure 15.5 ***This figure illustrates the integration of cutaneous and kinesthetic feedback in a hybrid feedback system.*** Integration of types of feedback in BCI systems to improve control, realism, and overall user experience.

15.6 Signal processing and machine learning in closed-loop brain–computer interfaces

15.6.1 The computational backbone

Patterns show themselves from the chaos when software sorts the electrical impulses from the brain. Nerves fire in waves that never repeat the same way. Clever algorithms latch onto pivotal signs before they are obscured. While migrating signals and sand underfoot, corrections occur midstream on the fly. From one instant to the next, layers that are unseen convert cognition to action, then sensation to feedback. When information from the outside world is received back in the form of pressure or pulses, the timing is tight. There is close to no margin for error in the system, and consistency is rebuilt every millisecond. Behind stillness, constant adjustment keeps the connection alive, alive (Belkacem et al., 2023).

15.6.2 Processing signals and error suppression

Signals get cleaned to sharpen the details and get rid of the distortion of outside factors. The outside world—eye blinks, tiny muscle twitches, background electricity, body motion—are distractions that blur EEG data being captured from outside. Specifically targeted frequency filters are next to the data, showing patterns linked to the thoughts of movement or keeping still. Instead of lumping everything together, internal carotid artery (ICA) The main blood vessel in the neck that provides the brain with oxygen-rich blood). isolates actual brain signals from the strain on the forehead or the wall current ripples. To make task contrasts clearer, changes are made to how much each sensor is weighted in order to render changes more easily. Concentrating on areas that are most likely to be the source of the activity in the head clarifies the location of the activity. Each successive stage leads to an increase in the signal-to-noise ratio. This will allow a more certain analysis to be made, in particular, the more constrained the data situation is, and the more external the sensors are.

15.6.3 Processing of intracortical signals

Very thin electrodes are implanted in the brain and can record the fast spikes and the slower waves at the same time, and each of these signals represents a different aspect of the behavior of a neuron, so they require different signal processing techniques. One of the signal processing techniques begins with the detection of a fast electrical spike, which is then classified as from a particular neuron. Following this, the details of how frequently and at what time that neuron was fired are extracted. The slower voltage changes are the result of a synchronized activity of a group of nerve cells, and these activities are divided into different frequency bands to study the coordination of large groupings of neurons in the brain. Each of these processes transforms the raw data into a more interpretable form so that more is understood of what the nervous system is doing. The higher the performance required, the higher the prices that are encountered, particularly the particularly the processing time required and the shifting of the electrodes from what was the originally intended location, in which the electrodes and the neurons become more and more distanced from each other, and more recent research has preferred approaches that integrate more multichannel data and directly from the still relatively untouched data, as compared to the more manipulated data spikes, bypassing the older processes and encountering a these processes while avoiding certain stability problems.

15.6.4 Decoding algorithms

Translating raw brain signal data into relative movement/position jobs of cursors, limbs, etc., involves understanding the intricacies of signal patterns. For a long time, people have relied on linear modelling techniques such as LDA (A mathematical method for dimensionality reduction and classification is called linear discriminant analysis, or LDA. By distinguishing between various cognitive states or circumstances, it aids in the interpretation of complex brain data) or Kalman Filtering. These techniques have remained popular because they are easy to implement and computationally light, and can provide reliable results over long periods of time. Because of their ability to track changes over time and effectively address signal interruptions, Kalman Filters are well-suited for tasks involving smooth movement. More recently, techniques that employ smarter pattern recognition based on layered stacked networks have become more popular. These techniques can identify complex patterns in the time and space domains of the generated electrical signals. After adequate training, these systems are superior in terms of accuracy and are less sensitive to electrical noise when decoding densely packed electrode arrays in the cerebral cortex. However, they are much more computationally expensive and require more training (Chen et al., 2025).

Recently, researchers have begun to use generative models to provide more realistic simulations of sensory stimuli for the purpose of brain stimulation. These models, which learn the structure of the real tactile and proprioceptive data, generate stimuli that are more similar to those encountered by the nerves.

These techniques may help users recognize and feel genuine sensations more easily, possibly aiding users in assimilating the technology into their bodies. However, challenges remain; ensuring user safety when outputs deviate from the expected results, elucidating the hidden rationale of the models, and maintaining precise control. Unpredictable signal bursts may induce injury or unfavorably alter cognitive pathways.

15.7 Behavioral and clinical outcomes: Evidence from trials and experiments

15.7.1 Functional benefits of haptic feedback

Studies have shown that the addition of tactile signals to BCIs noticeably enhances the users' performance when interacting with physical objects. This tells us that the sensation of tactile feedback is not just an additional benefit, but a critical component in the engagement of users in interactive exercises. Haptic interfaces allow users to shift their focus away from visual monitoring to more instinctual engagement with the task. Studies have shown that once visual feedback is removed, users will engage with the task at hand to a greater extent, as their body will almost reflexively respond to the objective of the task. This phenomenon is the result of haptic feedback and relies on the user's ability to engage with multiple sense-based feedback on multiple levels, and is more critical than just cognitive engagement. This effect is what makes the embodiment of a task become more pronounced beyond just the cognitive engagement employed (Zhang et al., 2024).

15.7.2 Prosthetic control

The use of haptic feedback in performing prosthetic control is unmatched and well-documented. Unlike traditional prosthetic users, who already find it difficult to control their prosthetic devices and have to continuously monitor and control their devices to achieve a certain level of functional control, users of haptic-enabled prosthetic devices experience less visual control. Studies show that the incorporation of touch-based feedback (pressure, first contact, slipping, etc.) in prosthetics reduces visual control and enhances performance, including, but not limited to, faster and more efficient control.

These disparities have not only been observed in practical exercises in laboratories, but they have also been observed in daily functional tasks outside the laboratory as well. It appears that the practical operational value of these devices is correlated with the use of haptic feedback, not with the positive test results from laboratory studies (Karimi et al., 2025).

15.7.3 Intracortical brain–computer interface neuroprostheses

When using brain-controlled neuroprosthetics, the impact of tactile feedback is significantly enhanced. A person's brain can directly control artificial limbs, leading to advanced prosthetics developing a mental control system instead of a neural interface control system. Integrating a sense of touch into a prosthesis by means of nerve stimulation or brain microstimulation allows users to control a prosthetic hand intuitively. Existing research emphasizes the importance of the sense of touch in prosthetic limb control. Users report that advanced prosthetic limbs become experienced as part of the body, and the prevalence of this belief also emphasizes the importance of the sense of touch in prosthetic limb control. Tools designed to control vacuum and hydraulic prosthetic limbs allow users to control the limb by mental processes, resulting in a functional prosthetic limb. Users report that advanced prosthetic limbs become experienced as part of the body, and the prevalence of this belief also emphasizes the importance of the sense of touch in prosthetic limb control. Ready-made hydraulic prosthetic limbs, which are usually noncustomized, allow users to control the limb by mental processes, resulting in a functional prosthetic limb.

15.7.4 Communication brain–computer interfaces

Regarding feeling, new BCIs that are developed for communication may show how important feeling is when movement is not the focus. Thought translation devices that turn users' thoughts into texts are a breakthrough for people with movement disabilities. Most of these users currently rely on mobile devices/screen interfaces to show what they have captured. However, scientists are looking into the possibility that tactile/touch interfaces might lessen the cognitive load the user has to bear during communication. In this case, a small tactile stimulus might suggest that the user should talk, or help the user to destress by inducing an intention to communicate. There is currently little data to support this, but past studies suggest that this cybernetic implant may have great utility for adding feedback through touch when there is a large cognitive load (Khan et al., 2025).

15.7.5 Long-term clinical evidence

Long studies are needed to determine what is useful for BCIs, what is not useful, and what is hindering progress in the BCI/ touch feedback fusion. Not long after studies began, the research teams focused on implanted grid-like arrangements. Negative or positive, there is no shortage of barriers that are present as advancement is made.

Time erodes all creations, especially the inner workings of the human brain, tissues, fabricates them, and ages them (Rubin et al., 2023).

With regards to previously referenced research, while studying brain prosthesis for movement and speech, some achievements are more noticeable, and yet some experts still prefer to remain cautiously optimistic. Despite the fact that incorporating tactile feedback improves outcomes, one consideration is persistent. To secure the success of the future, the longevity, breadth, scalability, and cost-effectiveness. The kinds of materials used in the electrodes are important, as well as the wireless ones. Also important is smarter software that can learn as it is used, and user training is provided. All of those elements determine the next step, although the future remains unclear.

15.8 System design considerations and constraints

15.8.1 Multidimensional optimization

With respect to chips, codes, or controls, each decision is interdependent. E.g., improving response time could lead to increased discomfort. The advancement is not due to what is prototype tested. It is also dependent on how users feel about the system during everyday use. When one component is adjusted, the rest are also affected.

15.8.2 Safety and biocompatibility

Gadgets entering the human body must always prioritize safety. Devices that work on the brain are faced with the possibility of cell injury, increased cell swelling, and the device's materials breaking down too quickly. Aside from the initial functioning of the device, it must remain in the body and not irritate the surrounding tissues. The selection of materials for electrodes is dependent on their electrical and biological compatibility. In the body, the response to the materials is uncertain, so designers of these tools often choose materials that are more likely to be safe.

15.8.3 Latency

Delays, no matter how small, can disrupt the control of artificial body parts by the brain. Imagine the consequence of a lag in the feeling that is involved in movement. Consider a situation where you are grabbing an object; if the feeling involved in the grabbing act arrives late, it becomes impossible to be accurate or precise in the act. Delays exist in each of the steps involved in movement: the sensing of the movement, the neural signals, the interpretation of the signals, the instruction (signal) to the muscles, and the actual movement of the muscles involved. There are always delays that are small in every stage, but they accumulate and cause more than expected control. Delayed response results in loss of control and reduced accuracy in the movement.

This involves balancing processing load, network constraints, and power usage—not racing mindlessly, but rather keeping pace with the natural flow. Speed is located within tight logic, clever turns, and deep wiring responsiveness.

15.8.4 Bandwidth and resolution

The amount of data that can flow in either direction is dependent on both the width and the precision of the signal. When impulses travel quickly across multiple pathways, reactions are quick, and the feedback becomes more responsive. However, these systems typically require more energy and processing. Slower systems are more easily integrated and processed, but they provide less control over the system and reduce the range of sensing. Selecting components requires balancing the real need against the costs of potentially disruptive bodily effects, energy consumption, and complex wiring.

15.8.5 Portability and wearability

Once gadgets move outside of laboratories, easy and flexible movement becomes key. Bulky pieces, dangling wires, or constant battery changes will move users off quickly. Therefore, materials that can bend with the body are becoming more common, and instead of tangled pathways, wireless signals will leap across the void. Additionally, less than a few watts will allow microchips to run for a long time on a single battery. As technology evolves, the ability to draw power from the environment becomes a reality. Therefore, to ensure comfort when wearing equipment or clothing, and to ensure a positive experience, it will be necessary to minimize waste and optimize it at every step (Li et al., 2025).

15.8.6 Human factors

The main area of interaction will be the body; touch feedback will be studied and adapted across a wide diversity of users. Thus, the placement of vibrating devices or electrodes will be fully within the design consideration. Moreover, people, in particular, vary and differ—the response of the skin varies, and routine activities do differ—thus, the design of the response device must be flexible to the user. Custom designs will not be optional and will be built in.

15.8.7 Regulatory pathways

Approval of medical implants takes a considerable amount of time, which results in teams organizing their plans with an emphasis on the stringent safety protocols and the prolonged review and approval processes. Testing can occur over many years and is often conducted postlaunch of the medical device. In contrast, fitness trackers and subdermal devices bypass many review processes and, therefore, can roll out updates on a more frequent basis than medical devices. Updates to fitness trackers and subdermal devices do not have the ability to treat or improve medical conditions. Quicker approval processes and regulations on medical devices dictate how the budget is allocated, the features that are prioritized, the metrics used to evaluate the device, and the overall regulatory process. As a result, designers are forced to nest their innovative ideas within the existing regulatory framework. Innovators must find an appropriate balance between regulatory constraints and innovative ideas.

15.9 Ethical, legal, and social implications

15.9.1 Unprecedented ethical challenges

The potential use of brain-interface technology raises complicated questions regarding the legal, moral, and practical ramifications that far exceed the regulation and use of ordinary consumer products and medical devices. As devices start to interact with people's thoughts, feelings, and emotions, the ethically relevant aspects of design begin to multiply. If we do not begin to address the issues associated with these types of products before they are developed, we will create brain–skin interfaces that are devoid of any ethical consideration and may conflict with people's moral expectations and theoretically justified social claims.

15.9.2 Privacy and mental privacy

The issues surrounding ethics have brought forth concerns regarding privacy. Initially, BCIs were designed to capture basic plans of movements or simple control instructions. Newer models, however, have begun to uncover more sophisticated patterns such as silent speech, fleeting emotions, and more complex cognitive processes. We are witnessing, for the first time, raw thought traces that have emerged from brain activity. This presents a never-before-considered level of privacy and brings forth questions regarding the privacy of one's own mind. The primitive biometric systems, such as fingerprints and iris scans, have almost no comparison to these mental signatures. Brain data is direct to the individual and the subjective experience of being alive. When the brain data is not predictable, obtaining consent becomes complicated. This creates the need for clear regulations regarding the ownership of neural data, the control and access to such data, and the purpose for which such data are used. As technology continues to integrate with cloud and smart systems, the need for these regulations becomes even more important.

15.9.3 Autonomy and agency

The ability to comprehend the situation, provide meaningful consent, and intervene in critical moments is particularly relevant with systems that adapt and learn independently. The user's comprehension keeps the control with the user, but a lack of transparency regarding the workings of a tool gives the user the sense that the technology is controlling them instead of the other way around. This creates a perception that the tech is driving them.

15.9.4 Equitable access

The ethics of the situation raise some moral questions around fair access and fair distribution. Advanced neural interfaces that can simulate a realistic sense of touch are incredibly expensive to develop, implant, and maintain. If there are no purposeful actions from governments, these devices will only be available to a handful of patients in affluent healthcare systems. This scenario is likely to amplify existing inequities in healthcare, creating a hierarchy in access to restorative or augmentative technologies. Access to these technologies is determined by reimbursement policies, the funding available, and whether the clinical trials were inclusive. The phrase "fair use" almost always relates to the legal rights granted to an individual; however, in the context of these technologies, it is more relevant to the people who will be excluded. The choices that are made will create gaps in the future.

15.9.5 Identity and embodiment

The use of BCIs that restore lost senses provides a new look at personal identity and control. Our self-perception is influenced by our sense of feeling, and fake touch signals might cause a disconnect from one's physical body. Someone might feel a robotic hand is completely theirs, and some might feel discomfort due to a signal mismatch. Because the feeling of altered touch can cause disruptions to one's mental sense of the body, individual responses can vary greatly. Emotional responses to stimuli are equally important as the restoration of neural pathways. Instead of only addressing whether a device can be operated safely, researchers and engineers should look for changes in mood, self-confidence, and coping mechanisms. The impact of sensation on self-identity is intricate and demands greater consideration than highly advanced technological components. Synthetic components will likely change the perception of an object being real and will require ongoing support. The most important factors are the emotional components and how they affect self-perception.

15.9.6 Regulatory evolution

The rules surrounding brain-linked technology are constantly shifting, but they tend to stay behind the curve. While safety assessments and regulations are important, they are often at the surface level. Other major concerns are how personal thoughts and cognitive processes will be transformed and utilized, as well as the changes to our identities and the people we become. Currently, there are advocates in the research and government sectors for the integration of multidisciplinary teams of clinicians, engineers, ethicists, lawyers, and users. These teams will consider how to use the technology in an ethical manner. Once technologies are released, there will be a need to monitor and assess them in the postmarketing context to evaluate their safety and effectiveness. Engaging and monitoring communities to determine how to assess technology will help to establish and maintain equity in the design and use of such technologies.

15.10 Technical and scientific challenges

15.10.1 Continuing obstacles

Despite advancements in interface technology and systems that simulate the sense of touch, the application of these systems outside of the lab continues to be a challenge. Whether it be system signal capture, system response adaptability, or any number of system interaction and trial complexity, these challenges all add to fine-tuning the systems for successful application. Progress on these challenges requires the integration of specialists from disparate fields. While both the human body and machinery eventually succumb to wear, the human body also possesses unique capabilities that the machine cannot replicate. Many designs that are successful on the conceptual level often fail when subjected to operational realities. While the human body has the ability to adapt, it is the only system that possesses such an ability, and these designs quickly reveal their limitations when subjected to operational pressures. In the conceptual design phase, boundaries are often defined by the failures of the design. Only once these boundaries are defined can the design be tested on a human body.

15.10.2 Long-term recording stability

Maintaining a consistent connection to brain activity over a prolonged period of time is problematic, particularly when the technology being used to do so is implanted within the brain. Implantable electrode grids trigger host response (scarring) in the form of accumulations of glial cells, which cause the loss of nearby neurons. As a result of this neuronal loss, the recorded signals become fainter, more difficult to interpret, and ultimately more unreliable. Additionally, over time, some of the electrode channels become less responsive, and the material properties of the electrodes used for stimulation/detection change. For example, as the probes become stiffer relative to the brain's compliance, they tend to indent the brain, which creates the potential for a variety of changes to the position of the electrodes as the brain undergoes movement. In this regard, while the use of more modern materials, flexible electrodes, and coatings, as well as improved surgical techniques, results in better performance over time for chronic implantation, the ability to maintain recording quality over a decade or more is definitely a gap that affects the true clinical potential for BCIs.

15.10.3 Shortcomings of noninvasive brain–computer interface

When signals are obstructed by bone or skin, they lose intensity. The head obstructs signal clarity; EEGs do not capture crisp data. As a result, the data are inconsistent, and the level of detail needed for a response is lost. Movement can be a source of interference. Whether it be the subject relocating or outside factors like people or objects moving. Breathing, blinking, and even eye movements can all distort measurements. More sophisticated signal processing systems can attempt to alleviate some of these limitations. However, any further advancements will be limited by the laws of physics. As such, it is necessary to assign these systems to functional tasks that require slower rates of response. Due to the location and methods used for data capture, performance limitations will always be present.

15.10.4 Naturalistic sensation evoking

Researchers are still perplexed by the ability to feel genuine sensations through artificial means, particularly when the stimulation of the brain is directly involved. Although brain stimulation has been shown to elicit feelings related to normal touch, pressure, or body awareness, it is challenging to achieve this without undesirable side effects. What works perfectly for a given individual may provoke discomfort or abnormal sensations in another individual due to the variances in brain wiring. As a result, the individual differences make designing and implementing a universal system nearly impossible, forcing the system to be tailored for each individual, and the excessive customization takes significant time and expert labor prior to anything being functional.

Over time, everything that involves an individual's thought processes and control strategies will start to change with the individual's mood, their concentration, and their energy levels. When the tech itself is doing updates, time is very important. If an individual moves too fast, their actions will become greater than their understanding. Results are likely to be greater if updates are synchronized with gradual drops in the clarity of the signal. People lose trust when things seem to operate under an absence of control, shifting unpredictably. People require time to make adjustments; however, changes tend to happen too rapidly. After a brief moment in time, a situation that seems familiar can become a lot more strange as the rhythm between the user and the technology falls out of sync. It is also the case that when the system is too rigid, the system will also become worse, because small changes will be ignored, and performance will fall. There is a need for a system to adapt to these changes, but not in a contradictory way. There is always a tolerance in the system that will stop it from changing too much. This is why the brain-technology interface hasn't improved in years. What is needed more than a system to adapt is a system to flex, because a system that adapts to too many changes will require people to learn a lot and serve no real function.

15.10.5 Multimodal integration

Integrating several senses into one system increases design complexity. The combination of visual, auditory, and tactile signals requires harmony. If signals conflict or some arrive too late, confusion occurs, resulting in slower reactions or loss of immersion. The way people react to different signals at varying senses means that designers need to offset inputs across different senses to ensure synchronicity, optimal intensity, and the right dominance. The need to feel natural and instantaneous creates design challenges.

15.11 Emerging directions and technologies

15.11.1 Advances in material sciences

New ways of integrating materials are impacting how machines interface with nerves. Standard materials are stiff and do not move like soft human tissues. This can lead to irritation and discomfort, and even make the nerves transmit weaker signals. To address this, scientists are developing flexible rubber-like sensor sheets that can mimic the behavior of nerve and skin layers. The rubber-like materials are softer and reduce stress to the cells, and reduce micro movements, which contribute to the minimization of discomfort, while maintaining monitoring activity. Similar to previous designs, flexible robotic muscles and rubber touch feedback tools offer smooth skin-like interactions over large areas. The comfort of the system is increased with the improved fit, and skin-friendly materials reduce chafing and wear, which is especially important for long-term use in medical devices or in artificial limbs. The use of materials for comfort and ease of use is important, especially over extended periods of time (Karimi et al., 2025).

15.11.2 Electronics miniaturization

Recent advances in materials have made it possible to make components of BCI systems smaller and to make them operate wirelessly. The absence of wires that penetrate the skin has significantly reduced infection risks and equipment failures. The ability to transmit data wirelessly increases user mobility and independence. Miniaturization of electronic components reduces power consumption, which results in longer device operation without overheating or the need for frequent recharge.

The reduction in size of components means that more channels can be accommodated in smaller spaces, enabling more sophisticated touch feedback configurations. Systems, even when densely packed, can perform more functions without increasing the difficulty of operation.

15.11.3 Novel neural stimulation

Incorporating them into humans would involve gene modification, which is an austere undertaking. Issues concerning risk, approval, and ethics impede advancement toward practical application. However, even without a potential biological application, the implications will help the development of better nonlight-operated devices.

15.11.4 Inner speech decoding

Decoding inner speech goes beyond the simple control of movement with BCIs. Only a few years ago, researchers discovered the patterned brain activity corresponding with the attempt or rehearsal of speech. Currently, these brain activities can be converted to text or even synthesized speech. Inner speech decoding is a game-changer for the speech disabled, particularly for those with severe paralysis. Envision a technologically advanced inner speech decoder that allows for the seamless control of the system. Imagine an interface that doesn't just respond to the user, but also provides a tactile sensation to control the rhythmic and responsive timing of the system. Picture a system that is fluid and responsive. Over the years, the array of inner speech decoders will be less of a machine and more of a voice. It will be seamless and responsive to the user's will. It will be more activated as the user's intention redirects the system, and it will be responsive to the user's will. Such a system will be able to provide a responsive, fluid experience that mimics normal human speech, complete with natural rhythms, pauses, and disruptions.

15.11.5 Integration of immersive technologies

Now imagine a world of immersive technology and the ability to interface it to control your own inner thoughts.

The brain's signaling domains, movement, while mimicking touch feedback, creates steps that appear automatic. It's not just sight and auditory stimuli. There is tactile feedback as pressure is felt under the finger, and the surface counter as the user pushes. All of these are controlled by the user's intention. When the mind is in control of most actions and reactions, making use of the digital space feels less like work. Some users are attempting activities that are risky, while others are rebuilding lost skills, and others are in the digital space recovery, redefined as the risk perceptions are altered by the functionality that is felt through the machine's interface. Digital space is redefined as the risk perceptions are altered by the functionality felt through the machine's interface. It is this combination of elements that allows the technology to evolve beyond the rehabilitation of the human body to the use of sophisticated collaboration constraints between man and machine.

15.12 Research roadmap and recommendations

15.12.1 Translational priorities

Material integrations with haptic technology on BCIs will only advance with clear and precise objectives. Signal strength is just one component. Past prototypes and iterative models tell us that what will endure is more important than what may captivate attention momentarily. The everyday systems users influence iterative improvements. Engineering solutions and real-world use create a paradigm for cyclical wear from use. Designs survive iterative improvements and engineering solutions. Designs survive and stay when they endure use beyond the controlled conditions of laboratory environments adapted for real-world use.

15.12.2 Improving durability and biocompatibility

Brain implants that are durable and play nice with body tissue will allow for more widespread use of the implants. The materials used in implants will dictate the response of the body and how durable the implants will be. The correct materials, bolstered by innovative engineering solutions that leave no space for the body's moisture, will aid in promoting implants that seamlessly interact with biological systems. Reducing the need for surgeries will create a reduction in costs, an increase in the public's trust, and a willingness to shift more towards the more implanted options.

15.12.3 Feedback based on user modality

Developing sensitivity in prosthetic devices should always begin with what the nervous system already knows. Feedback of the type matched to natural sensations works best, especially with emerging prosthetic devices and new rehabilitation technologies. Examples include devices that provide tactile sensations through skin electrical signals, joint motion, and pressure data using feedback and control systems. Several studies show that with the right type of feedback, the users of these devices demonstrate improved performance in carrying out particular everyday tasks. More complex technologies in the devices do not necessarily improve their outcomes. Familiar feedback systems, without adding complexity, provide the basis for considerable functional improvement. When feedback is provided in rehabilitative devices in response to the movements and the system nerves respond, the nervous system responds, and the brain adjusts to the movements more rapidly as well. Saving advanced components for later is more effective than adding more complex technologies to devices (Karimi et al., 2025).

15.12.4 Algorithms that are adaptive and stable

Developing reliable and smooth adaptive control of the encoding and decoding processes is one of the important problems. Since brain signals are adaptive, the system should change to optimal control with the least shifts in the control parameters. When a system is disrupted, the outcomes of its work may be reduced, resulting in reduced trust and reduced mastery of the system. The development of adaptive systems for control is one of the priorities for control systems in the future. The control system should be designed to change predictably to meet the natural learning and control systems, rather than overriding them.

With the right mix, the burden of constant learning and saving cognitive energy can make using the product much more pleasant.

15.12.5 Policy recommendations

Integrating the comprehensive psychosocial outcomes framework encourages the use of outcome measures that are more focused on and kind to the lived experience of users, as opposed to the split measures commonly used in more clerical and cerebral processes. In addition to the commonly used counts of the numbers of correct and incorrect responses with the measures of time taken, and the measures of user satisfaction, the user experience is "measurable" in ways that sip and puff users may report feelings of control, loss of control, and rage, and in extreme cases, the overall experience may be classified as deeply satisfying. "Controllers," "joy stick," and "sip and puff" navigators are portals, and loss of control is just one of the potential psycho–socio–emotional impacts. With controlled movement also come the psychosocial impacts of control, and for the user, the experience may go beyond the functional to the deeply valuable.

15.12.6 Interoperable standards

What is the largest challenge? Each laboratory has its own method of recording findings, which creates difficulty in correlating results or sharing insights. Common standards for the formation, testing, and evaluation of laboratory findings would clear confusion and promote collaboration. Imagine dozens of transnational teams working with the same standards, leading to increased efficiency, ease of collaboration, and fewer bottlenecks. Commonly overlooked is the inclusion of fairness criteria in the standards, ensuring that they do not overlook patient rights. When safety and ethical considerations are uniformly incorporated, trust in the standards increases.

15.12.7 Interdisciplinary collaboration

What unifies these goals? Continuous collaboration among diverse professionals. Innovations in BCIs with haptic feedback are the result of cross-disciplinary collaboration rather than individual contributions. Neuroscience understands the firing of neurons; engineering creates the devices and writes the code; medicine provides pathways to practice; and ethics and the public address the impact and the daily reality, which also fosters the development of better instruments. When the invention also meets a practical need, combined with clear guidelines and a focus on real-world issues, the outcome is a useful solution that is delivered in an unanticipated timeframe.

15.13 Conclusion

Consider, if you will, what it means to communicate directly with a machine, as if it were an extension of yourself. This kind of experience can arise from an emerging category of BCIs that integrate the use of touch-based signals. These interfaces can create a control-response loop that is similar to the way natural limbs function. Control of an artificial limb starts with a thought, and sends a command to the device, resulting in a touch-based response, and ultimately a feedback control command from the user. This kind of interaction over an extended period supports the user's ability to perform smooth and controlled movements and integrate artificial limbs as their own. Previously, prosthetic limbs would move in response to a user's commands without feedback or response to the user. Today's prosthetic limbs can respond to the user and integrate with the user's neural control system to achieve a level of function similar to skin.

Fluidity and comfort of interactions are critical for the seamless operation of a BCI system. The inclusion of tactile and/or sensory stimuli as a component of a BCI system can alleviate the need for a user to visually monitor the system or interface and allow for a more instinctive and natural interaction with the BCI system. Many studies have demonstrated the existence of a positive correlation between the inclusion of some form of sensory feedback (stimuli) and the user's cognitive and physical response—dominance of the user's physical motor response, increased cognitive load, greater sense of control, and increased overall user engagement with the BCI system. It is possible to employ various stimulating devices to provide feedback, including peripheral nerve stimulation and, in some cases, stimulation of brain tissue.

All the hardware advancements made the said changes possible. Newer materials paired with smarter layout designs and miniaturized circuitry enable tools used for anything from scalp EEG's to implants for deep brain neuromodulation. A sense of touch feedback systems, soft and flexible sensors, and wireless systems enable users to move freely and comfortably, and most importantly, to keep the tools on for extended periods. The neural interface's brain–computer connection instantaneous and adaptive response to neurophysiological changes and user-specific response patterns in neural activity, harnessing improvements in engineering together with data processing algorithms to seamlessly adapt to neural activity and personalize responses based on the neural activity patterns. Real improvement is evident, not from the individual components of the interface, but from the components working in synergy for the improvement of the tools and the improvement of the quality of life of the users (Zhang et al., 2024).

Equally important is how computers have improved to keep a function steady and continuous over a period of time. Neurons change over time; the new methods used in decoding have to adapt effortlessly, and this is how users can stay in control without losing control. The engagement of advanced models, while increasing the richness of the available experiences, raises important questions about control and transparency. It becomes possible to have realistic feedback from a generative model, but the system must be monitored closely. A system needs to provide sufficient user trust to maintain its functionality over a long period of time, but more importantly, the system needs to continue to function to maintain its activity.

Real-time technologies necessitate increased scrutiny (Zhang et al., 2024). For new technologies to meaningfully transform the practice of medicine and everyday life, enduring systems, while offering realistic, safe, and enjoyable sensations, some new elements to the systems will be needed. With new and different forms of use, the training of systems should permit multiple comfortable uses of the system. Technologies blind people to the technology while providing them with a form of control over their privacy. When technologies begin to practice thinking, people become a different person without knowing it. Before technology creates harm, governance to mitigate those harms must be implemented. Technologies blind people to the technology while providing them with a form of control over their privacy. When technologies begin to practice thinking, people become a different person without knowing it. Before technology creates harm, governance to mitigate those harms must be implemented.

Teams made of professionals from different areas will be key to socially and ethically integrating the brain–machine interface (BMI) technology. Collaboration with people from outside medical or technical domains at each step is essential. This includes engineering safe and operational BMIs. The key is to go beyond increasing the speed of BMIs and to intelligently integrate the designed-to-be-safe and operational BMIs with the socio-psychological reality surrounding their use. The potential for BMIs goes beyond restoring previous functions or even integrating technology with the human body. Bidirectional (or closed-loop) BMIs will significantly alter human interaction with technology. Touch feedback systems, mental commands, and virtual reality environments will further enhance the already significant potential of BMIs. Transformative technological advancements must meet the needs of people to be considered progress. This is certainly not the end of the journey, but the beginning of more. The future will not seek to simplify the nervous system or the human brain. Instead, it will respect the existing boundaries and gently expand the capabilities of the human brain.

References

Belkacem, A. N., Jamil, N., Khalid, S., & Alnajjar, F. (2023). On closed-loop brain stimulation systems for improving the quality of life of patients with neurological disorders. *Frontiers in Human Neuroscience, 17*. https://doi.org/10.3389/fnhum.2023.1085173.

Chen, S., Chen, M., Wang, X., Liu, X., Liu, B., & Ming, D. (2025). Brain–computer interfaces in 2023–2024. *Brain, 3*(1). https://doi.org/10.1002/brx2.70024.

Karimi, M., Yeganeh, N., Makarov, I., Sverrisson, A. Ö., Gunnarsson, K. F., Briem, K., Brynjólfsson, S., Kristjánsson, Á., & Unnthorsson, R. (2025). Haptic feedback systems for lower-limb prosthetic applications: A review of system design, user experience, and clinical insights. *Bioengineering, 12*(9). https://doi.org/10.3390/bioengineering12090989.

Khan, S., Kallis, L., Mee, H., El Hadwe, S., Barone, D., Hutchinson, P., & Kolias, A. (2025). Invasive brain–computer interface for communication: A scoping review. *Brain Sciences, 15*(4), 336. https://doi.org/10.3390/brainsci15040336.

Li, J., Chen, G., Li, G., Xiao, L., Jia, R., & Zhang, K. (2025). Flexible brain electronic sensors advance wearable brain–computer interface. *npj Biomedical Innovations, 2*(1). https://doi.org/10.1038/s44385-025-00029-7.

Rubin, D. B., Ajiboye, A. B., Barefoot, L., Bowker, M., Cash, S. S., Chen, D., Donoghue, J. P., Eskandar, E. N., Friehs, G., Grant, C., Henderson, J. M., Kirsch, R. F., Marujo, R., Masood, M., Mernoff, S. T., Miller, J. P., Mukand, J. A., Penn, R. D., Shefner, J., … Hochberg, L. R. (2023). Interim safety profile from the feasibility study of the BrainGate neural interface system. *Neurology, 100*(11), E1177. https://doi.org/10.1212/WNL.0000000000201707.

See, A. R., Choco, J. A. G., & Chandramohan, K. (2022). Touch, texture and haptic feedback: A review on how we feel the world around us. *Applied Sciences., 12*(9), 4686. https://doi.org/10.3390/app12094686.

Zhang, H., Jiao, L., Yang, S., Li, H., Jiang, X., Feng, J., et al. (2024). Brain–computer interfaces: The innovative key to unlocking neurological conditions. *International Journal of Surgery, 110*(9), 5745–5762. https://doi.org/10.1097/js9.0000000000002022.

SECTION IV

Validation, emerging healthcare integration, and future challenges

16. Clinical validation and safety standards of haptic medical robotics 303
17. Revolutionizing healthcare: The convergence of medical robotics and digital therapeutics 317
18. Securing the future: Legal and privacy challenges in haptic medical systems for disabled children 353

CHAPTER 16

Clinical validation and safety standards of haptic medical robotics

Pushan Kumar Dutta[a] and Bhupinder Singh[b]
[a]School of Engineering and Technology, Amity University Kolkata, Kolkata, West Bengal, India
[b]Sharda School of Law, Sharda University, Greater Noida, Uttar Pradesh, India

16.1 Introduction

The landscape of surgical practice has undergone a transformative evolution with the emergence of haptic medical robotics, a technological frontier that promises to redefine precision, training, and patient care. By integrating advanced tactile feedback mechanisms into robotic systems, this innovative field addresses critical challenges in surgical interventions, particularly in minimally invasive procedures where traditional sensory perception is limited. The genesis of medical robotics can be traced to the late 20th century, when researchers and clinicians recognized the potential of robotic systems to transcend the limitations of traditional surgical techniques. Initial developments focused on enhancing surgical accuracy and enabling minimally invasive procedures, allowing surgeons to manipulate instruments with unprecedented precision through control panels and visual interfaces. As artificial intelligence and software technologies advanced, these robotic systems expanded their applications across surgical, rehabilitation, and prosthetic domains (Khan et al., 2020). However, the promise of haptic medical robotics is not without significant challenges. The integration of tactile feedback technology demands rigorous clinical validation and adherence to stringent safety standards. Unlike conventional medical technologies, robotic systems must not only demonstrate functional effectiveness in controlled environments but also prove their reliability and safety in dynamic clinical settings.

This necessitates comprehensive experimental evaluations, compliance with international standards, and ongoing dialogue among researchers, clinicians, and regulatory bodies. The critical importance of haptic feedback lies in its ability to restore the sense of touch, a fundamental aspect of surgical interaction that is often compromised in robotic-assisted procedures (Rani, 2022). By providing surgeons with nuanced tactile information, these systems enable finer adjustments, potentially reducing surgical risks and improving patient outcomes. Moreover, the technology extends beyond surgical applications, showing significant potential in rehabilitation robotics and advanced prosthetic design. This chapter explores the multifaceted landscape of haptic medical robotics, examining its

Haptic Feedback Control in Medical Robotics
DOI: https://doi.org/10.1016/B978-0-443-27777-1.00005-6

historical development, current technological capabilities, clinical validation processes, and the complex regulatory framework that governs its implementation. By critically analyzing the challenges and opportunities in this emerging field, we aim to provide a comprehensive understanding of how haptic technologies are reshaping medical interventions and setting new standards for precision, safety, and patient care. As robotic systems continue to proliferate in healthcare settings, the need for rigorous clinical validation and safety standards has become increasingly critical to ensure their effectiveness and reliability in real-world applications (Ahirwar & Khan, 2022). The integration of haptic feedback in robot-assisted surgeries addresses the challenge of reduced sensory perception typically encountered during minimally invasive procedures.

16.1.1 Latency and sensory complexity

Haptic medical robotics, the use of robotic technology that incorporates haptic feedback systems, is transforming the way we approach precision and analogue tasks in the medical field. Such robotic systems industries, like telesurgery, rehabilitation, and prosthetics, heavily depend upon real-time haptic feedback that provides tactile sensations similar to human touch for surgeons and clinicians. But two important criteria must be overcome before these systems can be clinically validated in controlled settings and meet rigorous safety standards: latency and sensory complexity. Such factors affect the efficacy, safety, and regulatory approval of haptic medical robotics; therefore, they are often left out of clinical studies, safety guidelines, and industry standards (Han et al., 2020). Latency, the time it takes from the presentation of a command to its action, is a significant concern in haptic medical robotics. Even a minor delay, which in a clinical setting can lapse into errors of time, affecting the quality of surgery and patient prognosis, can cause significant disruption. Haptic system latency comes from several sources, such as network transmission delays, the time to process control algorithms, the time for sensors to respond, and the time for the actuator to execute (Pramanik et al., 2020). Latency matters depending on the application. such as in robotic-assisted telesurgery, where surgeons are not directly present at the location and work in a different location, latency can drastically affect the real-time control of robotic arms, increasing the chance of unintended movements, as well as other surgical complications, etc.

A significant contributor to latency is the network infrastructure employed by teleoperated medical robotics. Traditional internet networks impose time delays on communication because the data packets must travel long distances before they can be received and processed at a different endpoint, causing interruptions in haptics feedback loops. These delays can result in challenges for real-time responsiveness in haptic medical robotics, but low-latency communication technologies like 5G networks and edge computing are being investigated as means to reduce these latency times, to improve real-time response for haptic systems. Also, accounts of computational lags in the robotic control system need to be minimized by optimizing processing algorithms and hardware. Recent advances in prediction, through machine learning and elsewhere, make it possible to tightly couple robotic systems with the human to anticipate the next inputs, and in a sense, offset the inevitable lag. The third critical issue is the perceptual latency threshold, the amount of delay a human operator can stand without losing control or feeling discomfort. Haptic feedback systems ought to experience latency below 10 milliseconds (ms), as previously studied, for seamless operator-robotic device interaction. But in real system applications, it is not so easy to achieve such low latency due to system complexity and external network factors. Thus, clinical validation of haptic medical robots must include tests to assess performance under varying amounts of latency to define safe operating parameters and to verify the robustness of the proposed systems (Nazari, 2020).

16.1.2 Varying sensory inputs and multisensory feedback

Haptic medical robotics sensory complexity, this is the hallmark of medical robotics that gives you an exact amiable experience that feels just like human touch. Unlike classical robotic systems that depend only on visual and auditory feedback, haptic medical robots combine force, vibration, texture, and temperature sensing to create a more realistic and full-bodied interaction. This complexity of sensation is important for applications like robotic-assisted surgery (RAS), which needs the delivery of high-fidelity tactile feedback to distinguish between tissues, blood vessels, and bones. A key issue for delivering true-to-life over haptic feedback involves enabling hyper-realistic sensor systems. Haptic sensors need to be extremely sensitive to measure even small variations in force and pressure to enable immediate feedback. Modern haptic medical robotics often uses piezoelectric and capacitive sensors to achieve this precision. Yet, sensor reliability and consistent performance in varying clinical profiles are a critical challenge. Environmental influences like temperature, sways, and dampness can render sensor readings variable, causing intermittent well-being concerns over the haptic comments.

Using a sense of involvement in the task allows the operator to engage the robotic system more usefully based on multimodal feedback (tactile, kinesthetic, and visual). For example, haptic medical robotics are now being combined with augmented reality (AR) and virtual reality (VR) interfaces to give surgeons a 3D view of internal structures while feeling immediate force feedback (Sahi & Kaushik, 2022). Much like the combination of stereo vision, binocular parallax, and optical focal length corrects distance perception, this "stereo vision" creates depth perception and improves precision and safety in complicated procedures. Nonetheless, combining more sensory modalities adds complexity to the system due to requirements for real-time synchronization between haptic, visual, and auditory inputs. Any of these feedback loops, delayed in time or inconsistent with each other, could lead to a misinterpretation of sensory information and deteriorate the clinical outcome.

16.1.3 Validation of safety and clinical standards

Haptic medical robotics will also require stringent clinical validation and adherence to safety standards to guarantee patient safety and system reliability. Established regulatory bodies, including the U.S. Food and Drug Administration (FDA), the European Medicines Agency (EMA), and the International Organization for Standardization (ISO), have also outlined guidelines for medical robotic systems, mandating extensive testing before linking them to clinical approval. Clinical trials should systematically assess the potential effect of latency and sensory complexity on operational safety, guaranteeing that the robotic system reliably executes its tasks in various contexts. Protocols for hazard identification, risk assessment, and failure mitigation in haptic medical robotics are defined in safety standards, such as ISO 13482 for personal care robots and ISO 14971 for risk management in medical devices. Registration and obtaining market approval are required, along with IEC 60601, medical electrical equipment-safety, which regulations will apply. It requires extensive human factor studies, usability tests, and failure mode analysis to ensure compliance with these standards as the risks of delayed or inaccurate haptic feedback can have functional and human risks (Oukhatar et al., 2021). Artificial intelligence (AI)-based validation methods are also being used to improve the reliability of haptic medical robots. By utilizing machine learning models that process real-time data from sensors and user interactions, professionals can identify irregularities, forecast system failures, and enhance response times. These insights are enhancing the entire sensory experience for operators, enabling AI-assisted testing, which allows the validation process to be faster as it informs on decreasing latency. Latency and sensory complexity are the primary concerns within the clinical verification and validations of haptic medical robotics. Advanced communication networks, predictive algorithms, and optimized processing minimize latency for real-time responsiveness—vital for precision-driven medical uses. On the other hand, the combination of high fidelity haptic feedback and multimodal interactions helps solve the complexity of senses contributing to the effectiveness of robotic-assisted procedures. To ensure compliance with safety standards, top with extensive testing and validation with the aid of AI will be key to promoting the use of haptic medical robotics in clinical practice as regulatory frameworks evolve. With the right innovations addressing these challenges, the future of haptic medical robotics will be a landscape of safer, more efficient, and ultimately immersive solutions for healthcare (He & Eastburn, 2022).

16.2 Clinical validation in haptic medical robotics

By restoring a sense of touch, these systems enable surgeons to make finer adjustments, ultimately leading to safer and more effective interventions. However, the path to clinical adoption is fraught with challenges, including the need for experimental evaluations that accurately replicate clinical conditions, adherence to stringent safety regulations, and compliance with established international standards such as those set by the International Organization for Standardization (ISO). Notably, controversies surrounding the clinical validation processes highlight the complexities of ensuring that robotic systems operate reliably in dynamic surgical environments. Issues such as study heterogeneity, small sample sizes, and insufficient reproducibility in research findings pose significant obstacles to establishing a universal framework for safety and efficacy in haptic medical robotics. These challenges necessitate ongoing dialogue among researchers, clinicians, and regulatory bodies to foster transparency and collaboration in the development of robust guidelines and protocols (Mujawar et al., 2020). Haptic medical robotics stands at the forefront of surgical innovation, combining advanced technology with critical safety and validation measures to enhance patient care. As the field continues to advance, addressing the associated technical, ethical, and regulatory challenges will be essential for fully realizing the benefits of these transformative systems in clinical practice.

16.2.1 Use cases of haptic medical robotics

Haptic medical robotics is a transformative technology enabling high-precision touch feedback and real-time interaction in modern healthcare. They include force, vibration, and kinesthetic feedback, which enable increasingly accurate, precise, and efficient procedures that significantly benefit both the medical practitioner and the patient. Haptic medical robotics is used in RAS, telemedicine, rehabilitation, prosthetics, medical education, and diagnostics. These robots—when integrated with artificial imagination (AI), machine learning (ML), and advanced sensor technologies—have transformed the way healthcare professionals can assist diagnosis, treatment, and recovery. The acquisition of force information in connecting the surgeon's hand with the surgical instruments can even bring haptic properties to the operated tissues, thus providing the required biofeedback and assistance via these tailored robotic instruments, improving surgical precision, enhancing utility in telehealth, and increasing patient recovery rates by making haptic medical robotics crucial for the future of medicine (Udoh et al., 2023).

16.2.1.1 Robotic-assisted surgery

Haptic medical robotics has one of its most crucial applications in the field of RAS, wherein robotic systems accompany professional surgeons in carrying out minimally invasive surgeries with better control and precision. They also utilize haptic technology to give surgeons real-time perceptions of the forces they are exerting on the various tissues they contact, so they can "feel" the tissues even when they are away. Systems such as surgical systems allow surgeons to maneuver robotic arms with precision, lessening the chance of complications and optimizing patient outcomes. The haptic technology in robotic surgery is especially useful in delicate surgeries like cardiovascular, neurological, and orthopedic surgeries, which require precise movements. In laparoscopic surgeries, for instance, haptic feedback provides surgeons with the ability to feel the tension and resistance of internal tissues, thus reducing the risk of accidental tissue damage. AI-enabled haptic robotics can modulate force feedback according to tissue properties, improving surgical safety and efficiency (Chelliah et al., 2021).

16.2.2 Surgical and tele-surgery

The other area where haptic medical robotics has made a significant impact is telesurgery, where it enables remote surgical interventions by providing surgeons with robotic systems that offer real-time haptic feedback. This application is especially important in rural or underserved areas where there may not be a wide range of specialized surgeons. The low-latency 5G networks and AI-driven predictive models work to cut transmission delays between surgeon and robotic system, so they can be in concert at every moment. In fact, the MIRA (Miniaturized In Vivo Robotic Assistant) telesurgery project has shown that a tele-operated robotic system was able to operate on the patient from a distance in another country, giving us all hope of providing febrile surgical means for us all through haptic medical robotics (Materon et al., 2021). Further, this technology is being tested in military and space health services where RASs are performed on astronauts or injured soldiers in extreme conditions without human intervention.

16.2.3 Therapeutic intervention and rehabilitation

The important application of haptic medical robotics is in rehabilitation and physical therapy, where it can aid patients recovering from injuries, strokes, or neurological conditions. Haptic feedback has been developed to be used in robotic rehabilitation devices, allowing patients to receive personalized resistance, moving trajectory, and in-time somatosensory feedback, and thus enhancing their recovery efficiency. An example of this is the Lokomat robotic exoskeleton, which helps patients with spinal cord injuries or stroke regain their walking abilities by giving controlled movements, which can also be augmented with haptic feedback and adjustments controlled by AI. The therapeutic tool functions on the same principle as haptic gloves and robotic arms for use in hand and upper-limb rehabilitation, providing stroke patients the opportunity to perform movements in a controlled manner, with the level of force that they can generate gradually varying. Haptic robotics is also being deployed into the realm of neurological rehabilitation, which combines the use of robotic systems that elicit nerve responses with specific pressure and vibrations to help those with neurological conditions like Parkinson's disease and cerebral palsy recover motor function (Kumar & Kumari, 2021). These robots monitor the patient's progress during treatment and in real-time so that the therapist can modify the therapy protocols with the help of AI, based on the analysis of performance.

16.2.4 Control of prostheses and artificial limbs

Haptic medical robotics is a branch of robotics, it is aimed to help people who have lost their limbs through the use of prosthetic anatomical limbs. Certain advanced bionic prosthetics, which can be equipped with haptic sensors and AI-controlled mechanisms, enable users to "feel" pressure, texture, and temperature with their artificial limbs (Gaobotse et al., 2022). The LUKE Arm, created by DARPA, is a bionic prosthetic equipped with hybrid haptic sensors that transmit the sense of touch to the nervous system of the user, allowing controlled grip precision and manipulation. Haptically augmented brain–computer interfaces (BCIs) enabling amputees to manipulate a prosthetic limb using the power of their minds without losing touch with the environment empower amputees to use their neural signals to control their prosthetics while receiving real-time data on what they are feeling, increasing their mobility and functionality. This has been accomplished with the recent introduction of soft robotics and biointegrated haptic sensors that enhance the versatility of prosthetic limbs by replicating the flexibility and responsiveness of human skin. These breakthroughs are laying the groundwork for complete, AI-driven prosthetic limbs that can restore quasinatural senses for those who lost their limbs.

16.2.5 Medical education and simulation

Haptic medical robotics is a common application in medical education and training, as this allows students and professionals to perform complex procedures in a risk-free, controlled environment. Force feedback in haptic-enabled surgical simulators enables the trainee to sense cutting, suturing, and other tissue manipulation, providing critical proprioceptive feedback that can feel unsafe in a live patient (Vivekananthan et al., 2022). The LapSim and Simbionix simulators, for instance, utilize haptic technology to simulate laparoscopic and endoscopic procedures, ensuring that surgeons refine their skills in the training room, rather than the operating theatre. These simulators also help to enhance hand-eye coordination and decrease training time, and improve surgical precision, leading to better outcomes for patients. The platforms that utilize VR for haptic medical training incorporate VR and haptic interfaces to mimic real-world surgical settings, offering a multisensory educational experience. Through constant analysis of trainee performance, AI algorithms offer a strategic edge by delivering real-time feedback to enhance surgical competency and decision-making skills (Pradhan et al., 2023).

16.3 Haptic feedback and robot-assisted surgery

The integration of haptic technology in medical robotics marks a significant advancement in the field of surgical practices and clinical training (Manogaran et al., 2021). Haptic devices, which provide tactile feedback, have been reported to enhance procedural performance and clinical skills training, allowing surgeons to navigate surgical tools more accurately and avoid critical regions during procedures. This evolution stems from the need to improve surgical outcomes and reduce the risks associated with traditional surgical methods. In the late 20th century, as robotic systems began to emerge in medicine, researchers and clinicians recognized the potential for robotics to transform surgical techniques. The development of sophisticated mechanical robotic systems was initially focused on enhancing the accuracy of surgical tasks and facilitating minimally invasive procedures. These robotic systems enabled surgeons to manipulate instruments via control panels while viewing operations on screens, providing superhuman capabilities that were previously unattainable. This advancement not only improved surgical precision but also expanded the possibilities for remote treatments and objective diagnoses. The historical progression of robotic surgery can be traced back to the introduction of various robotic systems in operating rooms during the 1990s.

Over the years, innovations in AI and software integration have further propelled the capabilities of medical robots, enabling their application across surgical robotics, microrobotics, prosthetic rehabilitation, and hospital care (Suma, 2021). As medical robotics gained traction, the necessity for rigorous clinical validation and safety standards became apparent. Researchers began to develop standardized assessment tools for robotic surgical skills, ensuring the reliability and effectiveness of these advanced technologies in clinical settings. Emphasis is placed on the technical and regulatory challenges hindering widespread deployment, such as inconsistent evaluation metrics, small sample sizes, and difficulties in replicating clinical environments during testing. Notably, research initiatives such as those from the Haptics and Medical Robotics (HAMR) Laboratory at Johns Hopkins University have contributed significantly to advancing the design and validation of safe, effective haptic systems in surgical robotics (Jenkins, 2022).

16.4 Safety standards for haptic medical robotics

As the field continues to evolve, the incorporation of haptic feedback technology is seen as a critical component in enhancing the interaction between surgeons and robotic systems. This not only aids in the training of new surgeons but also addresses the technical challenges associated with RAS, ensuring that the integration of these technologies into clinical workflows is both safe and effective (Malik & Kumar, 2022). Clinical validation in haptic medical robotics clinical validation in haptic medical robotics is essential to ensure that the technology not only functions effectively in controlled settings but also produces reliable outcomes in real-world clinical environments. Clinical applicability can be defined as the extent to which results from published studies are likely to reflect expected outcomes when interventions are applied in actual medical practice.

Importance of Good Clinical Practices: A series of Good Clinical Practices (GCPs) establishes the validity of medical research and provides guidance for implementation, which is crucial for developing new medical technologies, including surgical robotics. For instance, before a clinical trial is approved, a new drug must undergo extensive preclinical studies involving in vitro, in vivo, and ex vivo testing. Similarly, haptic technologies in surgical robotics require rigorous experimental evaluations in environments that closely mimic the clinical settings in which they will ultimately be used (Kumar et al., 2022). Conducting tests on cadaver or animal models is encouraged to enhance the validity of findings, although these tests are not strictly required.

Challenges in experimental evaluation: The challenges faced in the clinical validation of haptic medical robotics are ensuring that experimental settings adequately replicate the clinical environment. Often, validation occurs in experimental settings that may not fully reflect real-world conditions, which can undermine the robustness of the results obtained. The need for a systematic approach to evaluate the criteria for experimental evaluation and data selection is highlighted in current research to enhance the replicability of findings and develop reliable methodologies for surgical robotics.

Haptic feedback and RAS: The integration of haptic feedback into RAS poses additional challenges and is a critical area of focus for research and development. Understanding how human perception of touch can be translated into robotic interactions is fundamental for enhancing the usability and effectiveness of these systems (Han, 2023).

16.5 Challenges in experimental evaluation

Research efforts in this area aim to address the technical difficulties associated with creating reliable haptic feedback mechanisms that can be effectively employed in surgical procedures, thus contributing to the overall safety and effectiveness of robotic surgeries (Raghav et al., 2024).

Safety standards for haptic medical robotics: The development and implementation of haptic feedback systems in medical robotics must adhere to rigorous safety standards to ensure the reliability and effectiveness of these technologies during surgical procedures. Various International Organization for Standardization (ISO) standards outline the necessary safety requirements that apply to machinery and robotic devices used in medical contexts.

Key ISO standards: ISO 13855:2010 and ISO 13855:2010 address the safety of machinery by specifying the positioning of safeguards with respect to the approach speeds of parts of the human body. This standard aims to prevent accidents by establishing parameters that ensure operators remain safe while interacting with machinery (Banerjee et al., 2020). ISO 18497:2018 and ISO 18497:2018 focus on the safety of highly automated machinery, particularly in agricultural machinery and tractors, but the principles can be adapted for medical applications. This standard underscores the importance of safety in automated systems, emphasizing risk assessments and appropriate safety measures. ISO 13851:2019 and ISO 13851:2019 provides guidelines for the design and selection of two-hand control devices, which are crucial for ensuring that operators can safely interact with robotic systems. This standard helps to mitigate the risks associated with unintentional machine activation during medical procedures. ISO 13482:2014 and ISO 13482:2014 outline safety requirements specifically for personal care robots, which can include robotic surgical systems. It emphasizes the need for careful design considerations to protect patients and operators alike, fostering the development of safe robotic technologies.

Importance of compliance: Compliance with these ISO standards is essential not only for regulatory approval but also for enhancing clinical applicability, which refers to the extent that results from controlled studies can be expected to reflect real-world outcomes (Barnes & Zvarikova, 2021). Ensuring that haptic devices meet established safety criteria is critical in maintaining the integrity of surgical procedures and minimizing potential risks during operations.

16.6 Evaluation metrics and reproducibility

The ongoing testing and validation are necessary to demonstrate that haptic feedback systems provide reliable and safe performance in clinical settings. This includes adhering to good clinical practices (GCPs) and conducting thorough experimental evaluations, often in environments that closely mimic actual clinical conditions (Desai & Shende, 2021). By following these rigorous standards, developers can help assure both the safety and efficacy of haptic medical robotics in real-world applications. *Applications of haptic medical robotics*: Haptic medical robotics plays a pivotal role in enhancing surgical procedures and rehabilitation practices by providing surgeons and patients with better interaction and feedback mechanisms. These applications are particularly significant in minimally invasive surgeries, rehabilitation robotics, and upper-limb prosthetic devices.

Minimally invasive surgery: Minimally invasive surgery (MIS) has revolutionized the field of surgery, offering patients reduced trauma to muscles, nerves, and tissues, alongside benefits such as decreased bleeding, pain, and scarring, and shorter hospital stays. However, the complexity of MIS requires surgeons to undergo specialized training to effectively manipulate the robotic devices involved. A major challenge in RAss has been the loss of direct haptic feedback, which allows surgeons to feel the tissue texture and resistance during procedures (Hurley & Popescu, 2021). The integration of haptic feedback systems into surgical robots has addressed this challenge, enabling surgeons to regain a sense of touch, which is essential for making precise adjustments in real-time.

Rehabilitation robotics: In the realm of rehabilitation, haptic technologies are applied in robotic devices designed for physical therapy. These rehabilitation robots can assist patients in performing therapeutic exercises while providing feedback that helps maintain their capabilities within safe physiological limits. For instance, devices like Achilles are designed specifically for ankle rehabilitation, helping patients recover through guided movements that are tailored to their individual progress.

Upper-limb prosthetics: Upper-limb prosthetic devices are another area where haptic technology has made significant contributions. The development of prosthetics equipped with haptic feedback systems enhances the user's ability to perceive touch and force, thereby improving the functionality of the prosthetic and the overall user experience (Goel et al., 2021). This advancement helps individuals better interact with their environment, leading to a more natural and effective use of their prosthetic limb.

The development and implementation of haptic medical robotics face numerous challenges and limitations that can impact clinical validation and safety standards. Technical complications are one of the primary issues, as substantial technical complications are involved in executing automatic actions in medical interventions with the required reliability. The complexities associated with robotic systems necessitate a thorough understanding of their operational environment, which is not always achievable. For instance, experiments often take place in "open environments," complicating the evaluation of their effectiveness in more constrained laparoscopic settings (Kumar et al., 2022).

Study heterogeneity and sample size: Another significant limitation is the heterogeneity across studies and small sample sizes. This variation makes it difficult to draw generalized conclusions and hampers the reproducibility of research findings. The scarcity of comprehensive, long-term studies further complicates the establishment of consistent safety standards and clinical validation for these technologies.

Evaluation metrics and reproducibility: In many cases, the criteria for evaluation are not clearly defined, which can lead to ambiguities in assessing the performance of robotic systems. Some experiments lack sufficient documentation to enable other researchers to reproduce the results effectively. This is critical, as reproducibility is a cornerstone of scientific research and essential for validating findings in the context of clinical applications (Huang et al., 2021).

16.7 Legal and ethical concerns

The integration of haptic feedback technology in surgery also raises legal, regulatory, and ethical issues. As robotic systems become more prevalent in surgical settings, it is crucial to develop appropriate frameworks to address these concerns effectively. There is a pressing need for collaboration among academic institutions, scientific journals, and funding organizations to address the challenges posed by these technologies, particularly in ensuring their safe and effective deployment in clinical practice.

Latency and sensory complexity: Latency remains a significant challenge in haptic feedback technology, as delays between a surgeon's actions and the corresponding feedback can disrupt precision and adversely affect surgical outcomes. The complexity of sensation involved in providing effective haptic feedback requires further investigation to enhance the reliability and intuitiveness of these systems in real-world applications (Malik et al., 2021).

16.8 Future directions

The future of haptic medical robotics holds significant promise, yet it is accompanied by notable challenges that must be addressed to ensure its successful deployment and clinical applicability. Key areas for development include improving the accuracy and intuitiveness of haptic interfaces, which necessitates a comprehensive understanding of human touch and interaction dynamics in medical contexts. As these technologies evolve, it is crucial to conduct rigorous testing to ensure the safety and reliability of haptic devices, particularly in high-stakes medical procedures. There is an increasing need to establish robust credentialing guidelines for healthcare institutions that incorporate advanced surgical robots into their clinical practices. This is essential to ensure that both the technology and its operators meet the requisite safety standards for patient care. The interplay between human-centered artificial intelligence (AI) and transparent machine learning (ML) also presents opportunities for innovation in haptic medical robotics, particularly when integrated with anatomical models that enhance training and operational efficacy (Kumar et al., 2021).

16.9 Conclusion

The significant direction is the enhancement of clinical validation processes. Current experimental setups often fail to replicate real clinical environments, which can lead to discrepancies between research outcomes and practical applications. To overcome this limitation, there is a call for open research practices where scientists share their methodologies and findings transparently. Such initiatives will contribute to the reproducibility and reliability of results, thereby bolstering confidence in haptic technologies within the medical fields. As haptic technology continues to expand its reach in healthcare, from medical education to rehabilitation, collaborative efforts among academic institutions, funding organizations, and scientific journals will be essential in addressing the challenges posed by this rapidly advancing field. By fostering interdisciplinary research and innovation, stakeholders can better harness the potential of haptic technology to improve patient outcomes and enhance the overall quality of medical care.

References

Ahirwar, R., & Khan, N. (2022). Smart wireless nanosensor systems for human healthcare. In *Nanosensors for futuristic smart and intelligent healthcare systems* (pp. 265–292). Informa UK Limited. https://doi.org/10.1201/9781003093534-15.

Banerjee, A., Chakraborty, C., & Rathi, M., Sr. (2020). Medical imaging, artificial intelligence, internet of things, wearable devices in terahertz healthcare technologies. In *Terahertz biomedical and healthcare technologies* (pp. 145–165). Elsevier. https://doi.org/10.1016/B978-0-12-818556-8.00008-2.

Barnes, R., & Zvarikova, K. (2021). Artificial intelligence-enabled wearable medical devices, clinical and diagnostic decision support systems, and internet of things-based healthcare applications in COVID-19 prevention, screening, and treatment. *American Journal of Medical Research, 8*, 9–22.

Chelliah, R., Wei, S., Daliri, E. B. M., Rubab, M., Elahi, F., Yeon, S. J., & Oh, D. H. (2021). Development of nanosensors based intelligent packaging systems: Food quality and medicine. *Nanomaterials, 11*, 1515.

Desai, D., & Shende, P. (2021). Integration of internet of things with quantum dots: A State-of-the-art of Medicine. *Current Pharmaceutical Design, 27*, 2068–2075.

Gaobotse, G., Mbunge, E., Batani, J., & Muchemwa, B. (2022). Non-invasive smart implants in healthcare: Redefining healthcare services delivery through sensors and emerging digital health technologies. *Sensors International, 3*, 100156.

Goel, S. S., Goel, A., Kumar, M., & Moltó, G. (2021). A review of internet of things: Qualifying technologies and boundless horizon. *Journal of Reliable Intelligent Environments, 7*, 23–33.

Han, B., Tomer, V., Nguyen, T. A., Farmani, A., & Singh, P. K. (2020). *Nanosensors for smart cities* (pp. 1–566). Elsevier. https://doi.org/10.1016/C2018-0-04422-9.

Han, X. (2023). A novel assimilated navigation model based on advanced optical systems (AOS), internet of things (IoT) and artificial intelligence (AI). *Optical and Quantum Electronics, 55*, 655.

He, L., & Eastburn, M. (2022). Smart nanosensor networks for body injury detection. In *2022 IEEE international conference on smart internet of things (SmartIoT)* (pp. 15–19). Suzhou: IEEE.

Huang, J., Wu, X., Huang, W., Wu, X., & Wang, S. (2021). Internet of things in health management systems: A review. *International Journal of Communication Systems, 34*, e4683.

Hurley, D., & Popescu, G. H. (2021). Medical big data and wearable internet of things healthcare systems in remotely monitoring and caring for confirmed or suspected COVID-19 patients. *American Journal of Medical Research, 8*, 78–90.

Jenkins, T. (2022). Wearable medical sensor devices, machine and deep learning algorithms, and internet of things-based healthcare systems in COVID-19 patient screening, diagnosis, monitoring, and treatment. *American Journal of Medical Research, 9*, 49–64.

Khan, T., Civas, M., Cetinkaya, O., Abbasi, N. A., & Akan, O. B. (2020). Nanosensor networks for smart health care. In *Nanosensors for smart cities* (pp. 387–403). Elsevier. https://doi.org/10.1016/B978-0-12-819870-4.00022-0.

Kumar, P. M., Hong, C. S., Afghah, F., Manogaran, G., Yu, K., Hua, Q., & Gao, J. (2021). Clouds proportionate medical data stream analytics for internet of things-based healthcare systems. *IEEE Journal of Biomedical and Health Informatics, 26*, 973–982.

Kumar, P. M., Khan, L. U., & Hong, C. S. (2022). Notice of violation of IEEE publication principles: Affirmative fusion process for improving wearable sensor data availability in artificial intelligence of medical things. *IEEE Sensors Journal*. https://doi.org/10.1109/JSEN.2022.3153410.

Kumar, S., & Kumari, P. (2021). Flexible nano smart sensors. In S. Thomas, T. A. Nguyen, M. Ahmadi, A. Farmani, & G. Yasin (Eds.), Nanosensors for smart manufacturing (pp. 199–230). Elsevier. doi:10.1016/B978-0-12-823358-0.00011-3.

Malik, A., & Kumar, A. (2022). Assimilation of blockchain with internet of things (IoT) with possible issues and solutions for better connectivity and proper security. In *New trends and applications in internet of things (IoT) and big data analytics* (pp. 187–207). Springer International Publishing.

Malik, P. K., Sharma, R., Singh, R., Gehlot, A., Satapathy, S. C., Alnumay, W. S., & Nayak, J. (2021). Industrial internet of things and its applications in industry 4.0: State of the art. *Computer Communications, 166*, 125–139.

Manogaran, G., Alazab, M., Song, H., & Kumar, N. (2021). CDP-UA: Cognitive data processing method wearable sensor data uncertainty analysis in the internet of things assisted smart medical healthcare systems. *IEEE Journal of Biomedical and Health Informatics, 25*, 3691–3699.

Materon, E. M., Gómez, F. R., Joshi, N., Dalmaschio, C. J., Carrilho, E., & Oliveira, O. N., Jr (2021). Smart materials for electrochemical flexible nanosensors: Advances and applications. In *Nanosensors for smart manufacturing* (pp. 347–371). https://doi.org/10.1016/B978-0-12-823358-0.00018-6.

Mujawar, M. A., Gohel, H., Bhardwaj, S. K., Srinivasan, S., Hickman, N., & Kaushik, A. (2020). Nano-enabled biosensing systems for intelligent healthcare: Towards COVID-19 management. *Materials Today Chemistry, 17*, 100306. https://doi.org/10.1016/j.mtchem.2020.100306.

Nazari, A. (2020). Nanosensors for smart cities: An introduction. In *Nanosensors for smart cities* (pp. 3–8). Elsevier. https://doi.org/10.1016/B978-0-12-819870-4.00001-3.

Oukhatar, A., Bakhouya, M., & El Ouadghiri, D. (2021). Electromagnetic-based wireless nano-sensors network: Architectures and applications. *Journal of Communications, 16*, 8.

Pradhan, M. R., Mago, B., & Ateeq, K. (2023). A classification-based sensor data processing method for the internet of things assimilated wearable sensor technology. *Cluster Computing, 26*(1), 807–822. https://doi.org/10.1007/s10586-022-03605-3.

Pramanik, P. K. D., Solanki, A., Debnath, A., Nayyar, A., El-Sappagh, S., & Kwak, K. S. (2020). Advancing modern healthcare with nanotechnology, nanobiosensors, and internet of nano things: Taxonomies, applications, architecture, and challenges. *IEEE Access, 8*, 65230–65266.

Raghav, A., Singh, B., Jermsittiparsert, K., Raghav, R., & Yadav, U. (2024). Artificial intelligence in environmental and climate changes: A sustainable future. In B. Singh, C. Kaunert, K. Vig, & S. Dutta (Eds.), *Maintaining a sustainable world in the nexus of environmental science and AI* (pp. 485–506). IGI Global. https://doi.org/10.4018/979-8-3693-6336-2.ch019.

Rani, P. (2022). Nanosensors and their potential role in internet of medical things. In *Nanosensors for futuristic smart and intelligent healthcare systems* (pp. 293–317). CRC Press.

Sahi, K. S. S., & Kaushik, S. (2022). Smart Nanosensors for healthcare monitoring and disease detection using AIoT framework. In *Nanosensors for futuristic smart and intelligent healthcare systems* (pp. 387–400). CRC Press.

Suma, D. V. (2021). Wearable IoT based distributed framework for ubiquitous computing. *Journal of Ubiquitous Computing and Communication Technologies, 3*, 23–32.

Udoh, E. E., Hermel, M., Bharmal, M. I., Nayak, A., Patel, S., Butlin, M., & Bhavnani, S. P. (2023). Nanosensor technologies and the digital transformation of healthcare. *Personalized Medicine, 20*, 251–269.

Vivekananthan, V., Khandelwal, G., Alluri, N. R., & Kim, S. J. (2022). E-skin for futuristic nanosensor technology for the healthcare system. In *Nanosensors for futuristic smart and intelligent healthcare systems* (pp. 133–157). CRC Press.

CHAPTER 17

Revolutionizing healthcare: The convergence of medical robotics and digital therapeutics

Mohammad Shahnawaz Shaikh[a,b]
[a]Department of AI and Big Data, Faculty of Engineering and Technology, Marwadi University, Rajkot, Gujarat, India
[b]Lincoln University College, Petaling Jaya, Selangor Darul Ehsan, Malaysia

17.1 Introduction

Artificial intelligence (AI) and robotics are an emerging trend in the field of modern healthcare, which is promoting changes in the future of medicine. Medical robotics and digital therapeutics are overturning conventional healthcare models, where precision-mediated surgery is only the starting point in applications of the technologies. The technological advances in sensor technology, machine learning (ML), imaging, and autonomous systems have made it possible to create robotic platforms beyond the capability of increasing the surgical precision, and improve patient outcomes, decrease the time required to recover, and increase access to care. The innovations are exceptionally vital in a global environment characterized by the growing need for high-quality medical interventions, the aging population, and inequality in healthcare opportunities (Farid et al., 2024).

Medical robotics has evolved from being restricted to helping in monotonous or restricted surgical procedures to assuming a pivotal role in other intricate procedures in neurosurgery, orthopedics, urology, and otolaryngology. The combination with an AI allows these systems to assist in decision-making in real-time, foresee complications, and customize the approach used in the surgery to the patient, respectively. Similarly, the remote care, data-driven interventions, and constant engagement of patients that digital therapeutics are already facilitating are most notably being applied in chronic disease management and posttraumatic rehabilitation (Dupont et al., 2021).

Albeit with huge potential, there are obstacles to the implementation of robotic and digital solutions, including high costs, lack of haptic feedback, complexity of the system, and ethical issues like autonomy, data privacy, and human responsibility of the clinician. We believe that this chapter will allow us to develop an in-depth understanding of the state of the art of medical robotics and digital therapeutics, including their technical underpinnings and current clinical applications, as well as case studies, hence

Haptic Feedback Control in Medical Robotics
DOI: https://doi.org/10.1016/B978-0-443-27777-1.00016-0

our intended contribution to this book. The innovations that are incorporated in the chapter to illustrate the sheer potential of the disruptive nature of the technologies are AI-assisted surgery systems, magnetic resonance imaging (MRI)-compatible robots, minimally invasive microbots, and neurological disorders rehabilitative platforms (Chen et al., 2024; Safahieh et al., 2024).

17.1.1 Motivation and objective

The rapid progress of medical robotics and AI has revolutionized surgical innovation, offering increased accuracy, reducing the patient's trauma, and improving the results of recovery. However, challenges such as high cost, technical boundaries, and moral concerns obstruct widespread adoption. This chapter is motivated to detect the ability to remove these obstacles and to detect the ability to integrate the latest techniques, such as AI, advanced sensors, and autonomous systems, to replace surgical practices and improve global access to high-quality care. The primary objective is to analyze the transformational effects of medical robotics, which focuses on innovations such as AI-operated systems, tele-robotics, and MRI-compatible robots. By evaluating their advantages, boundaries, and future capacity, the purpose of the study is to propose innovative solutions to increase surgical accuracy, remove challenges, and pave the way for safe, more efficient, and patient-focused healthcare.

17.2 Literature survey

The improvement in technology in computers and robotics has led to such medical procedures as tele-surgery being done at a distance. In 2001, a feat was made to illustrate that robotic surgery across the Atlantic Ocean is a possibility and that tele-robotics has possibilities that will transcend geographical boundaries (Farid et al., 2024). In 2008, the United States medical robotics market was estimated to be worth about 648 million dollars and was estimated to increase to around 1.5 billion dollars by the year 2014, with surgical robot systems taking up a significant portion of the market share (Gupta et al., 2023). Some benefits come along with robotic surgery, i.e., high precision, decreased risk, better depth perception, and the ability to carry out intricate procedures. Among the most outstanding advantages of telesurgery is the capacity to enable life-saving procedures in far-flung locations, which minimizes or limits the amount of delay in procedures and travel (Chen et al., 2024; Safahieh et al., 2024). These advantages notwithstanding, there is still limited acceptance of seeking medical help with the condition, even when it is misrepresented or distorted, most likely in the media or represented fictionally. To solve this problem, there is a need to raise more awareness, educate, and change the thinking of people toward medical tele-robotics and medical innovative thinking (Katsimperis et al., 2025; Shaikh et al., 2025a;Yang et al., 2024).

At the same time, the educational theories proposed the significance of experiential learning in which students develop knowledge by interacting with technological artifacts (Nelson et al., 2024). Learning environments that are based on robotics technology, such as modular robot toolkit applications, have been shown to be able to help students experience levels of engagement (Felizi et al., 2021; Malik & Brinjikji, 2022). These platforms also allow learners to design, construct, test, and debug robotic systems, and as a result, they help learners to gain experiential knowledge. Moreover, the means of learning involving physical action as part of the training process, such as kinesthetic or cinestatic strategy, have long been used as part of the medical training process and military training (Chen et al., 2023; Shaikh et al., 2025e; Wang et al., 2025). When combined with robotics, these techniques aid in enforcing the kinesthetic memory, contributing additional understanding to the concepts, and eventually making the learning process more engrossing and productive (Tariq et al., 2024).

It is possible to divide medical robotics depending on the role of the robot in executing the tasks, active or passive. The advent of minimally invasive surgery (MIS) has doubled the role and extent of medical robotics to a new level. MIS techniques have significant benefits over conventional surgery as they allow the use of small ports to implant instruments into the body, which causes less trauma to the patient and their postoperative complications (Van Kessel et al., 2023; Yoo et al., 2023). Problems that surround such procedures as poor visualization access and the complexity of the maneuvering of instruments are already being solved through computer-aided processes and robots. There is also future work on developing less invasive robotic surgery, even intrabodily ingestible robotic capsules that are capable of collecting physiological data via diagnostic activities autonomously and administering drugs to specific sites (Shaikh et al., 2025f). These trends represent the first evolution of the way medical conditions are diagnosed and treated, putting the focus on the harmony of robotics, automation, and intelligent therapeutics (Lantzsch et al., 2022).

17.3 Significant advances in medical robotics

Disruptive technological advancements in the field of medical robotics during the past few years have pushed medical robots into a new paradigm, as shown in Fig. 17.1. They even go so far as to improve procedural accuracy and use these developments to support innovative, patient-centered, and intelligent care models. AI and ML are improving the surgical process; for instance, they can assist with preoperative planning (e.g., 3D modeling from scans), real-time decision-making, and postoperative care (e.g., prediction of complications). Additionally, it uses natural language processing to improve surgical documentation. Emerging methods like explainable AI and federated learning are improving data security, trust, and training. Surgical robots equipped with haptic (touch) sensors and high-resolution images can currently give such safety and precision.

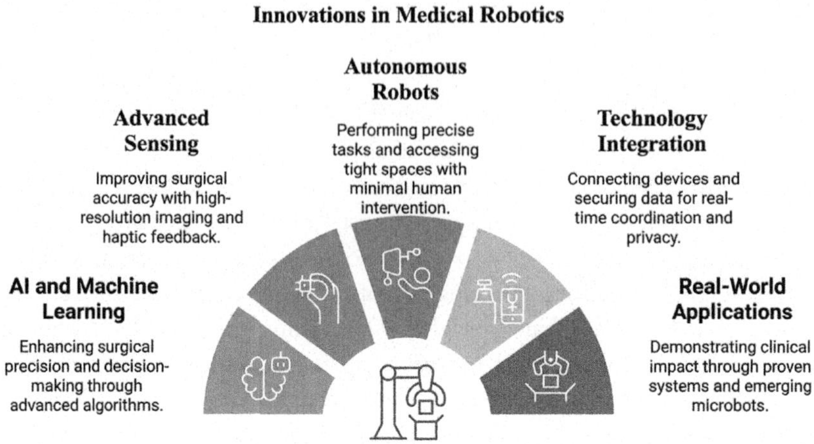

Figure 17.1 Emerging medical robotics innovations.

Augmented reality (AR) and virtual reality (VR) technologies are combined with real-time visual data (such as tumor boundaries) to project the visual data as real-time data during intricate procedures. When deployed as a robot, the smart tissue autonomous robot (STAR) can perform suturing procedures with a great degree of precision, sometimes even better than human surgeons. Semiautonomous systems modify tools or positions. Specialized robots, such as microbots, can perform minimally invasive surgeries in confined locations. Medical robots and other operating room equipment exchange real-time data over the Internet of Medical Things (IoMT). Blockchain technology is being utilized to improve privacy and integrity in the safe exchange of surgical data. Certain systems, like da Vinci and ROBODOC, have had a real therapeutic impact since they are more accurate and recover more quickly. Microbots are emerging for the aim of targeted medication delivery at the nanoscale, which presents new possibilities for noninvasive treatment (Ansari et al., 2024; Balasubramanyam et al., 2024; Gupta et al., 2023; Haig et al., 2023; Kumar & Shaikh, 2026; Long et al., 2022).

"Medical robotics" is developing quickly thanks to ground-breaking discoveries that are altering the surgical field. In addition to improving surgical procedures' precision and effectiveness, this advancement is allowing patients' treatment to completely embrace novel strategies. The most significant advancements in medical robotics are highlighted in Fig. 17.2, with an emphasis on advanced sensor technologies, autonomous systems, and AI (Acosta et al., 2022).

17.3.1 Artificial intelligence and machine learning

In medical robotics, "artificial intelligence (AI) and machine learning (ML)" are driving revolutionary advancements that are revolutionizing preoperative planning, intraoperative guidance, and postoperative care. An accurate 3D model can be created in prospective

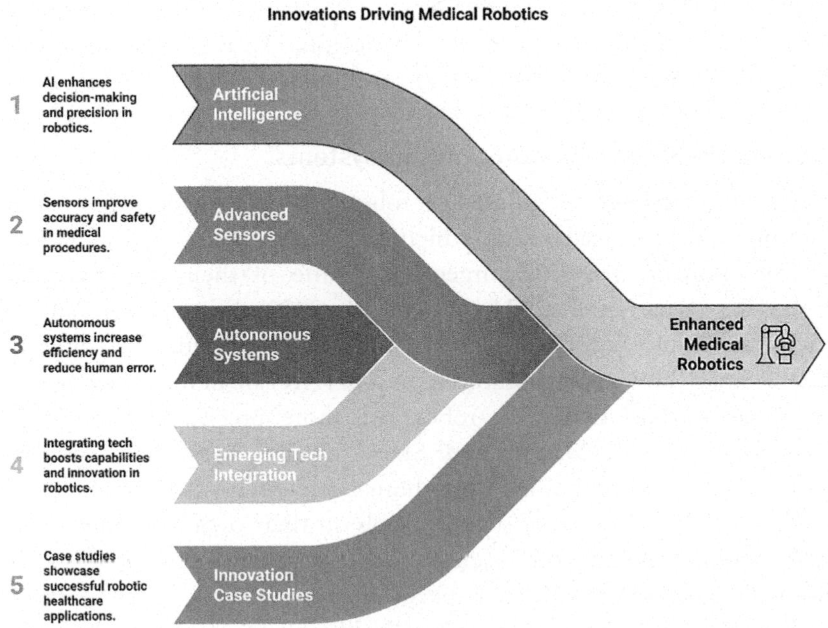

Figure 17.2 Innovations driving medical robotics.

planning by an AI system that analyzes patient-specific data, such as computed tomography (CT) and MRI scans, allowing surgeons to assess problems and adjust their strategy. AI predicts outcomes and identifies issues after surgery, allowing for prompt intervention that lowers readmission rates to the hospital and enhances recovery (Shaikh et al., 2025f; Van Kessel et al., 2023). To improve postoperative treatment, for instance, the AI model has been used to forecast acute renal damage and surgical site infection (Lantzsch et al., 2022). Natural language processing also lessens administrative load, increases accuracy, and automates surgical paperwork. AI-enhanced robot training, federated learning, and explaining AI are some of the developing ideas that promise to improve surgical robotics by enhancing training, guaranteeing data confidentiality, and enhancing transparency in AI judgments (Haig et al., 2023).

17.3.2 Advanced sensor technologies

Medical robots' incorporation of cutting-edge sensor technology has significantly improved surgical systems' capabilities, allowing for increased precision, security, and versatility. High-resolution cameras and imaging methods included in contemporary robotic systems give doctors a unique perspective of the surgery site. For instance, real-time tissue and organ reconstruction made possible by 3D mapping enables surgeons to accurately and easily navigate complex anatomy (Haig et al., 2023). Furthermore, even in minimally invasive procedures, surgeons may "feel" the roughness and resistance of tissues thanks to haptic feedback provided by embedded forces and pressure sensors in robotic weaponry. The robotic system that overlays critical information, like the tumor margin or blood flow pattern, also uses enhanced AR and VR technology (Long et al., 2022).

17.3.3 Autonomous and semiautonomous systems

One of the most intriguing areas of medical robotics is the development of autonomous and semiautonomous surgical robots, which can carry out particular tasks with little assistance from humans, relieving surgeons of some of their cognitive burden and enhancing results (Ansari et al., 2024; Gupta et al., 2023). For repetitive and extremely exact activities like suturing, tissue cutting, and implant placement, fully autonomous robots like the Smart Tissue Autonomous Robot (STAR) are being developed; preclinical experiments have shown that these robots are more precise than human surgeons (Balasubramanyam et al., 2024; Kumar & Shaikh, 2026). Conversely, semiautonomous systems work alongside surgeons by fusing human judgment with robotic accuracy. For example, they can alter postures or stabilize tools on their own depending on real-time feedback, freeing up surgeons to concentrate on crucial decision-making (Acosta et al., 2022). Minimally invasive operations are becoming more feasible because of task-specific robots, such as snake-like robots that can reach difficult-to-reach places like the sinuses or gastrointestinal tract, or microrobots that can navigate the bloodstream for targeted therapy (Liu et al., 2025).

17.3.4 Integration with emerging technologies

Increasingly, medical robotics is being combined with other state-of-the-art technologies to increase its impact through a synergistic effect (Desikan et al., 2025). Data interchange between robotic systems and other operating room devices is made possible via the "Internet of Medical Things (IoMT)" (Chen et al., 2026; Shaikh et al., 2025b). To improve accuracy and safety, robotic instruments, for instance, might interact with anesthetic machines or patient monitors to modify their actions in response to real-time patient vitals. To address important concerns about patient privacy and data integrity, blockchain technology is being investigated to guarantee the safe exchange of surgical data between robotic systems, hospitals, and research organizations (Desikan et al., 2025). Table 17.1 summarizes the clinical implications of medical robotic systems, exhibiting an overview, primary use cases, key advancements, and clinical significance of various advanced medical robotic systems.

17.4 Robotic-assisted surgery

Modern medicine now offers more options due to robotic-assisted surgery, which makes procedures that were previously thought to be extremely difficult or dangerous possible.

Table 17.1 Clinical implications of medical robotic systems.

Robotic system	Overview	Primary use cases	Key advancements	Clinical significance
Versa surgical platform (da Vinci's inspiration)s	One of the most advanced robotic systems is used in precision-guided surgery.	Minimally invasive procedures in general surgery, gynecology, and urology	Motion control and AI-assisted imaging improve surgical skill and cut down on operating room time.	Motion control and AI-assisted imaging improve surgical skill and cut down on operating room time.
The OrthoNav System, which is comparable to ROBODOC	Orthopedic robotic platform with specialized features for self-alignment and bone resurfacing.	Complete joint replacement, including knee and hip replacement	Uses self-governing planning algorithms to ensure accurate implant placement.	Improves postoperative functional outcomes and reduces the likelihood of complications.
Microrobots from MagnetoBot	External magnetic manipulation is used to maneuver micro-scale devices for internal navigation.	Drug delivery to specific targets and vascular treatments	Preclinical research validates minimal tissue disturbance and site-selective medication payload delivery.	Promises highly localized, noninvasive therapeutic uses in oncology.

The research findings and minimally invasive procedures, backed by case studies, as well as their applications in remote telesurgery, are outlined in Fig. 17.3 (Singh et al., 2025).

17.4.1 Surgical robotics

The goal of using robotics in surgery is to provide the surgeon with new, incredibly versatile tools that increase the range of patient treatment options, not to replace them. For the medical industry to use technology in a safe, moral, and responsible manner, health care workers must have the proper education and training (Gupta et al., 2023). There is ample objective and quantifiable proof that using robots in surgery offers clinical benefits over current manual methods. The expenditures of utilizing robotic technology must be justified by both technology suppliers and robotic systems end users (Kumar & Shaikh, 2026). Supporting, laboratory, rehabilitative, and surgical applications are the four primary types of robot applications in the medical profession that require a structural classification (Balasubramanyam et al., 2024). An appropriate framework was used to

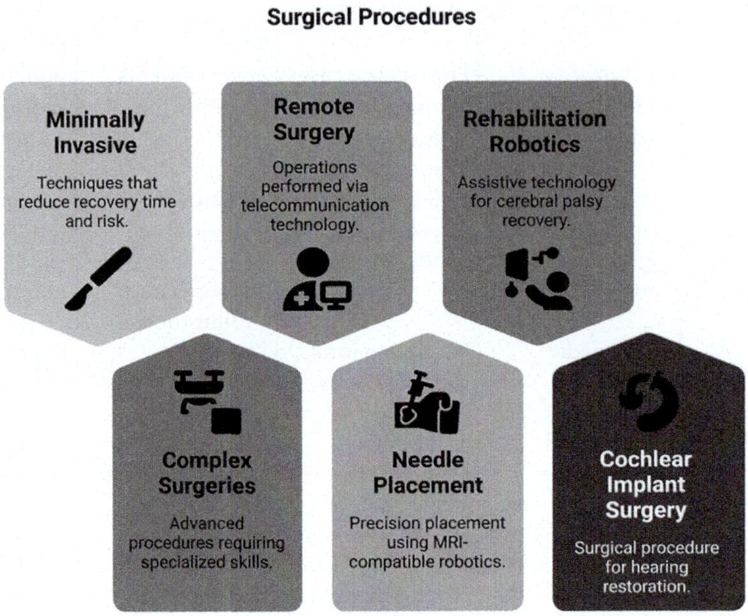

Figure 17.3 Robotic-assisted surgical procedure.

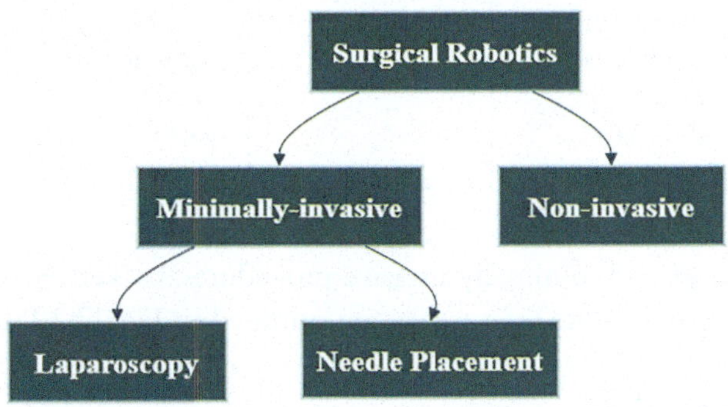

Figure 17.4 Surgical robotics framework.

describe the various medical specialties and surgical techniques where surgical robots can be employed, organizing our study as shown in Fig. 17.4. The surgical techniques were selected as drivers because they provide the foundation of our methodical review of the literature.

17.4.2 Minimally invasive procedures

Robotic technologies have improved accuracy and decreased patient trauma, revolutionizing minimally invasive surgeries like laparoscopic and endoscopic treatments. Surgeons

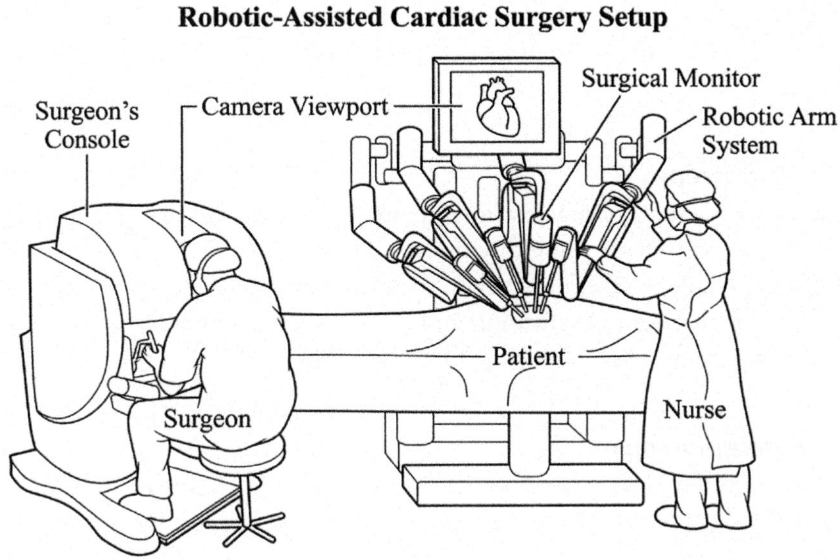

Figure 17.5 Robotic-assisted cardiac surgery setup.

can execute complex procedures with minimal harm to surrounding tissues thanks to devices like the da Vinci Surgical System. For example, studies have demonstrated substantial advantages such as decreased blood loss, shortened hospital stays, and quicker recovery times, making robotic-assisted laparoscopic cholecystectomy (gallbladder removal) the norm. Robotic-assisted laparoscopic procedures enhanced recovery times by 30% and decreased postoperative complications by 25%, according to a 2022 study published in Surgical Endoscopy. Both patients and doctors favor robotic-assisted minimally invasive surgery because of its many advantages, which include less trauma, smaller scars, a decreased risk of infection, and a faster return to regular activities (Ficuciello et al., 2019; Qian et al., 2019; Weerarathna et al., 2023). "Minimally invasive surgery (MIS)," a practical, economical surgical method with minimal consequences, is being used more and more for many surgical procedures. The surgeon performs minimally invasive surgery by making tiny skin incisions and introducing a laparoscope and specialized instruments into the patient's abdomen as shown in Fig. 17.5. This allows the surgeon to use video of the operating field to perform internal anatomy operations without creating large incisions (Balasubramanyam et al., 2024; Gupta et al., 2023; Kumar & Shaikh, 2026).

17.4.2.1 Noninvasive

Surgery is always changing, from open operating rooms to endoscopic methods that use robots to perform noninvasive surgeries. The next course of action is noninvasive surgery.

When a smart pill is swallowed, it gathers health information, diagnoses the condition, and, if required, releases the appropriate medication to treat it (Weerarathna et al., 2023).

17.4.2.2 Laparoscopy

During laparoscopic surgery, tiny instruments are placed into the abdominal wall incisions that have already been made using tiny trocars (Acosta et al., 2022). This kind of surgery shortens hospital stays, recuperation times, and the need for any kind of treatment restart. It also lowers the chances of transfusion and blood loss, postoperative pain, and intra and postoperative problems (Acosta et al., 2022).

17.4.2.3 Needle placement

Needles are inserted during one of the most frequent minimally invasive surgical procedures (Desikan et al., 2025; Liu et al., 2025). Its success depends on precisely aiming the needle point at a particular soft tissue target (Chen et al., 2026; Shaikh et al., 2025b). Soft tissue displacement brought on by the patient's respiration, postural changes, or tissue tensions employed during insertion may decrease success (Wang et al., 2025). Certain treatments demand precise needle insertion in the millimeter range, even in the micromillimeter range (Chen et al., 2026). Hand tremors can be reduced with robotic help, freeing up the surgeon to concentrate more on supervision and planning the procedure (Singh et al., 2025).

17.4.3 Complex and precision surgeries

Robotic devices surpass human constraints to accomplish remarkable results in intricate and precise surgery. Robots like the ROSA system offer submillimeter accuracy for delicate jobs like deep brain stimulation and epilepsy surgery in neurosurgery and spinal surgeries. Robotic-assisted spinal fusion operations had a 98% success rate, with shorter operating times and better patient outcomes, according to a 2021 study published in Neurosurgery. Systems such as CorPath GRX allow for accurate suturing and stent placement in cardiac and vascular procedures. A 2023 study published in the Journal of the American College of Cardiology demonstrated a 20% decrease in postoperative complications for robotic-assisted coronary artery bypass grafting. Robots like ROBODOC and MAKO guarantee accurate implant placement and alignment during orthopedic and reconstructive procedures. A 2022 study published in The Lancet showed that robotic-assisted knee replacements had a 15% longer implant lifespan and fewer revisions. These developments demonstrate how robots are revolutionizing precision, safety, and patient outcomes in a variety of intricate surgical domains (Antonelli et al., 2008; Qian et al., 2019; Ward et al., 2021).

17.4.4 Remote and tele-surgery

Patients in underserved or remote places can now receive remote surgery and consultations thanks to robotic technology that is removing geographical restrictions. Surgeons can operate on patients remotely thanks to tele-surgery, as demonstrated by the first transatlantic treatment carried out in 2001 with the Zeus robotic system. According to a 2023 study published in Nature Digital Medicine, 5G networks can lower latency to less than 100 ms, increasing the viability of real-time telesurgeries. Furthermore, robotic technologies provide remote consultation and real-time telepresence, enabling seasoned surgeons to mentor less experienced peers. As an illustration of the possibility of international cooperation, a robotic-assisted tele-surgery was successfully carried out in India in 2022 under the supervision of a surgeon in the United States. By increasing access to knowledge and guaranteeing top-notch care wherever one is, these developments are revolutionizing the healthcare industry (Ahmad et al., 2025; Shaikh et al., 2025c; Yang et al., 2020). Table 17.2 shows various surgical robotic systems along with their explanation and possible uses.

17.4.5 Needle placement by magnetic resonance imaging-compatible robotics

This section aims to introduce a patient-mounted, MRI-compatible robot for percutaneous diagnostic and interventional techniques, starting with arthrography as an initial application. The robot will provide stable needle guidance, automatic registration between MRI and patient coordinates, and reduce the needle passes, radiation risk, and the process will reduce time compared to traditional manual methods. Currently, arthrography, used to assess joint conditions, requires contrast injection after MRI imaging, which requires a two-phase workflow associated with fluoroscopy or ultrasound, which increases the patient's anxiety, radiation risk, cost, and disability. The proposed robot enables a single-step process within the MRI suit within the MRI suit, addressing these challenges while working effectively in the strong magnetic field of the MRI environment. Major stages include securing robots, obtaining preoperable images, registering the coordination system, determining the needle trajectory, and guiding the needle for manual insertion by the radiologist. This innovation is particularly important for pediatric care, where it is important to reduce radiation risk (Ahmad et al., 2025; Guo et al., 2025b; McDonald & Ranzani, 2021; Yang et al., 2020). Fig. 17.6 showing MRI compatible robot workflow.

17.4.6 Rehabilitation robotics for patients with cerebral palsy

A nonprogressive neurological condition called Cerebral palsy, caused by a developmental brain injury, is the most common lifelong physical disability in childhood, affecting around 300–500 children annually. Typically diagnosed by age 2, it often results in spasticity, motor skill delays, and issues with posture, gait, balance, and muscle tone. More than half of affected children experience gait disorders due to ankle and foot

Table 17.2 Types of surgical robot systems, their descriptions, and possible uses.

Types of Surgical Robot Systems	An Explanation and Possible Uses
Passive robots: not powered, mechanically fixed, static devices.	
Prerobots: Iron Intern (automated medical products), Omni-lapo tract (Omnitract), surgassistant (solos endoscopy), robotrac system (Aesculap), Bookwalter retraction system (Codman), trocar sleeve stabilizer (Richard Wolfe), first assistant (Leonard Medical)	The initial generation of passive robots, known as prerobots, is mostly employed as retractors or adjuncts to existing robotic systems.
Semiactive synergistic robots possess a certain amount of self-moving capability	
Active constraint robot: Acrobot (Imperial College, London), Steady-Hand robot (Johns Hopkins), LARS: Laparoscopic Assistant Robotic System (Johns Hopkins and IBM) NeuroMate (Integrated Surgical Systems)	Semiactive, four-DOF synergistic robot Only used in experimental surgery and for safe and accurate bone cutting in advance of hip resurfacing and knee replacement. Intended for a microcontroller to do stereotactic neurosurgery using images.
Automated endoscopic system for optimal positioning: AESOP 3000 (computer motion)	3-DOF camera-holding robot The first lap robot approved by the USFDA. An arm that is voice-activated, utilized in heart valve, neuro, ENT, and microsurgery
Master-slave robotic systems *Advanced robotics and telemanipulator system* for MIS: ARTEMIS (Karlsruhe Research Center) ZEUS (Computer Motion), da Vinci, and da Vinci S (Intuitive Surgical)	Master-slave manipulators for thoracic and abdominal MIS. For lap surgery, either urological or nonurological

malalignments, increasing the risk of falls, joint deformities, and early arthritis. While traditional therapies aim to improve gait through strength and range-of-motion exercises, they often fall short in addressing foot stability and muscle strength. Robotics has emerged as a promising adjunct to therapy, though the field remains in its early stages (Ali et al., 2024; Dupont et al., 2021; Goodrich & Schultz, 2007).

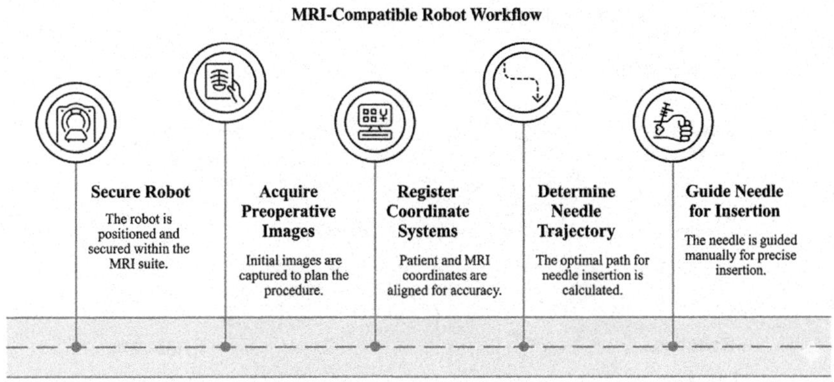

Figure 17.6 MRI-compatible robot workflow.

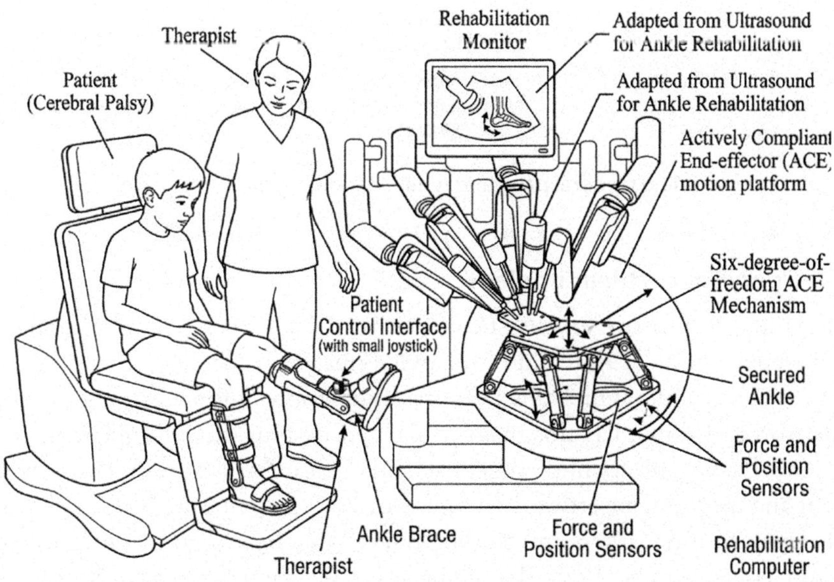

Figure 17.7 Rehabilitation robotics for cerebral palsy therapy.

The actively compliant end-effector motion platform, a six-degree-of-freedom device initially designed for ultrasound scanning, has been adapted for ankle rehabilitation in cerebral palsy patients as shown in Fig. 17.7. In an approved trial, the actively compliant end platform measures ankle range of motion and gathers data to enhance therapy, aiming to motivate patients and improve their mobility in future studies (Ali et al., 2024). Fig. 17.8 illustrates various components related to cerebral palsy and rehabilitation therapy.

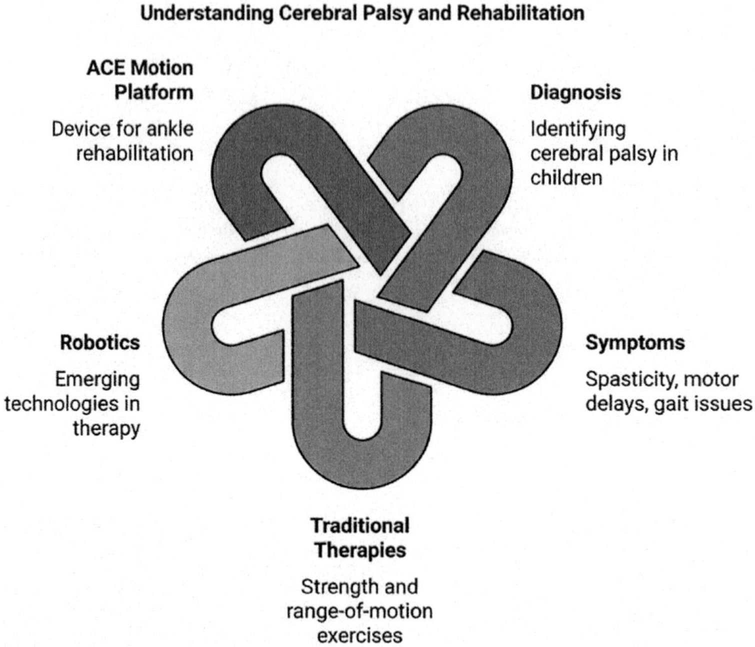

Figure 17.8 Understanding cerebral palsy and rehabilitation.

17.4.7 Surgery for cochlear implant

Cochlear implant (CI) surgery has been performed on over 320,000 people globally since its inception in the 1960s and is still a vital treatment for sensorineural hearing loss. The surgical technique, which involves mastoidectomy, tympanostomy, cochleostomy, and facial recess drilling, has not mainly changed despite the advancement of implant technologies. Due to the cochlea's small size and intricate structure, precise cochleostomy placement and electrode insertion are crucial for both patient outcomes and good device function. By mapping markers and guiding trajectories, image-guided technologies, such as CT and cone beam CT-based systems, have demonstrated promise in enhancing surgical precision. Additionally, new techniques such as optical tracking systems and stereotactic frames are improving implantation accuracy (Patil et al., 2011; Singh et al., 2023). Fig. 17.9 shows an anatomical illustration of CI surgery.

This study assessed the viability of utilizing cutting-edge imaging methods, such as augmented reality overlays of crucial anatomy, in conjunction with the "daVinci robotic system" for cochlear implantation. Successful bilateral cochlear implantation without damage to vital tissues was shown in two cadaveric case studies. Postoperative imaging verified that the electrodes were positioned correctly, and intraoperative image guidance and augmented reality correctly depicted the cochlea, round window, and facial nerve.

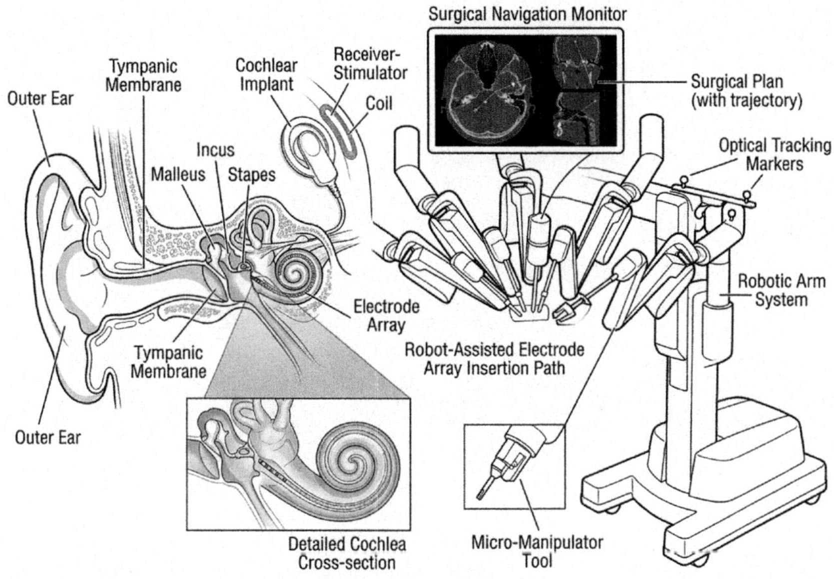

Figure 17.9 Anatomical illustration of cochlear implant surgery.

This method demonstrates how robotic and image-guided technologies can be used to increase the precision and safety of CI procedures (Fig. 17.10) .

17.4.8 Orthopedic surgery

The orthopedic field has been leading in the use of robots, especially in cases that demand high precision, like joint replacement, spinal correction, and fracture fixation. One of the most significant advancements in the field to date was the development of the Masor robotic system by the Israel Institute of Technology as an offshoot of the Miniature Robot for Surgery project. This small robotic, bone-mounted unit was a breakthrough in the paradigm of computer-assisted intervention in the orthopedic world, as shown in Fig. 17.11. The Masor is a semiautonomous surgical assistant, part of which can be remotely controlled and the rest can be operated locally by the surgeon. In contrast to the classical robotic arms that need substantial working zones, Masor can be applied to the patient's bone directly, which gives it the best stability and accessibility to the surgical field. Such closeness allows the accurate operation of instruments that drill bones, cut them, and put implants in them (Shaikh et al., 2025d; Osman et al., 2025).

Among the most outstanding effects of Masor, its dependency on image-based preoperative planning, which involves the assistance of computers, must be mentioned. The patient has a virtual three-dimensional model created by the computer processing of detailed imaging data obtained by a CT or MRI scan. The surgeons may then specify

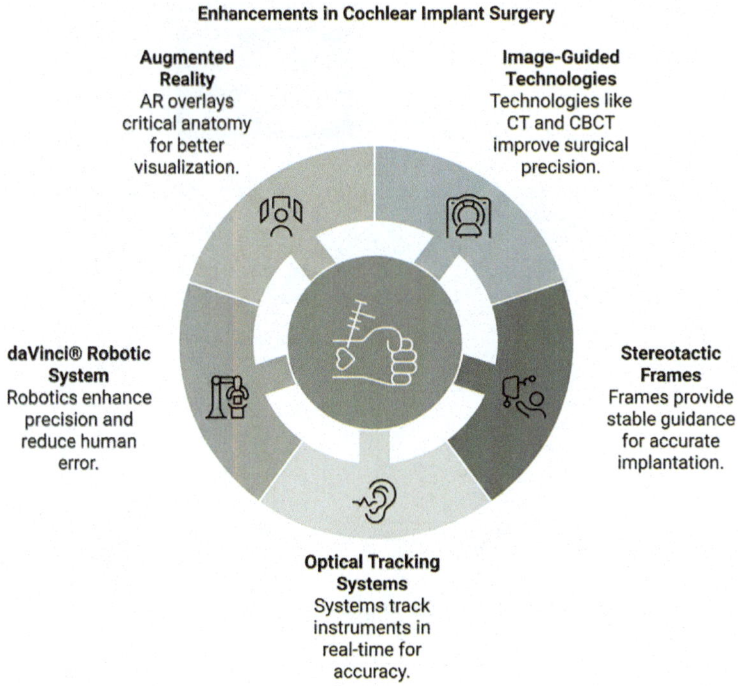

Figure 17.10 Enhancement in cochlear implant surgery.

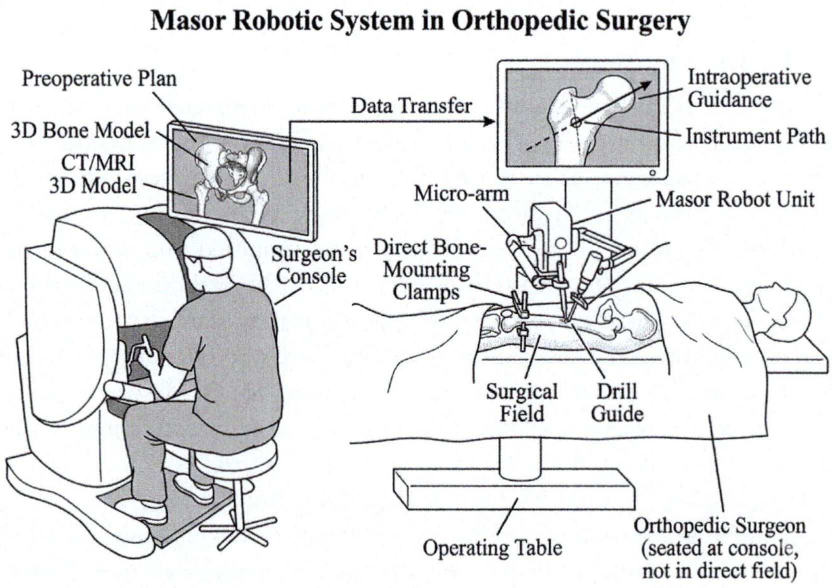

Figure 17.11 Masor robotic system in orthopedic surgery.

the best paths to follow and target points, and then the robot system would be able to limit the path of the instrument in a predetermined line. Eradicating the openness of manual processes, Masor increases the accuracy of the surgical incisions, reduces the disconnection of the soft tissues, and enhances the location of the implants, important determinants in the success of the long-term after-effects of orthopedic surgeries. Also, the robotic system considerably decreases the radiation exposure during the operation by minimizing the interval fluoroscopy, which also saves time in the operation process since the system allows performing more operations (Boutros et al., 2022; Guo et al., 2025a;Shaikh et al., 2025d).

With future development of robotic platforms in the orthopedics domain, the combination of options such as AI-based planning, intraoperative navigation, and real-time feedback systems appears to boost accuracy and safety as well. The example of Masor and other tools shows the potential of miniaturization and patient-specific robotic technologies to transform the future of orthopedic surgery by delivering more predictable, reproducible, process-oriented surgical outcomes (Yang et al., 2023).

17.4.8.1 Total knee arthroplasty

Retinal vein occlusion (RVO): The occlusion of retinal veins leads to the condition known as RVO. It is a major source of vision loss worldwide in both older adults and in people with comorbidities that involve the heart vessels. The disorder causes hemorrhages and macular edema, as well as ischemic retinal injury because of the inefficient drainage of the blood. The standard interventive approaches toward the management of RVO, in the form of intravitreal injections of antivascular endothelial growth factor agents or corticosteroids, are less aimed at curative periodic responses and exist more as a means of managing the symptoms on a case-to-case basis since the traffic of the concerned vasculature is very tiny and fragile to require a definite reaction. New developments in robot-assisted microsurgery are coming up as ways to counter these constraints. Robot-assisted retinal vein cannulation (RVC) is one of the more notable examples of such an experimental procedure, as it is a very accurate task that involves inserting a microneedle into the blocked vein to use clot-lotion agents and introduce them directly into the blockage. The method has great potential to recover the venous circulation and maintain retinal functionality, which could potentially give a curative approach, as opposed to palliative care (Boutros et al., 2022; Zhang et al., 2024).

The robotic systems used in RVC incorporate filters of tremor, motion scaling, and haptic feedback, and are thus capable of submicron resolution, which surpasses the human ability in manual dexterity by a large margin. Such systems enable the ophthalmic surgeons to work in the ultimate delicate surroundings of the posterior segment of the eye with little injury to the surrounding tissues. Moreover, the concept of using real-time intraoperative optical coherence tomography imaging improves navigation and positioning, so that the cannula is well positioned in the target vessel. Even though the method of RVC is at an experimental stage, the initial feasibility trials of robot-assisted RVC show that a safe, repeatable, and effective intervention is possible. Following the further development of robotic technologies, there may be an increase in the number of uses in the field of ophthalmic microsurgery; this will give rise to new ways of treating retinal diseases that were considered untreatable before because of conventional approaches (Shaikh et al., 2025d).

17.4.9 Neurosurgery

Neurosurgery is one of the most challenging branches of medicine; it needs very high accuracy, manipulation skills, and spatial orientation, as the brain and spinal cord contain complicated structures and are expressions of essential body functions. The bone cutting, cranial drilling, and spinal decompression are particular procedures that are remarkable in the treatment of brain cancer, spinal fracture, intracranial hemorrhage, and neurodegenerative disorders. These interventions are highly risky because most of the time, damage to the dura mater, nerve roots, or the adjacent neural tissue might arise and cause serious complications such as paralysis, loss of function, or even death (Zhang et al., 2024).

Robotic neurosurgery is becoming an inevitable tool to eliminate these risks and improve the outcomes of the surgery. New robotic generations with high-resolution 3D imaging, stereotactic navigational assistance, and indoctrination using AI can localize pathological targets together with mapping crucial brain structures in real time. Such systems have intelligent algorithms that help identify the mechanical anatomical limits, dynamically alter the tool path, and cut-off speed and pressure set according to tissue response. As an example, force sensors and adaptive control loops can be used to avoid overshoot as the tool nears a delicate tissue, so that tasks such as cortex incision, biopsy, or the insertion of an electrode can be done with submillimeter precision. In addition to this, robotics platforms improve skills, thus providing a surgeon with tremor filtration, increased dexterity within tight environments, and stable instrument positioning that is essential in deep-seated lesions or when the surgery is minimally invasive. There is also preoperative planning during which volumetric data from MRI or CT scans are planned and then incorporated during the intraoperative period to direct navigation and the alignment of the instruments (Henn et al., 2023; Yang et al., 2023).

These developments minimize intraoperative risks, decrease surgical times, and play a significant role in better patient outcomes, particularly in intricate surgeries like deep brain stimulation, spinal fixation, and brain tumor removal close to eloquent areas. With the further advancement of robotic neurosurgery, the incorporation of augmented reality, intraoperative imaging, and real-time AI analytics is likely to add to the ongoing revolution in terms of the safety, accuracy, and efficiency of interventions to the nervous system (Kulikov et al., 2023; Shaikh et al., 2025d). Fig. 17.12 illustrates that various advanced medical procedures can be treated with medical robotics.

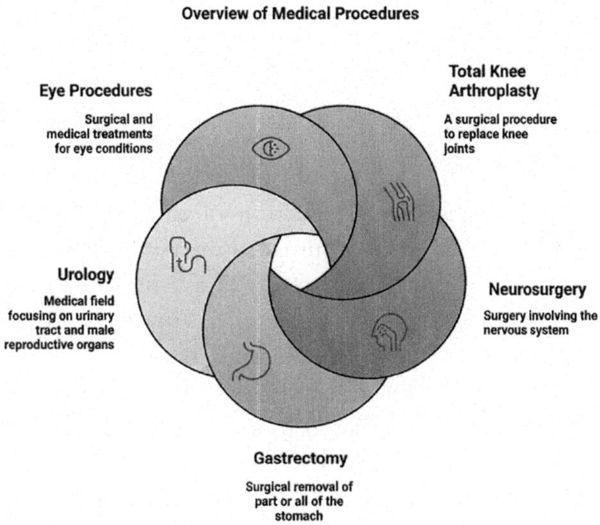

Figure 17.12 Overview of advanced medical procedures.

17.4.10 Gastrectomy

Da Vinci Surgical System has transformed the minimally invasive surgical arena with the combination of better optical control, 10x magnification, and a high-resolution 3D view of the field of surgery. These features also enable a surgeon to see even small anatomical details that are extremely essential during a surgical procedure involving extensive dissection of tissues and sharp sutures, like that of gastrectomy. Gastrectomy, especially in the context of oncological procedures, requires extensive accessing of various abdominal quadrants, namely the upper, middle, and lower quadrants. The rigid nature of the instrument arms and limited torque and motion of the traditional laparoscopic and earlier-generation robotic systems was cumbersome and, in many cases, demanded that the system or patient be broken and physically manipulated.

The most recent innovation in the da Vinci Xi System has been related to such limitations by introducing serpentine (flexible) and rotational robotic arms, allowing smooth and easy access regardless of the quadrant without repositioning. An overhead boom with dynamic docking and a significant improvement in instrument reach and flexibility of placement of ports results in a more comfortable and efficient surgical flow, as is also possible with the Xi platform. These developments have made more complicated and longer surgeries like total gastrectomy, lymphadenectomy, and multivise resections more precise and require less time for the operation. In sum, the introduction of multiquadrant accessibility, advanced optics, and smart robotic instrumentation to the Xi platform marks an enormous sea change in robotic-assisted gastrointestinal surgery, where robotic surgery makes surgery safer, more holistic, and patient-personalized (Dupont et al., 2021; Goodrich & Schultz, 2007).

17.4.11 Urology

One of the first examples of introducing robotic technology in the urological operating room was the TURP (Transurethral Resection of the Prostate) robotic system. The system was assembled with a six-axis Unimate Puma robot, which is integrated with an aspirator and an endoscopic liquidizer to create a high-tech robotic frame that is able to dissect precise tissue in the tight space of the urethra and prostate. The degrees of freedom permitted the six degrees of freedom to experience better maneuver and fine motor skills, which is important to minimize collateral damages to the surrounding tissues. Early clinical trials were conducted to show the clinical safety of robot-assisted TURP, its reproducibility, and efficacy, which was an added advantage in that it may lead to sufficient hemostasis during the process. These data reported fewer intraoperative hemorrhages and increased sight during the resection, the two parameters that are most significant when it comes to the safety and effectiveness of procedures existing in urology.

Subsequently, based on such early innovations, the first robot-assisted prostate biopsy system was developed, which can be considered one of the great achievements in the framework of diagnostic urology. The system assisted in automating the direction and placement of biopsy needles with submillimeter accuracy, which makes it possible to sample suspected cancerous areas based on preoperative imaging information. This kind of precision not only enhances the accuracy of diagnosis but also reduces complications and remedial procedures. These early systems had laid the groundwork toward the development of the next generation of urological robots and opened the door toward robot-assisted radical prostatectomy, image-guided interventions, and automated platforms of therapy. Continued development of robotic technology will allow an increase in the safety of the procedures, minimize patient morbidity, and improve long-term outcomes in urologic practice (Ali et al., 2024; Dupont et al., 2021).

17.4.12 Eye procedures

When the retinal veins become occluded, the resulting disorder is known as RVO. It is a major source of vision loss worldwide in both older adults and in people with comorbidities that involve the heart vessels. The disorder causes hemorrhages and macular edema, as well as ischemic retinal injury because of the inefficient drainage of the blood. The standard interceptive approaches toward the management of retinal vein occlusion (RVO), in the form of intravitreal injections of antivascular endothelial growth factor agents or corticosteroids, are less aimed at curative periodic responses and exist more as a means of managing the symptoms on a case-to-case basis since the traffic of the concerned vasculature is very tiny and fragile to require a definite reaction. New developments in

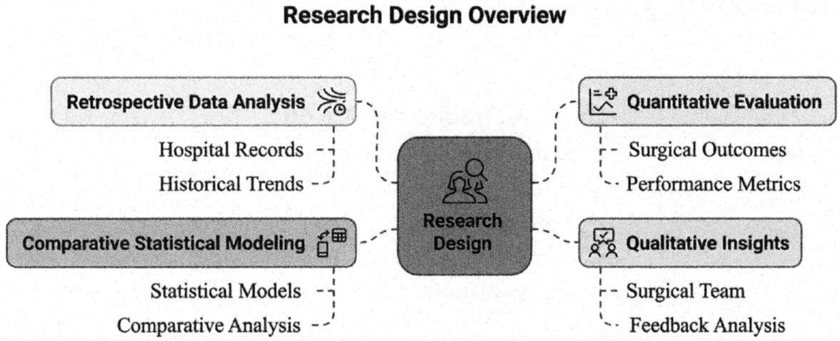

Figure 17.13 Research design overview.

robot-assisted microsurgery are coming up as ways to counter these constraints. Robot-assisted RVC is one of the more notable examples of such an experimental procedure, as it is a very accurate task that involves inserting a microneedle into the blocked vein, with the aim of using clot-lotion agents and introducing them directly into the blockage. The method has great potential to recover the venous circulation and maintain retinal functionality, which could potentially give a curative approach, as opposed to palliative care.

The robotic systems used in RVC incorporate filters of tremor, motion scaling, and haptic feedback, and are thus capable of sub-micron resolution, which surpasses the human ability in manual dexterity by a large margin. Such systems enable the ophthalmic surgeons to work in the ultimate delicate surroundings of the posterior segment of the eye with little injury to the surrounding tissues. Moreover, the concept of using real-time intraoperative optical coherence tomography imaging improves navigation and positioning, so that the cannula is well positioned in the target vessel. Despite the fact that the method of RVC is at an experimental stage, the initial feasibility trials of robot-assisted RVC show that a safe, repeatable, and effective intervention is possible. Following the further development of robotic technologies, there may be an increase in the number of uses in the field of ophthalmic microsurgery; this will give rise to new ways of treating retinal diseases that were considered untreatable before because of conventional approaches (Singh et al., 2023).

17.5 Methodology

17.5.1 Research design

The study methodology integrates observational techniques from qualitative and quantitative methodologies to assess the impact of AI-powered robotic surgery systems on surgical precision and patient recovery outcomes, as shown in Fig. 17.13.

Table 17.3 Selection of patient groups and their features.

Parameter	Group A (Traditional, n = 60)	Group B (Robotic-Assisted, n = 60)
Surgical procedures	Laparoscopic hysterectomy, radical prostatectomy, cholecystectomy	Same as Group A
Age range	35–70 y	35–70 y
Procedure type	Elective, nonemergency	Elective, nonemergency
Surgical history	No complex prior surgeries	No complex prior surgeries
ASA classification	Physical Status I-II	Physical Status I-II
ICU admission	Not admitted	Not admitted
Comorbidities (Exclusion)	No malignant conditions, no unstable vitals, no immunocompromised states	Same as Group A

Patients who underwent robotic-assisted surgery and those who received AI-enhanced robotic platforms were the two patient groups that formed the foundation for data analysis through retrospective evaluation using comparative statistical modeling. In addition to evaluating system operational excellence and operating room job execution during procedures, the method includes numerical evaluation of clinical indicators.

17.5.2 Patient selection and categorization

A tertiary care hospital's surgery data were used to choose 120 participants for the study, as summarized in Table 17.3. A total of 120 patients were treated in either Group A, which included 60 patients who underwent traditional laparoscopic surgery, or Group B, which included 60 patients who underwent robotic-assisted surgery using the da Vinci Xi Surgical System with AI characteristics. Enrollment is open to patients aged 35–70 who require a laparoscopic hysterectomy, cholecystectomy, or radical prostatectomy as a nonurgent operation. Patients with ASA Physical Status I or II who had no history of complex surgeries and no chronic systemic disorders were included in the chosen patient cohort. All study participants had to be free of immunosuppressive disorders, malignant or failing organs, or emergency department admissions.

17.5.3 Robotic and artificial intelligence system integration

Several AI modules were controlled by the da Vinci Xi robot-based surgery platform to offer real-time assistance with tool management and clinical judgments throughout procedures, as summarized in Table 17.4. Convolutional neural networks are used by the AI system to improve image segmentation and screen for anatomical landmarks. Bayesian models are used to evaluate intraoperative risk, and reinforcement learning algorithms

Table 17.4 Robotic-AI surgical platform details.

Component	Function
AI vision system (convolutional neural network-based)	Enhances real-time camera feed for anatomical landmark detection
Risk stratification engine (Bayesian model)	Estimates the probability of bleeding, infection, or nerve damage
Motion planning (RL)	Suggests optimal tool paths to minimize tissue trauma
AI-based haptic feedback emulator	Compensates for loss of tactile feel in robotic controls

are used to select the best tool path. Because robotic control systems inherently have performance limitations, the AI system and its haptic feedback emulator worked to simulate tactile feelings. When adjusting to anatomical or procedural changes in patients, AI components kept a collaborative system that helped doctors perform precise minimally invasive procedures. The da Vinci Xi Surgical System, which used the following AI components to function, was used to perform the robotic surgery.

17.5.4 Surgical procedures and workflow

Preoperative planning, intraoperative navigation, and postoperative monitoring were the three primary phases of the medical process. Patients first had preoperative 3D imaging using CT or MRI scans, which AI systems then converted into surgical simulation models and the identification of important risk areas. Surgeons were able to undertake adaptive path planning and intraoperative blood vessel and nerve cluster safety monitoring thanks to real-time imaging and AI-based predictions. Following surgery, surgeons used Internet of Things connections to assess the patient's vital signs, and AI algorithms identified possible dangers of infection or delayed wound healing. For both audit and instructional purposes, intraoperative performance metrics and video records were kept on audit systems (Ansari et al., 2024; Gupta et al., 2023; Kumar & Shaikh, 2026; Long et al., 2022). Fig. 17.14 shows a step-by-step surgical pipeline for robotic-assisted surgery.

17.5.5 Variables and data collection

The study made use of the hospital's electronic medical records and surgical logs as shown in Table 17.5. In addition to postoperative infection rates and bleeding, the study monitored the length of the operation, blood loss during the procedure, nerve injury occurrences, recovery durations, and length of hospital stay. The documentation contained feedback from the surgical team on the usability of the system as well as metrics pertaining to system efficiency, such as robotic calibration mistakes and AI-alert reaction rates. Before receiving institutional consent for the ethical processing of information, the obtained data were anonymized. Operational logs and hospital electronic medical record

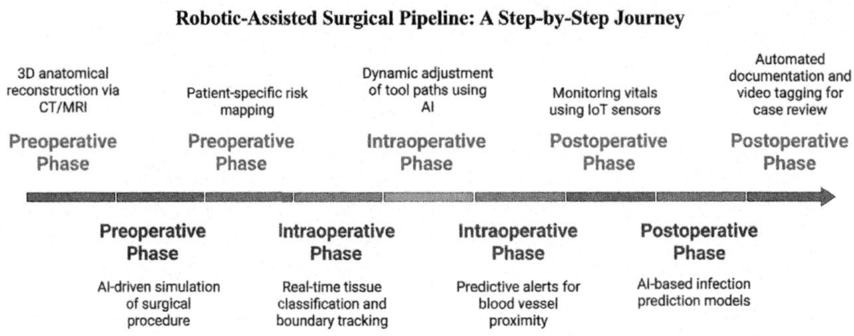

Figure 17.14 Robotic-assisted surgical pipeline.

Table 17.5 Measured variable for an important domain.

Domain	Measured variables
Surgical process	Duration, precision errors, and blood loss
Patient recovery	Hospital stay, wound healing, mobility
Safety	Complications (infection, bleeding, and nerve damage)
System efficacy	System uptime, tool calibration errors, AI-alert responses

systems provided information on the variables that were identified for this investigation (Acosta et al., 2022; Balasubramanyam et al., 2024; Liu et al., 2025).

17.5.6 Statistical analysis

The collected data was analyzed using statistical programs based on Python and SPSS version 25.0. Means, standard deviations, and range values for surgery and recovery variables were calculated as part of a descriptive statistical analysis. Because researchers looked at their continuous variables, the independent t-test showed differences between the two groups for both recuperation and operation time length. The researchers examined categorical outcomes, particularly complications, using chi-square statistical methods. By examining surgical parameters along with patient age, blood loss, and operation time, the study employed multivariate regression to investigate several aspects that influence quick recovery. With an R^2 value of 0.76 and a significant correlation at $P < .001$, the model's analysis results showed that AI-assisted surgery had a significant impact on favorable recovery outcomes. Fig. 17.15 illustrates various components related to the statistical analysis of surgical outcomes.

17.5.7 Validation and reliability

10-fold cross-validation was used in the regression modeling to evaluate its correctness and dependability. With a value of 0.87, Cronbach's Alpha validated the excellent reliability of the survey and feedback data that were gathered. After reviewing the methodology

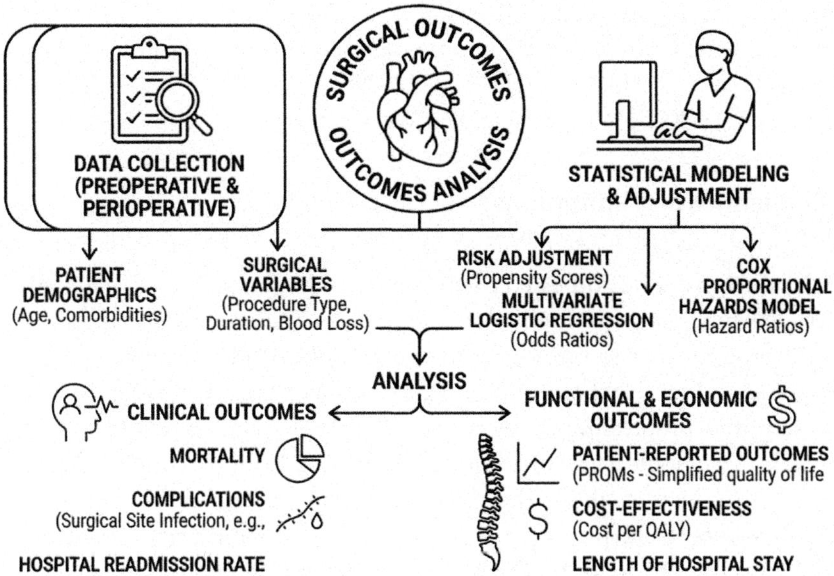

Figure 17.15 Surgical system statistical analysis.

and interpreting the research findings, three prominent surgical practitioners validated the study.

17.6 Difficulties and obstacles

A lot of challenges and restrictions need to be addressed before robotic-assisted surgery can become popular and successful. Access may be restricted by technological challenges such as the high cost and maintenance of robot systems, especially in settings with low resources.

Despite improvements in force sensors, the current technologies' limited touch responsiveness and sensory awareness also hinder surgeons from "feeling tissues." The integration of the robot system into the existing surgical workflows presents additional operational and logical obstacles. Concerns regarding patient safety, consent, and responsibility for errors are taken into account, making legal and ethical ideas equally important. Data privacy and security are significant issues that necessitate strong cybersecurity measures since robotic systems collect and transmit sensitive patient data. To provide surgeons with the skills they need to do their profession effectively, specific programs are required because education and training remain significant obstacles. Although simulation-based learning and VR platforms are emerging as helpful resources for learning new skills, efforts are still ongoing to bridge the gap between clinical practice and technological advances. These problems need to be fixed if we are to see the full potential of robot-assisted surgery to enhance patient outcomes and revolutionize healthcare (Ahmad et al., 2025; Antonelli et al., 2008; Ficuciello et al., 2019; Qian et al., 2019; Shaikh et al., 2025b, 2025c; Singh et al., 2025; Ward et al., 2021; Weerarathna et al., 2023).

17.6.1 Challenges and limitations

Robotic-assisted surgery leads to many challenges and boundaries, as summarized in Table 17.6, which must be addressed to ensure its widespread adoption and effectiveness. Technical challenges include high cost and maintenance of robot systems, which can limit access, especially in low-resource settings. Additionally, limited touch reaction in current systems and sensory perception obstruct surgeons from "feeling tissues," although the force sensor is progressing. Integrating the robot system into the existing surgical workflows also faces logical and operational difficulties. Ethical and legal ideas are equally important, with concerns about the patient's safety, consent, and liability informed in the event of errors. Data privacy and safety are also major issues, as robotic systems collect and transmit information of sensitive patients, which requires strong cybersecurity measures. Training and education remain significant obstacles, requiring special programs to equip surgeons with skills to operate effectively. Simulation-based learning and VR platforms are emerging as a valuable tool for skill acquisition, but decreasing the difference between technological progress and clinical practice remains a task in progress. Addressing these challenges is necessary to fully understand the ability of robot-assisted surgery in improving the patient's results and change healthcare.

17.7 Case studies of innovation in action

The "da Vinci Surgical System" remains a flagship example of robotic-assisted surgery, with over 10 million procedures performed globally. Its latest versions integrate AI and advanced imaging technologies to enhance precision and reduce operative times, solidifying its role in modern surgery. ROBODOC, an autonomous robotic system, is revolutionizing orthopedic surgeries by enabling precise bone cutting and implant placement, demonstrating improved accuracy and faster patient recovery compared with traditional methods. Meanwhile, researchers are pioneering magnetic microbots, which can be guided through the body using external magnetic fields. These microbots have shown promise in preclinical trials for targeted drug delivery to areas like tumors, minimizing collateral damage and offering a new frontier in minimally invasive treatments. Together, these innovations highlight the transformative potential of medical robotics in improving surgical outcomes and patient care. Table 17.7 exhibits a case study analysis of various advanced medical robotic systems incorporating their respective applications, key achievements, and relative impact.

Table 17.6 Summary of key insights and challenges in surgical robotics.

Category	Metric/Observation	Sources
Medical specialties	Supporting, laboratory, rehabilitative, and surgical applications	Wang et al. (2025)
	High adoption in minimally invasive surgery (MIS)	Ansari et al. (2024), Gupta et al. (2023), Kumar and Shaikh (2026), Balasubramanyam et al. (2024)
Surgical techniques	Laparoscopy: Shortened hospital stays, lower blood loss, reduced pain	Acosta et al. (2022)
	Needle placement: Accuracy requirement in the millimeter range	Chen et al. (2026), Shaikh et al. (2025b)
Patient benefits	Shorter hospital stays, fewer complications, reduced scarring	Yoo et al. (2023)
	Reduced blood loss, quicker recovery times	Acosta et al. (2022)
	Enhanced precision with robotic assistance	Singh et al. (2025)
Surgeon benefits	Ergonomic consoles reduce fatigue	Yoo et al. (2023)
	Haptic feedback allows for precise tissue manipulation	Yoo et al. (2023)
	Reduced hand tremor and improved focus on operation design	Singh et al. (2025)
System limitations	Design flaws, high costs, component failures	Van Kessel et al. (2023), Shaikh et al. (2025f), Lantzsch et al. (2022), Haig et al. (2023)
	Fixed axis of movement restricts the surgeon's freedom	Haig et al. (2023)
	Space and training requirements are significant	Chen et al. (2024)
Safety concerns	Human errors, software/hardware issues, and inaccurate instructions	Chen et al. (2024), Long et al. (2022)
	Risks include patient physical harm during close contact	Long et al. (2022)
Adoption trends	Increasing use of MIS procedures for economic and patient care benefits	Ansari et al. (2024), Gupta et al. (2023), Kumar and Shaikh (2026), Balasubramanyam et al. (2024)
	Needle placement techniques focus on precision and minimizing tissue displacement	Liu et al. (2025), Desikan et al. (2025), Chen et al. (2026), Shaikh et al. (2025b)

Table 17.7 Case study analysis: Pioneering robotic systems in healthcare.

System	Description	Applications	Key Achievements	Impact
da Vinci Surgical System	A flagship robotic-assisted surgical system with over 10 million procedures performed worldwide.	Minimally invasive surgeries (e.g., prostatectomies, hysterectomies)	Latest iterations incorporate AI and advanced imaging for enhanced precision and reduced operative times.	Improved surgical precision and patient outcomes.
		Complex procedures in urology, gynecology, and general surgery	Over 10 million procedures performed globally.	Reduced recovery times and hospital stays.
ROBODOC	An autonomous robotic system designed for orthopedic surgeries, specializing in precise bone cutting and implant placement.	Total hip and knee replacements	Showed improved accuracy in implant placement compared to traditional methods.	Faster recovery times and reduced postoperative complications.
		Orthopedic reconstructive surgeries	Autonomous operation reduces human error.	Enhanced patient mobility and long-term implant success.
Magnetic microbots	Microscale robots guided by external magnetic fields for targeted drug delivery and minimally invasive procedures.	Targeted drug delivery to tumors	Preclinical trials show successful navigation and drug delivery with minimal collateral damage.	Potential to revolutionize cancer treatment and reduce side effects of traditional therapies.
		Minimally invasive procedures (e.g., clearing arterial blockages)	Ability to navigate complex anatomical structures.	Enables precise, localized treatment with minimal invasiveness.

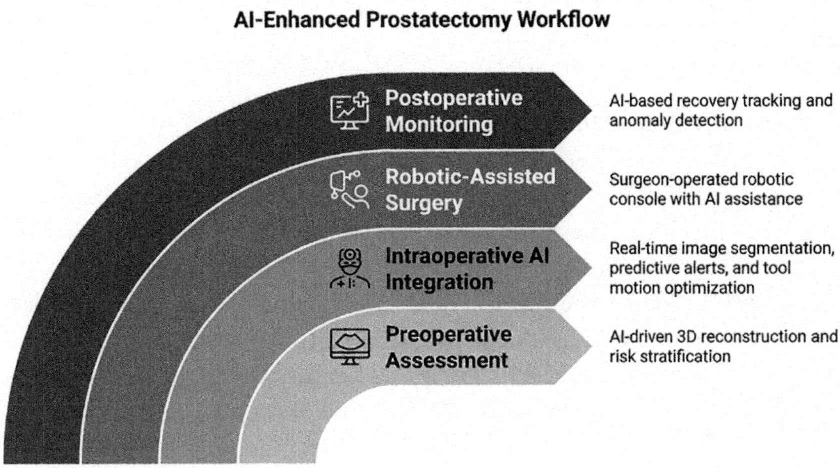

Figure 17.16 AI-assisted prostatectomy.

17.7.1 Case study: Robotic-assisted prostatectomy with artificial intelligence enhancement

The use of AI-modified robotic-assisted radical prostatectomy to treat a 62 -year-old patient with localized prostate cancer is examined in this study. The procedure was carried out by a da Vinci Xi robotic surgical system using its platform, which integrated AI-assisted equipment. The primary goal was to ascertain how AI in conjunction with robots improves surgical precision while lowering risks and accelerating patient recuperation (Dupont et al., 2021; Farid et al., 2024). Fig. 17.16 shows an AI-enhanced prostatectomy workflow.

17.7.1.1 Preoperative assessment

Due to his managed hypertension, the patient was classified as being in ASA Physical Status II with a Gleason score of 7(3 + 4) and a PSA level of 9.2 ng/mL. Both high-resolution pelvic MRI and contrast-enhanced CT imaging were included in a comprehensive preoperative imaging panel. These gathered datasets were processed by AI, enabling a sophisticated 3D reconstruction. A total of 90% of critical features, including the neurovascular bundle, bladder neck, and rectal wall, were accurately delineated by AI models trained on before surgical data. An AI-based risk stratification module was used to deliver this predictive analytics module, estimating a 5% chance of positive surgical margins, 8% chance of rectal damage, and 12% chance of intraoperative hemorrhage. The surgical team used these projections for preoperative planning, which led to the adoption of a nerve-sparing strategy that preserved functional outcomes without sacrificing oncologic safety (Chen et al., 2024; Dupont et al., 2021; Farid et al., 2024;

Katsimperis et al., 2025; Nelson et al., 2024; Safahieh et al., 2024; Shaikh et al., 2025a; Yang et al., 2024).

17.7.1.2 Intraoperative artificial intelligence integration

Unfortunately, AI has not been included in the robotic platform during the robotic-assisted radical prostatectomy (RARP) technique, so even if it offers situational awareness and operational precision, the robotic platform still functions as a stand-alone system. Among the main intraoperative AI features were. These alerts were given as anticipatory warnings about 1.5 s before the surgical instruments approached the major blood vessels in this instance, enabling prompt corrective action.

- Online image segmentation: AI algorithms identified critical anatomical structures on a line for their localization in the gland (prostate capsule, seminal vesicles, adjacent tissues).
- Reinforcement learning algorithms to optimize motion trajectories of the robotic arm during surgery to satisfy time and tissue strain trade-off, ergonomics, and surgeon fatigue. It took 2 h and 40 min for the treatment to begin to resemble a traditional laparoscopic prostatectomy, and the anticipated blood loss was 110 mL, significantly less than the 250 − 300 mL, typically involved in such a procedure (Chen et al., 2024; Dupont et al., 2021; Farid et al., 2024; Katsimperis et al., 2025; Nelson et al., 2024; Safahieh et al., 2024; Shaikh et al., 2025a; Yang et al., 2024).

Table 17.8 shows statistical results achieved from AI- augmented robotic-assisted prostatectomy, incorporating outcome metric, observed value, benchmark based on conventional RARP, and improvement percentage.

17.7.1.3 Clinical consequences

To improve the surgeon's situational awareness and ease surgical precision during the RARP, the AI capabilities were smoothly integrated onto the robotic platform. Among the main intraoperative AI features were:

- Online picture segmentation: AI algorithms identified key anatomical features on a line for the gland's location (prostate capsule, seminal vesicles, surrounding tissues).
- In this instance, these alerts were sent as anticipatory warnings around 1.5 s before the surgical instruments approached the major blood vessels, enabling prompt remedial intervention. To decrease tissue strain, the surgeon's labor, and the mechanical stress placed on his or her hands, the robotic arm trajectories were improved using reinforcement learning algorithms. It took 2 h and 40 min for the treatment to begin to resemble a traditional laparoscopic prostatectomy, and the anticipated blood loss was 110 mL, significantly less than the 250 − 300 mL typically involved in such a procedure (Chen et al., 2024; Dupont et al., 2021; Katsimperis et al., 2025; Nelson et al., 2024; Safahieh et al., 2024; Shaikh et al., 2025b; Yang et al., 2024).

Table 17.8 AI-Augmented robotic-assisted prostatectomy: Statistical results.

Outcome Metric	Observed Value	Benchmark (Conventional RARP)	Improvement (%)/Note
Operative time	2 h 40 min	3–3.5 h	~15%–25% reduction
Estimated blood loss (EBL)	110 mL	250 – 300 mL	~ 60% reduction
Intraoperative complications	None	5%–10% incidence	100% reduction in this case
Positive surgical margin	0%	10%–15%	Zero margin achieved
Hospital stay duration	3 d	4–5 d	~25%–40% reduction
Catheter removal	Day 7	Day 10–14	Accelerated by 3–7 d
Continence recovery (full control)	3 wk	6–8 wk	~ 50% faster recovery
Postoperative PSA level (1 mo)	< 0.02 ng/mL	< 0.1 ng/mL (target)	Exceeded oncological target
Pain Score (VAS)	10-Feb	4–5/10	50% reduction in reported pain
AI anatomical mapping accuracy	~ 90%	N/A (manual)	High-precision aid to planning
AI risk prediction accuracy	92% (based on outcomes)	N/A	High clinical reliability
Tool trajectory optimization (AI-driven)	Enabled	Manual	Reduced strain, increased precision

17.8 Conclusion and future work

Integrating medical robotics and digital therapeutics into the delivery system is radically changing the course of healthcare delivery using precision, potential risk reduction of procedures, and generalization of patients' treatment. The chapter has outlined the history of robotic-assisted interventions, starting with cases of less invasive surgery, da Vinci and ROBODOC, to the newest technologies in robotics, inventions with the assistance of AI in diagnostic disciplines, and the use of MRI-compatible robots and rehabilitative aids for neurological disorders. There are great clinical implications through these technologies, such as within surgical operations, shorter hospitalization, and quicker recovery of patients. Furthermore, AI implementation and improved sensor technology use have made it possible to conduct real-time decision-making, predictive analytics, and higher flexibility when it comes to performing complex medical procedures. However, in spite of these improvements, there are a number of barriers to mass adoption. High costs of implementation, technical intricacy, low levels of haptic feedback, privacy of data, and specialized training are still major drawbacks. In addition, the ethical and regulatory bodies also need to keep up with changes so that concerns of autonomy, responsibility, and consent in AI-mediated interventions can be reflected.

Going forward, the future of medical robotics and digital therapeutics sees a merging of intelligent, connected, and autonomous systems. It is the responsibility of a researcher to come up with more cost-effective and scalable robotic platforms that can be used in low-resource facilities. Improved haptics modes and virtual augmented reality interfaces will also help to close the gap between the human intuitive understanding and the robotic precise calculations. Moreover, explainable AI, federated learning, and blockchain-based data sharing schemes will be necessary to safeguard, guarantee transparency, and confidence in healthcare that is assisted by robots.

Collaboration between clinicians, engineers, as well as regulatory bodies also needs to be highlighted as an area of future work, as it is beneficial to standardize protocols, in addition to improving system interoperability, as well as education and training programs. These innovations are still in their infancy, but in the future, not only will they transform how clinical excellence is achieved but also democratize access to quality healthcare to make the vision of intelligent, patient-centered care a reality in the world.

References

Acosta, J. N., Jacobs, M. A., & Haspel, N. (2022). Multimodal biomedical AI. *Nature Medicine, 28*(12), 2389–2400. https://doi.org/10.1038/s41591-022-01981-2.

Ahmad, P. N., Ullah, I., M Aboelenein, N., Kumar Singh, S., Jiang, W., Al-Khasawneh. , M. A., & Daradkeh, Y. I. (2025). Deep neural network-based feature encoding for automated health monitoring using large AI models in online communication systems. *ACM Transactions on Internet of Things,* 7(1), 1–20.

Ali, S. I., Shaikh, M. S., Ponnusamy, S., & Chouhan, P. S. (2024). Technological collaboration, challenges, and unrestricted research in the digital twin: Digital twin technology. In S. Ponnusamy, M. Assaf, J. Antari, S. Singh, & S. Kalyanaraman (Eds.), *Harnessing AI and digital twin technologies in businesses* (pp. 380–399). IGI Global. https://doi.org/10.4018/979-8-3693-3234-4.ch028.

Ansari, S. T., Maik, V., Naheem, M., Ram, K., Lakshmanan, M., & Sivaprakasam, M. (2024). Hybrid image-guided navigation and robot-assisted spine surgery system. *arXiv preprint*, arXiv:2407.01578. https://arxiv.org/abs/2407.01578

Antonelli, G., Thor, F., & Dana, Y. (2008). *Handbook of robotics.* Springer.

Balasubramanyam, A., Ramesh, R., Sudheer, R., & Honnavalli, P. B. (2024). Revolutionizing healthcare: A review unveiling the transformative power of digital twins. *IEEE Access, 12*, 69652–69676. https://doi.org/10.1109/ACCESS.2024.3399744.

Boutros, C., Singh, V., Ocuin, L., Marks, J. M., & Hashimoto, D. A. (2022). Artificial intelligence in hepatopancreaticobiliary surgery: Promises and perils. *Artificial Intelligence Surgery, 2*, 213–223. https://doi.org/10.20517/ais.2022.32.

Chen, C., Zou, Q., Song, Y., Yu, M., Zhu, S., Song, S., & Li, X. (2024). Visual attention-based cognitive human-robot collaboration for pedicle screw placement in robot-assisted orthopedic surgery. *arXiv preprint.*

Chen, F., Chen, H., Yu, T., Wang, R., Wang, Y., Zhang, X., Li, J., Liu, K., Hai, D., Bao, X., Mo, Z., Yang, D., Wang, Z., Lin, Y., Xia, Q., Yang, G., & Shuai, J. (2026). AI-driven revolution of medical robotics across surgical innovation, rehabilitation intelligence, and multimodal healthcare delivery. *MedComm,* 7(3), e70597. https://doi.org/10.1002/mco2.70597.

Chen, X., Liu, C., Qiu, S., & Fu, J. (2023). Wearable sensor-based rehabilitation: Emerging solutions for post-stroke motor recovery. *Journal of NeuroEngineering and Rehabilitation, 20*(1), 57. https://doi.org/10.1186/s12984-023-01150-z.

Desikan, J., Singh, S. K., Jayanthiladevi, A., Bhushan, S., Rishiwal, V., & Kumar, M. (2025). Hybrid machine learning-based fault-tolerant sensor data fusion and anomaly detection for fire risk mitigation in IIoT environment. *Sensors, 25*(7), 2146.

Dupont, P. E., Nelson, B. J., Goldfarb, M., Hannaford, B., Menciassi, A., O'Malley, M. K., Simaan, N., Valdastri, P., & Yang, G.-Z. (2021). A decade retrospective of medical robotics research from 2010 to 2020. *Science Robotics, 6*, eabi8017.

Farid, Y., Azizi, M., Sefidgar, M., & Tavakkoli, M. (2024). Systematic review of medical robotic systems in reconstructive surgery. *Journal of Robotic Surgery, 18*, 88–97. https://doi.org/10.1007/s11701-024-01987-7.

Felizi, N., Tavakoli, M., Patel, R. V., & Atashzar, S. F. (2021). Robotics and AI for teleoperation, tele-assessment, and tele-training in the COVID-19 era. *Frontiers in Robotics and AI, 8*, 610677. https://doi.org/10.3389/frobt.2021.610677.

Ficuciello, F., Tamburrini, G., Arezzo, A., Villani, L., & Siciliano, B. (2019). Autonomy in surgical robots and its meaningful human control. *Paladyn Journal of Behavioral Robotics, 10*, 30–43. https://doi.org/10.1515/pjbr-2019-0002.

Goodrich, M., & Schultz, A. (2007). Human-robot interaction: A survey. *Foundations and Trends in Human-Computer Interaction, 1*, 203–275. https://doi.org/10.1561/1100000005.

Guo, C., He, Y., Shi, Z., & Wang, L. (2025a). Artificial intelligence in surgical medicine: A brief review. *Annals of Medicine & Surgery, 87*(4), 2180–2186. https://doi.org/10.1097/MS9.0000000000003115.

Guo, J., Liu, Y., & Leng, J. (2025b). Intelligent soft robots that couple soft sensing, actuation, computation, and power sources. *Matter, 8*(11), 102473. https://doi.org/10.1016/j.matt.2025.102473.

Gupta, A., Srivastava, D., Ghosh, A., & Mishra, V. (2023). Blockchain-enabled secure data transmission for surgical robots in IoMT environments. *IEEE Internet of Things Journal, 10*(4), 2103–2113. https://doi.org/10.1109/JIOT.2022.3199625.

Haig, M., Main, C., Chávez, D., & Kanavos, P. (2023). A value framework to assess patient-facing digital health technologies. *Value in Health, 26*(2), 198–207. https://doi.org/10.1016/j.jval.2022.11.006.

Henn, D., Trotsyuk, A. A., Barrera, J. A., Sivaraj, D., Chen, K., Mittal, S., Mermin-Bunnell, A. M., Chattopadhyay, A., Larson, M. R., Kinney, B. M., Nachbar, J., TerKonda, S. P., Reddy, S., Jeffers, L., Sacks, J. M., & Gurtner, G. C. (2023). Robotics in plastic surgery: It's here. *Plastic and Reconstructive Surgery, 152*(1), 239–249. https://doi.org/10.1097/PRS.0000000000010270.

Katsimperis, S., Tzelves, L., Feretzakis, G., Bellos, T., Tsikopoulos, I., Kostakopoulos, N., & Skolarikos, A. (2025). Innovations in robot-assisted surgery for genitourinary cancers: Emerging technologies and clinical applications. *Applied Sciences, 15*, 6118.

Kulikov, E. S., Fedorova, O. S., Tolmachev, I. V., Ryazantseva, U. V., Vrazhnov, D. A., Gubanov, A. V., Nesterovich, S. V., & Shmyrina, A. A. (2023). Russian-language repository of the open clinical data "SibMED Data Clinical Repository. *Bulletin of Siberian Medicine, 22*(2), 182–184. https://doi.org/10.20538/1682-0363-2023-2-182-184.

Kumar, B., & Shaikh, M. S. (2026). ZigBee based low latency IoT and AI integrated framework for real-time telehealth monitoring. *Emerging Science Journal, 10*(2), 1068–1083. https://doi.org/10.28991/ESJ-2026-010-02-024.

Lantzsch, H., Panteli, D., Martino, F., & Stephani, V. (2022). Benefit assessment and reimbursement of digital health applications. *Frontiers in Public Health, 10*, 933117. https://doi.org/10.3389/fpubh.2022.933117.

Liu, J., Wang, X., & Zhang, J. (2025). Investigating elderly individuals' acceptance of artificial intelligence (AI)-powered companion robots: The influence of individual characteristics. *Behavioral Sciences, 15*(5), 697. https://doi.org/10.3390/bs15050697.

Long, Y., Cao, J., Deguet, A., Taylor, R. H., & Dou, Q. (2022). Integrating artificial intelligence and augmented reality in robotic surgery using dVRK. *arXiv preprint*, arXiv:2201.00383. https://arxiv.org/abs/2201.00383

Malik, M. H., & Brinjikji, W. (2022). Feasibility of telesurgery in the modern era. *Neuroradiology Journal, 35*(5), 423–426. https://doi.org/10.1177/19714009221105005.

McDonald, K., & Ranzani, T. (2021). Hardware methods for onboard control of fluidically actuated soft robots. *Frontiers in Robotics and AI, 8*, 720702. https://doi.org/10.3389/frobt.2021.720702.

Nelson, B. J., Bendok, B. R., Turcotte, E. L., & Batjer, H. H. (2024). Remote magnetic navigation enables precision telesurgery. *Science Robotics, 9*(75), eado3187. https://doi.org/10.1126/scirobotics.ado3187.

Osman, E. I. A., Ismail, M. M. ElM. M, Mukhtar, M. A. H., Ahmed, A. U. B., Mohamed, N. A. A. E., & Ibrahim, A. A. A. (2025). Artificial intelligence and robotics in minimally invasive and complex surgical procedures: A systematic review. *Cureus, 17*(1), e81339. https://doi.org/10.7759/cureus.81339.

Patil, S., van den Berg, J., & Alterovitz, R. (2011). Motion planning under uncertainty in highly deformable environments. *Robotics: Science and Systems*. https://doi.org/10.15607/RSS.2011.VII.033.

Qian, L., Wu, J. Y., Maio, S., Navab, N., & Kazanzides, P. (2019). A review of augmented reality in robotic-assisted surgery. *IEEE Transactions on Medical Robotics and Bionics*, 1. https://doi.org/10.1109/TMRB.2019.2957061.

Safahieh, A. H., Nazemi, H., Arjmand, N., Azimi, P., & Khalaf, K. (2024). Advancements in robotic-assisted spine surgery. *Frontiers in Bioengineering and Biotechnology, 12*, 1416872.

Shaikh, M. S., Jain, N. K., Patel, N., Das, D., Mahesh, A., & Thakor, S. (2025a). Enhancing surgical precision and recovery: A review of medical robotics and AI integration in modern operating rooms. *EPJ Web of Conferences, 328*. Article 01016 https://doi.org/10.1051/epjconf/202532801016 .

Shaikh, M. S., Jain, N. K., Biswal, A., Patidar, P. K., Panwar, I., & Ali, S. I. (2025b). Sensors to insights revolution: Leveraging IoT, AI, and big data for next-gen patient monitoring. In *2025 12th international conference on emerging trends in engineering & technology - signal and information processing (ICETET-SIP)* (pp. 1–8). IEEE. https://doi.org/10.1109/ICETETSIP64213.2025.11156671.

Shaikh, M. S., Jain, N. K., Mungale, S., Das, R., Sarkar, B., & Wadekar, S. (2025c). Federated edge intelligence for privacy-preserving pre-eclampsia prediction in IoT-based maternal care. In *2025 12th international conference on emerging trends in engineering & technology - signal and information processing (ICETET-SIP)* (pp. 1–6). IEEE. https://doi.org/10.1109/ICETETSIP64213.2025.11156336.

Shaikh, M. S., Patidar, P. K., Vaghela, B. A., Pandwal, A. N., & Ali, S. I. (2025d). Advanced heart disease risk assessment using random forest algorithm: A comprehensive predictive model. *AIP Conference Proceedings, 3343*(1), 030007. https://doi.org/10.1063/5.0292473.

Shaikh, M. S., Sarkar, B., Nandi, P. K., Kumar, M., Patidar, P. K., & Wadekar, S. (2025e). The digital pulse: Reshaping clinical accuracy with sensor-based smart diagnostics. In *Proceedings of the 2025 world skills conference on universal data analytics and sciences (WorldSUAS)* (pp. 1–8). IEEE. https://doi.org/10.1109/WorldSUAS66815.2025.11199094.

Shaikh, M. S., Singh, Y., Patidar, P. K., Agrawal, N., & Vaghela, B. A. (2025f). The dawn of smart surgery: Medical robotics in the age of precision healthcare. In M. M. Ghonge, H. Liu, M. Khan, & T. A. Tran (Eds.), *Advances in emerging technologies and computing innovations. Sustainable artificial intelligence-powered applications*. Springer https://doi.org/10.1007/978-3-031-92854-3_25.

Singh, S. K., Virdee, B. S., Aggarwal, S., & Maroju, A. (2025). Incorporation of XAI and deep learning in biomedical imaging: A review. *Polytechnic Journal, 15*(1), 1–15.

Singh, S. K., Yang, L. T., & Park, J. H. (2023). FusionFedBlock: Fusion of blockchain and federated learning to preserve privacy in Industry 5.0. *Information Fusion, 90*, 233–240.

Tariq, A. R., Cao, J., Li, Q., & Zhen, X. (2024). Multimodal AI-enabled rehabilitation systems: A narrative review. *Sensors, 24*(2), 358. https://doi.org/10.3390/s24020358.

Van Kessel, R., Roman-Urrestarazu, A., Anderson, M., Kyriopoulos, I., & Field, S. (2023). Mapping factors affecting the uptake of digital therapeutics. *Journal of Medical Internet Research, 25*, e43310. https://doi.org/10.2196/43310.

Wang, Y., Han, X., Xin, B., & Zhao, P. (2025). Innovations in upper limb rehabilitation robots: A review of mechanisms, optimization, and clinical applications. *Robotics, 14*(6), 81. https://doi.org/10.3390/robotics14060081.

Ward, T. M., Mascagni, P., Madani, A., Padoy, N., Perretta, S., & Hashimoto, D. A. (2021). Surgical data science and artificial intelligence for surgical education. *Journal of Surgical Oncology, 124*(2), 221–230. https://doi.org/10.1002/jso.26496.

Weerarathna, N., Raymond, D., & Luharia, A. (2023). Human-robot collaboration for healthcare: A narrative review. *Cureus, 15*(11), e49210. https://doi.org/10.7759/cureus.49210.

Yang, G. Z., Kruger, J., Xu, X., & Fischer, P. (2024). Clinical applications of artificial intelligence in robotic surgery. *Journal of Robotic Surgery, 18*, 102–113.

Yang, G. Z., Nelson, B. J., Murphy, R. R., Choset, H., Christensen, H., Collins, S. H., Dario, P., Goldberg, K., Ikuta, K., Jacobstein, N., Kragic, D., Taylor, R. H., & McNutt, M. (2020). Combating COVID-19—The role of robotics in managing public health and infectious diseases. *Science Robotics, 5*(40), eabb5589. https://doi.org/10.1126/scirobotics.abb5589.

Yang, J. H., Goodman, E. D., Dawes, A. J., Goodman, E. D., Dawes, A. J., Gahagan, J. V., Esquivel, M. M., Liebert, C. A., Kin, C., Yeung, S., & Gurland, B. H. (2023). Using AI and computer vision to analyze technical proficiency in robotic surgery. *Surgical Endoscopy, 37*, 3010–3017. https://doi.org/10.1007/s00464-022-09781-y.

Yoo, J. H., Jeong, H., & Chung, T. M. (2023). Cutting-edge technologies for digital therapeutics: A review and architecture proposals. *Applied Sciences, 13*(12), 6929. https://doi.org/10.3390/app13126929.

Zhang, C., Hallbeck, M., & Salehinejad, H. (2024). The integration of artificial intelligence in robotic surgery: A narrative review. *Surgery, 176*, 552–557.

CHAPTER 18

Securing the future: Legal and privacy challenges in haptic medical systems for disabled children

Shashwata Sahu, Navonita Mallick and Hiranmaya Nanda
School of Law, KIIT Deemed to be University, Bhubaneswar, Odisha, India

18.1 Introduction

The quick rate of advancement of new medical technologies has provided new avenues for improving patient outcomes, especially for people with disabilities. The haptic medical system is remarkable for its distinctive capability to provide haptic feedback with real-time interaction between operators and medical devices. With increased usage in diagnosis, rehabilitation, and treatment therapies these days, haptic medical systems provide a more responsive and interactive experience of medicine (Haleem et al., 2022). Disabled Children, who tend to accumulate amplified disadvantage in getting persistent and versatile care, stand to gain through the important progress made by haptic technology by receiving interactive therapies that can be personalized to focus on their distinct physical and mental capabilities. In the digital health era, the nexus of disability and technology demands a delicate balance of risk and potential. The World Health Organization estimates that nearly 240 million children worldwide live with some form of disability, about 1 in 10 children, and some are out of reach of quality health care because of infrastructure, access, and adaptability constraints. The use of artificial intelligence (AI)-based haptic systems can potentially bridge these gaps. However, their use also raises challenging data privacy issues, consent, algorithmic prejudice, and access. For the child with disability who is most likely to have restricted ability to control and navigate online risks of exploitation, monitoring, and abuse of their data are in proportion.

While haptic health systems are becoming more mainstream in society, there remains an immense legal literature and policy void regarding directly addressing the singular vulnerabilities of disabled children in this technological landscape. Existing data protection law, as it evolves, tends to be impenetrable in being applied to advanced healthcare technologies involving AI, biometric monitoring, and remote monitoring (Habbal et al., 2024). Legislation like the General Data Protection Regulation (GDPR), the Children's Online Privacy Protection Act (COPPA), and India's Digital Personal Data Protection Act (DPDPA) puts in place crucial principles. However, it does not explicitly address

Haptic Feedback Control in Medical Robotics
DOI: https://doi.org/10.1016/B978-0-443-27777-1.00002-0

the stacked issues arising from pediatric haptic systems. India's Rights of Persons with Disabilities Act, 2016, is comprehensive but does not mention emerging digital health technologies. This research fills this regulatory and ethical deficit by investigating the privacy and legal issues of haptic medical technology for disabled children. It aims to fill the necessary gap in inclusive, rights-oriented, and future-oriented governance instruments that guarantee technological progress is not won at the cost of dignity, autonomy, and the protection of the most vulnerable.

18.2 Research objectives

The paper seeks to scrutinize the legal and privacy concerns that accompany haptic medical systems, especially their use in disabled children. The paper seeks to determine loopholes in existing legal systems and prescribe general solutions to bridge gaps in privacy threats and ethical concerns. The study additionally tries to address the intersectionality of international and domestic laws, i.e., GDPR, COPPA, DPDPA, and the Rights of Persons with Disabilities Act, 2016 (India), to facilitate a robust regulatory environment for such technology.

18.3 Research methodology

The research employs the doctrinal analysis of the legal statutes on data privacy, case laws, and international conventions on disability rights and medical technology. Global reports by organizations like UNICEF, WHO, and the United Nations are analyzed for empirical facts related to vulnerable aspects of disabled children in the online world. A comparison of privacy frameworks for nations like the European Union, the United States, and India is conducted to articulate the best practices and identify areas of substantial gaps.

18.4 Conceptual and theoretical framework

To scrutinize the legal and privacy issues of haptic medical systems within disabled children, the analysis should be placed against a robust conceptual and theoretical footing. The current chapter provides a foundation that becomes the interdisciplinary construct of the investigation, blending technological expertise, human rights rationality, theories of privacy, and bioethics (Chawki et al., 2024). Each subsection is accompanied by a critical analysis of how these frameworks intersect to inform on both the opportunities and challenges of deploying such new healthcare technologies.

18.5 Understanding haptic medical systems

Haptic medical systems are a form of assistive technology that allows users to perceive haptic feedback, simulating the sense of touch through vibrations, force feedback, or

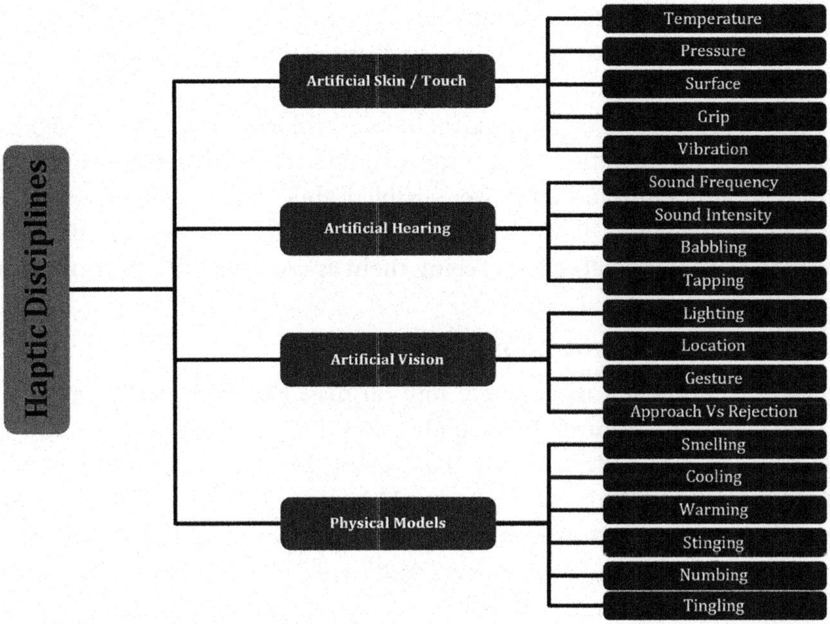

Figure 18.1 Domains of haptic disciplines in assistive technology.

motion control. They were initially created for virtual reality and robotics, and now they are increasingly entering healthcare, particularly in rehabilitative and therapeutic environments. During pediatric rehabilitation, haptic devices enable motor coordination, sensory stimulation, and therapy development in a lively and interactive atmosphere. In contrast to passive systems, haptic systems encourage children to work actively with them as care agents, improving cognitive and psychological outcomes during recovery (Kapoor et al., 2014). They usually consist of sensor integration, AI-driven algorithms, and live data capture capabilities. All these ensure personalized medical interventions depending on patient response. For disabled children with cerebral palsy, muscular dystrophy, or developmental delay, for example, these technologies can significantly augment physical capability and social interaction, provided they are integrated into ethically and legally possible systems (Fig. 18.1).

18.5.1 Human rights and disability models

The prevailing theoretical framework used to describe disability in legal scholarship today is the human rights model, articulated in the United Nations Convention on the Rights of Persons with Disabilities (UNCRPD). The model shifts the point of attention from impairment at the personal level toward institutional and societal obstacles to active engagement (Mladenov, 2013). In this view, disability is not caused by a medical condition but by exclusion from systems, environments, and services. UNCRPD articles 4, 9, and 25 collectively place on states' obligations to enable access to the right to assistive technologies, to obtain access to healthcare, and to uphold the dignity and autonomy of people with disabilities, including children (Lawson & Beckett, 2021). In haptic medical systems, the human rights model calls for developing technologies to maximize health gains and empower disabled children as right-holders with equal rights to new interventions. It also strengthens the call for accessibility, affordability, and adaptability in healthcare technology, seeing them as components of inclusive care.

18.5.2 Theories of data protection and consent

Digital health technologies increasingly rely on mass data processing, and privacy and consent are required for any ethical application. For children who cannot yet fully grasp the value of data collection, the theories of data protection must provide them with more protection. Both the European Union's GDPR and the United States' COPPA offer greater protection for children's data. The GDPR Article 8 proposes obtaining parental permission to process children's data until the age, and COPPA requires verifiable parent consent before gathering any data on children under the age of 13 (McGraw & Mandl, 2021). Such tools acknowledge children's growing capabilities and ensure consent mechanisms are transparent, accessible, and suitable. Besides these, more practical understandings of "privacy as contextual integrity" recognize that children's information should be collected and utilized only for purposes corresponding with their best possible well-being, and as consistently as possible with health care facility standards (Nissenbaum, 2011). Applied to haptic systems environments where information constantly flows on movement, behavior, and body response information, the concepts in each use take top priority as offering the actual protection that genuinely matters.

18.5.3 Bioethics and pediatric digital health

Bioethics provides a helpful framework to approach the use of haptic systems in the treatment of children, i.e., through the four traditional principles of autonomy, beneficence, nonmaleficence, and justice. Autonomy in disabled children can be balanced against the principle of capacity development, i.e., children should increasingly be engaged in decisions about health in proportion to increasing maturity and awareness. Beneficence and nonmaleficence require haptic systems to be developed to provide maximum beneficial therapy while avoiding physical, psychological, or digital harm (Varkey, 2021). For example, systems must not overstimulate, decrease dependency, and contain no manipulative design features. Justice issues are regarding fair access to such technologies, particularly to children from poor or rural backgrounds. In addition, developers and healthcare institutions must consider the ethical implications of algorithmic decision-making, data bias, and the commodification of children's biometric data (Vayena, 2003).

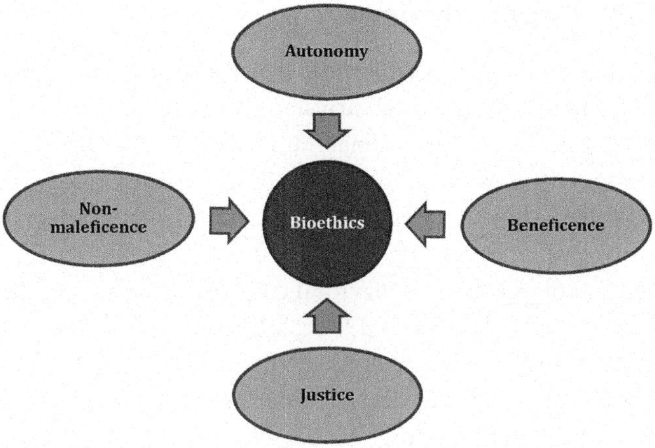

Figure 18.2 Core principles of bioethics in pediatric digital health.

A bioethical perspective guarantees that applying haptic medical systems is clinically effective, morally correct, and socially equitable (Fig. 18.2).

18.5.4 Justification for the interdisciplinary framework

Since issues regarding haptic medical systems are multifaceted problems, there can be no single-disciplinary perspective. A technical-only analysis may be oblivious to the agency and rights of disabled children, and a narrow legal analysis may be unable to grasp the real-time, data-dense technologies' ethical or pragmatic sense. Interdisciplinarity in this research brings technological ideas, human rights principles, data protection theory, and bioethical norms to address a wide understanding of the topic (Jafari et al., 2016). It provides for mindful balancing of opportunity and risk to accommodate disabled children's embodied experience, digital healthcare's flexibility, and the necessity of anticipatory regulation. Positioning the analysis here, the framework is intended to allow for developing inclusive, ethical, and legally sound solutions that benefit and empower the most vulnerable and stimulate innovation in pediatric rehabilitation.

18.5.4.1 Legal and policy landscape

The increasing use of haptic healthcare technologies for the treatment of children presents complicated legal issues, especially if one considers the vulnerable situation of the disabled child. They operate at the intersection of health, information, and digital innovation, each with overlapping but generally inadequate regulation regimes. The chapter establishes global and Indian legal instruments regulating haptic medical systems, followed by a comparative analysis of the area's crucial gaps in the law.

18.6 International legal instruments

Several international legal instruments provide general principles applicable to regulating haptic systems, particularly concerning privacy, children's rights, health standards, and disabilities. The most universal international framework for data protection is the GDPR, which has been adopted across the European Union since 2018. The GDPR treats children as a special data subject category and mandates parental consent in processing children's data below the age of 16 or otherwise as decided by national law. For haptic systems collecting physiological and behavioral data, transparency, data minimization, and purpose limitation, as defined in the GDPR, assume a foundational significance. The United Nations Convention on the Rights of the Child (UNCRC) secures each child the right to privacy (Article 16), health (Article 24), and safeguards from exploitation (Article 36). These apply when using AI-based haptic systems, with risks of digital profiling, biased algorithms, or data abuse. The COPPA of the United States places stringent demands on organizations handling children's details under 13 years (Richards & Hartzog, 2020). They mandate enforceable parental consent and institute transparency requirements for collection procedures. Although developed initially with internet-based settings in mind, its directives are increasingly configured for utilization on digital health equipment recording and processing children's data. The Health Insurance Portability and Accountability Act (HIPAA) regulates health data privacy in the United States. Even though HIPAA targets health providers and insurers, the HIPAA PHI requirements apply to pediatric healthcare technologies if and when they are enacted in clinical environments (Hoofnagle et al., 2019). The United Nations Convention on the Rights of Persons with Disabilities (UNCRPD) requires access to assistive technology on a nondiscriminatory basis. Article 25 assures equal access to healthcare, and Article 4 requests states to promote and develop technology that is accessible and, more particularly, encourages the use of haptic systems that were developed for disabled children.

18.6.1 Indian legal and policy framework

India's jurisdiction is arriving with changing health technologies, but is still decentralized in coverage and child-specificities. The Rights of Persons with Disabilities Act, 2016 (RPwD Act) is the enabling legislation safeguarding the rights of people with disabilities in India. Although it promotes access to medical treatment and encourages the utilization of assistive technologies (Section 25), it is not yet applicable to regulate new digital health technologies such as haptic systems or AI-rehabilitation. Digital Personal Data Protection Act (DPDPA), 2023, offers an overarching personal data protection regime. While being partly inspired by the GDPR, the DPDPA is noncommittal toward some provisions made about children with disabilities as well (Pujari, 2025). The DPDPA still mandates processing of children's personal data only based on their parents' verifiable consent; processing in opposition to children's interests is even prohibited under the

Table 18.1 Comparative legal frameworks for regulating haptic medical systems and child data privacy.

Jurisdiction	Key legal instruments	Child data protection provisions	Disability rights coverage	Technology-specific regulation
India	RPwD Act, 2016; DPDPA, 2023; IT Act, 2000	Parental consent under DPDPA lacks tech-specific child safeguards	Strong rights framework for healthcare access; lacks digital health provisions	No clear regulation for AI or haptic systems in child healthcare
European Union	GDPR; UNCRPD; UNCRC	Robust protections; GDPR mandates parental consent; rights-based child privacy focus	Comprehensive under UNCRPD; encourages assistive tech use	GDPR and the EU AI Act (proposed) apply indirectly to haptic and AI systems
United States	COPPA; HIPAA; ADA; UNCRC	Strict parental consent for data collection under COPPA; limited to commercial services	ADA ensures access to care; digital tech coverage is fragmented	Sectoral laws apply (health and online), but no haptic-specific regulatory framework

principle of "best interests of the child." The Information Technology Act, 2000, and rules for sensitive personal data or information (SPDI) do make some efforts toward addressing some of the privacy issues of digital health platforms. Still, the Act is not gifted with an end-to-end regulation regime for AI-based or sensor data-driven technology. The Clinical Establishments (Registration and Regulation) Act, 2010, has established minimum standards for healthcare facilities. While it standardizes treatment, no such extension is granted to software technology, wearables, or haptic devices utilized for home or remote care. India has also introduced initiatives like Ayushman Bharat Digital Mission (ABDM) to enable and enhance access to digital health records. However, no such initiatives yet that address child-centered haptic technologies specifically with protection or procedures.

18.6.2 Comparative legal matrix

The following is a comparative overview of legal provisions in India, the European Union, and the United States of America concerning haptic medical systems for disabled children (Table 18.1).

This comparative summary emphasizes the differential legal handling of haptic medical systems and related privacy rights of disabled children by courts. While the

European Union provides the most rights-based and harmonized model, supported by GDPR and UNCRPD, India's model, albeit transitioning, is uneven and lacks child-specific protection frameworks for new assistive technologies. The United States offers extensive privacy safeguards for children through COPPA and health data security by HIPAA, but not categorically for haptic rehabilitation tools. The table also reflects the necessity for harmonized, disability-informed, and child-informed legislation that adapts to the complexity of haptic systems.

18.6.3 Legal gaps in regulating haptic technologies for disabled children

Despite increasing global consensus on safeguarding children's rights in the virtual environment, current legal instruments inadequately address haptic medical systems, and there are wide regulatory loopholes. The loopholes are most visible in four general areas. First, it is a matter of inadequate definition and categorization. No legislation explicitly defines haptic systems; therefore, one does not know whether they are a medical device, an AI system, or consumer electronics. Lack of this explicit definition becomes extremely problematic if these systems are employed outside the institutional health environment, e.g., within the home or the school. Secondly, existing legislation, such as the GDPR and India's DPDPA, while recognizing increased vulnerability on the part of children, does not recognize disabled children who employ such technology for therapeutic reasons as being specially protected. Such children are exposed to specific and intersecting risks requiring special legal consideration. Third, current consent procedures are too simplistic and based mainly on self-consent or parental consent and do not account for children's developing capacities or promote group decision-making among children, guardians, and health care providers. In ever-changing haptic environments, static consent is inadequate, and dynamic, continuous consent procedures must be employed (Team, 2024). Fourth, there is virtually no institutional monitoring; no overseeing agency exists to monitor the ethical design, compliance with privacy, or outcome assurance of child-directed haptic systems. Without regulation, the potential exists for uncontrolled development and deployment of sensitive technology with weak controls or accountability mechanisms (Przegalinska, 2019). These issues need to be addressed by legal acknowledgement of haptic systems as a distinct field of the broader area of medical and assistive technologies, with enforceable technical and ethical guidelines, periodic audits, and real stakeholder engagement with groups like disability rights activists, child protection experts, and pediatric medical professionals.

18.6.3.1 Privacy and ethical challenges in haptic medical systems

Integration of haptic medical systems into the disabled child's care comprises a subtle meshwork of governance, ethical, and privacy concerns. The systems rely on ongoing interaction of data between the user and the device, generating enormous amounts of physiological, behavioral, and biometric data. While clinical benefits are vast, legal and ethical concerns, most critically to children, are a priority. This chapter deconstructs the most serious threats regarding data processing, ownership, AI decision-making, security, and governance from a child's point of view.

18.7 Data collection, storage, and consent issues

Haptic devices employed during the rehabilitation of children are heavily reliant on acquiring real-time information. Sensors from gloves, exoskeletons, or wearable sensors capture accurate data such as muscle tension, joint movement, heart rate, and response patterns. The data sets are critical to tuning therapeutic intervention, but ongoing generation and storage raise fundamental issues of consent, data retention period, and extent of secondary use (De Fazio et al., 2023). In the children's world, sound consent is complicated. Children cannot possibly comprehend what data is collected or what it is utilized for, and guardians, in turn, usually do not have the technical expertise to assess privacy risks. Moreover, current paradigms of consent are static and one time usage, which is less than ideal for haptic systems that dynamically adapt through machine learning and adaptive algorithms. Dynamic and real-time consent frameworks, where users or guardians can modify permissions in real time, are typically unused, leaving a governance gap (Dankar et al., 2019). Also, poor data minimization practices are prone to overcollection, with sensitive data stored in cloud storage indefinitely without retention schedules. Such practices violate core privacy principles under GDPR, DPDPA, and other child data protection regimes, exposing users to abuse and unauthorized use.

18.7.1 Ownership, profiling, and discriminatory risk

Ownership of the information generated by haptic systems is also an ongoing issue in all legal systems. In reality, data belongs to the service company or equipment maker, and little control or access is given to end-users and their relatives. Such a one-sided setup generates ethical issues, mainly when the information is utilized to build behavioral patterns or prediction algorithms without explicit approval or user notification. This kind of profiling, especially in children with disabilities, may inadvertently contribute to discrimination (Ramones & del-Rio-Guerra, 2023). Predictive models, for example, may incorrectly label delays in development or physical impairments as low responsiveness and issue lower or unsuitable therapeutic suggestions. When this is subsequently disclosed to third parties (e.g., insurers or schools), the consequences can impact access to services or opportunities and contribute to social exclusion. Discriminatory profiling also occurs in the scenario when AI systems operating on normative data do not account for the extremely high level of neurodiversity and bodily variation in disabled children. Such biases will often be latent and difficult to find, but could result in extreme long-term damage in automated decision-making where human correction is low.

18.7.2 Artificial intelligence bias, algorithmic opacity, and misuse in pediatric contexts

Most haptic systems incorporate AI and machine learning to customize exercises, recommend routines, or monitor progress. However, the decision-making algorithm behind them is usually transparent, a state with which most of us are acquainted as the "black box" problem. When decisions affect vulnerable groups like disabled children, transparency becomes an ethical and legal obligation. AI systems can add biases into training data that reflect adult motor patterns or nondisabled baselines, creating inaccurate measurements or inappropriately challenging tasks (Felin & Holweg, 2024). These algorithm misalignments reduce clinical performance and can hurt the child's confidence level, cause frustration, or develop physical exhaustion. Additionally, the use of AI in pediatrics also carries the threat of misuse. For example, AI models can be used for surveillance, behavior alteration, or experimentation without a proper ethical check. These demand the integration of explainability and accountability in algorithmic systems implemented in pediatric settings. In nations where AI legislation is unenforceable, like India, the absence of mandatory algorithm audits merely adds to such dangers (Bhargava et al., 2024). While the EU draft AI Act has provisions for high-risk systems like medical devices, haptic rehabilitation systems may remain in regulatory gray areas.

18.7.3 Security threats and cross-border data transfers

Haptic system threats are greater than unauthorized access or data breaches. Because sensitive health and behavioral information is contained, compromised haptic systems can put children at risk of several levels of harm, including identity theft and fraud, and emotional harm. Haptic devices usually function based on third-party cloud facilities and remote servers, most likely outside the user's domicile jurisdiction. That creates transnational data flows that are hard to subject to proper privacy enforcement. Data localization law is ad hoc or patchy, and therefore, it is hard for interested users to have redress or enforce domestic laws. Further, security on most devices is very weak (Hine et al., 2024). Encryption protocols are highly inconsistent, firmware updates are not necessarily on time, and device authentication controls cannot prevent intrusions. Children are especially at risk in this environment because they might be unable to identify phishing attacks, malicious program notifications, or unsolicited user interface changes. The absence of harmonized worldwide frameworks for ensuring cross-border movement of information in child health technologies is a governance gap of utmost priority (Li & Liu, 2021). It is because of this fact that bilateral data exchange agreements and harmonized cybersecurity standards for child health systems are imperative.

18.7.4 Ethical design failures and the absence of child-centric oversight

One of the strongest issues in haptic medical system design and deployment is the lack of ethical-by-design solutions for Children with disabilities. Though technologists and engineers design an enormous quantity of systems, they don't sufficiently involve pediatric therapists, child psychologists, or disability rights activists within their work. The outcome is that important considerations like the stimulation of feelings, sensitivity to culture, and usability are greatly neglected (Crandall & Karadoğan, 2021). Furthermore, boards of today's regulatory agencies are seldom composed of pediatric technology experts or members of child protective organizations. Ethical review committees that manage clinical trials of medical devices also cannot review digital therapeutics or AI systems, particularly when applied outside institutional hospital settings. A lack of independent regulation has enabled the market to grow without strong quality, safety, or ethical compliance checks. No specific agency exists in most countries, including India, that controls the design and use of haptic systems for child rehabilitation. Even if general health technology standards exist, these usually do not require child-specific risk analysis or usability testing on handicapped individuals (Pantanowitz et al., 2024). Closing these gaps will necessitate institutional transformation, such as the creation of interdisciplinary ethics committees, participatory design processes, and regulatory sandboxes with a pediatric digital health focus. These interventions will be needed to avoid institutionalizing unsafe or exclusionary technologies in child care environments.

18.7.5 Global insights and case illustrations

Evaluation of haptic medical systems' pragmatic implications for children with disabilities entails weighing international policy recommendations against execution at a jurisdictional level. International organizations such as UNICEF, WHO, and the United Nations stressed the paramount role of sustaining digital health innovation on a child rights and disability inclusion track. Concurrently, regional analyses of the EU, US, and India provide salient lessons about legal compliance, data management, and implementation matters. The chapter summarizes international sources' pertinent findings and displays contextual examples of haptic and AI-based healthcare tools implemented across jurisdictions.

18.7.6 Insights from UNICEF, WHO, and UN reports

UNICEF and WHO have been longstanding champions of child-centered approaches to digital innovation, particularly in health. In the 2021 UNICEF report, "Children and Digital Health," it is highlighted that although digital technologies hold the promise to shape access and quality of health, children's rights to privacy, protection, and participation are most commonly left behind in the process of developing and rolling out new technology (Cooray et al., 2021). The report highlights that data-intensive technologies, sensor-based technologies, and wearable devices are typically constructed in impermeable language and with low child-friendly protection features. The World Health Organization, in its 2021 Global Strategy on Digital Health, recommends that digital health products be inclusive, equitable, and accessible to marginalized groups such as disabled children. WHO also stresses incorporating digital literacy as inclusive and policy-compliant with international human rights instruments (Xie et al., 2025). The United Nations, for its part, has also argued that assistive technologies must be designed with the involvement of people with disabilities and their communities to avoid excluding either or tokenizing. All of these voices collectively argue that haptic medical systems for disabled children must be rights-based in their design, not as an afterthought.

18.7.7 Case study: GDPR compliance in pediatric haptic rehab (EU)

In the European Union, with the GDPR providing a robust legal framework, haptic systems have been implemented by some child rehabilitation centers to respond to strict data protection laws. A pilot in Germany, for instance, had AI-driven haptic gloves as part of a child's therapy session with cerebral palsy (Kourtesis, 2024). Apps in GDPR were mandated to have lawful grounds for data processing, use data minimization methods, and obtain verifiable parental consent. The project also used child-friendly privacy notices and interactive guardian consent dashboards. The developers also carried out Data Protection Impact Assessments (DPIAs) to evaluate risks on biometric data and algorithmic profiling. While the program was adequate in compliance with the overall standards and principles of GDPR, it also showed gaps in algorithmic transparency and the absence of pediatric-specific risk indicators (Innovation News Network, 2019). It validates how compliance can be made to function within digital pediatric care, provided institutional goodwill and regulatory clarity prevail.

18.7.8 Case study: COPPA and artificial intelligence health devices (USA)

In the United States, data operations among children under 13 years are governed by the COPPA primarily in the online space. Its provisions have, however, been extended to cover protection for AI-based health applications that children use, especially commercially available ones outside hospital environments (Zhang & Kollnig, 2024). Among the high-profile cases is an AI-driven therapeutic chatbot device with a haptic band worn by children with attention deficit disorders. It was sold as an in-home therapy device and collected behavioral and emotional information to change feedback patterns. After public outcry, the company was investigated for not providing adequate parental control settings and collecting data without end-to-end encryption. The U.S. Federal Trade Commission (FTC) recognized these activities as probable COPPA violations, which led to updating procedures for managing data (Khalid et al., 2024). The case illustrates how regulatory enforcement, though reactive, can be instrumental in correcting privacy misuse. However, the greater problem remains: COPPA does not explicitly apply to healthcare-specific devices or AI-haptic hybrids, leading to a patchwork regulatory landscape.

18.7.9 Case study: India's digital disability healthcare ecosystem

India's health technology innovation ecosystem has witnessed the emergence of digital technologies for rehabilitating disabilities, but haptic technologies are still in initial phases. Indian Institute of Technology (IIT) and All India Institutes of Medical Sciences (AIIMS) have shown a low-cost haptic interface's proof-of-concept for mobility training and neurorehabilitation. One successful pilot project in Odisha employed a tablet application and haptic enhancement to make physiotherapy possible in children with locomotor disabilities. Though promising in therapeutic returns, the project lacked a defined data protection policy. Informed consent was obtained through oral descriptions or hospital registration cards in bulk, without explicit data processing or storage disclosure (Giansanti, 2023). Children's data utilization impact assessments and institutional ethics reviews were not conducted. This case also showcases India's regulatory fragmentation, with the DPDPA and RPwD Act, but minimal enforcement or guidance on new assistive technologies. However, public health institutions, disability NGOs, and state health departments' entry into this market is evidence of opportunities for scalable, inclusive models with proper protection.

18.7.10 Lessons and best practices from comparative jurisdictions

The comparative analysis of the regimes has several vital lessons of prudence and hope for conclusions for haptic medical system regulation among disabled children. The European Union experience, in the first instance, illustrates the efficacy of child-centric compliance, where legislation such as the GDPR is framed in terms of a child's rights, and end-to-end solutions can be mapped in system design, for example, privacy dashboards and ethics audit tools (Duval, 2022). Second, the US approach focuses on enforcement: while it may be challenging to design the law COPPA perfectly for emergent technologies, aggressive enforcement methods like FTC investigations are required to discourage coercive practices. Third, Indian experience brings to the foreground that institutional integration is necessary; participation of education and health institutions at the early phases of haptic system design assures a fertile setting for participatory innovation. However, this promise has to be secured by formal policy machinery and effective ethical regulation (Marwick et al., 2024). Fourth, organizations such as UNICEF, WHO, and the UN agencies should not be ignored, as they offer globally agreed-upon standards that governments can adopt and integrate into national policy and law regimes. Finally, inclusive design generates trust. In all jurisdictions, one can see that where carers and children are included in the design, feedback, and decision-making, technology is most likely to be used effectively and ethically (Taylor, 2016). Together, these findings not only specify the difficulty of dominating haptic technologies in the care of child disability but also provide tangible templates for their successful integration into inclusive paradigms of care.

18.7.11 Regulatory blueprint for ethical haptic deployment

Creating an ethically strong and legally strong haptic medical system regulatory model will require paradigm-shifting the way the developing technology is currently being developed and regulated, disability and children's use being two particular areas of concern. This chapter presents a multi-layered solution that combines privacy-by-design, technological regulation, institutional regulation, and child-centered governance principles. It requires that the regulation must not be reactive but anticipatory, inclusive, and participatory, based on a rights approach that is reconcilable with the child protection and disability law.

18.8 Privacy-by-design and ethics-by-design standards

Privacy-by-design (PbD) is an intrinsic design philosophy for any technology handling sensitive data, particularly pediatric rehabilitation. Haptic medical systems must construct privacy as an inherent design aspect, not an add-on. It entails designing systems that collect low amounts of required data, are secure in processing, provide high-grained user control, and facilitate well-informed, dynamic consent regimes . For disabled individuals, Ethics-by-Design (EbD) must go further, ensuring the technology incorporates non-discrimination, dignity, and inclusion values. Designers must employ value-sensitive design practices like stakeholder involvement (e.g., caregivers and children) throughout the design process (Cummings, 2006). Emotional neutrality, accessibility options, and cultural sensitivity must be integrated into user interfaces and system behavior. EbD and PbD require that technical decisions be evaluated on performance and ethical alignment with the user's rights (Figs. 18.3 and 18.4).

Additionally, these two principles must be audited regularly and not taken for granted. Benchmarks like privacy impact assessments (PIAs) and ethical risk assessments must be integral to design, testing, and deployment. They must involve child rights specialists, disability rights activists, and healthcare workers in their overall review to enable a complete review.

18.8.1 Emerging technologies: Blockchain, artificial intelligence audits, federated learning

New technologies will extend trust, security, and conformance to haptic systems. Blockchain technology can, for example, be leveraged to produce irrevocable evidence of access to data and the consent timeline. Guardians can then have the ability to see when, where, and why data is being accessed, so much greater transparency and post-hoc control can be exercised. Smart contracts can be used to implement automatic recall or time-out of consent to counter time lapse, aging, or state modification. Another important mechanism is AI audits (Almarri & Aljughaiman, 2024). Given widespread

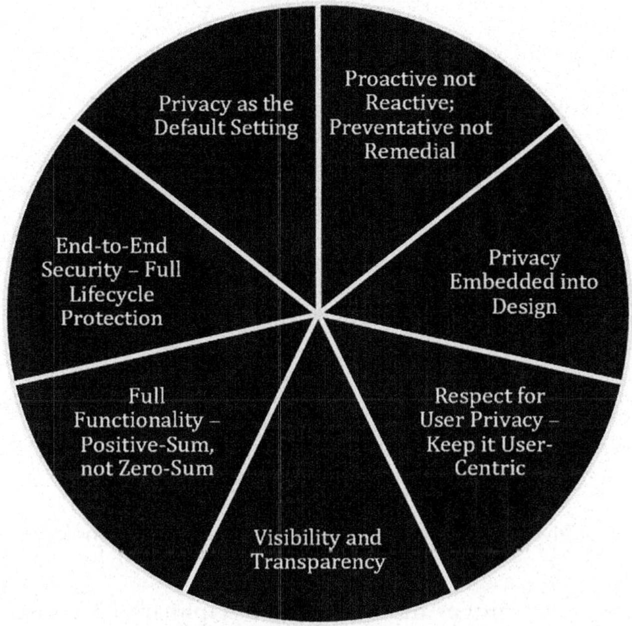

Figure 18.3 The seven foundational principles of privacy by design.

deployment of adaptive algorithms by most haptic systems, law would regularly require algorithmic verification to detect bias, attest fairness, and ensure performance on varied user segments. These would be conducted by third-party certifiers independent of open-source benchmarking tools created by developers. Where there is bias overfitting in nondisabled groups or mislabeled emotional reactions, developers would need to retrain models and keep tabs on patches. As a privacy-conscious approach to machine learning, Federated learning can best limit centralized data storage. With the federated architecture, algorithms are learned on decentralized devices without transferring raw data to central servers. Not only does this offer greater protection for data, but it also meets local data privacy regulations and minimizes potential cross-border data misuse (Nazir et al., 2023). In medicine, particularly in treating children, where data are highly sensitive, federated learning would be most useful. They must, however, be accompanied by open legal standards and technical specifications. Government certification of privacy-assuring technologies (PETs) employed in child health systems should exist.

18.8.2 Institutional mechanisms: Ethics committees and digital consent boards

Institutional governance has a crucial role in bridging principles of law to practice. Ethics committees need to be instituted at various levels, like national, regional, and organizational, to sanction, audit, and oversee the use of haptic technologies for children's

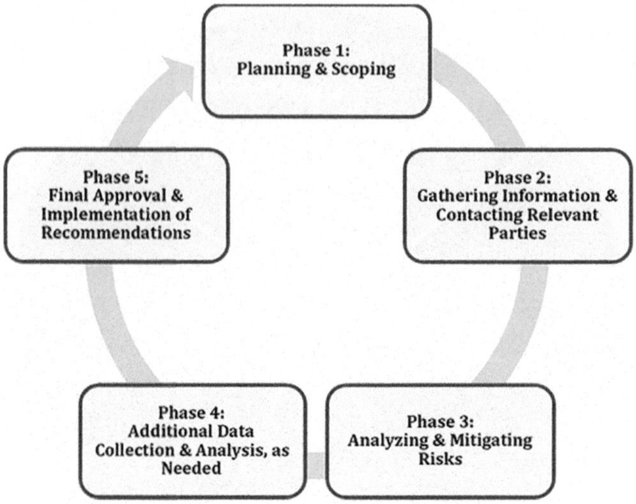

Figure 18.4 Phases of a privacy impact assessment (PIA) cycle with stakeholder engagement.

rehabilitation. These committees must be multidisciplinary in composition, including child psychology, disability rights, medical ethics, and the regulation of AI (Institute of Medicine (US) Committee on the Social and Ethical Impacts of Developments in Biomedicine et al., 1995). One institution that can be established is the Digital Consent Board (DCB). Unlike the traditional ethics review boards, a DCB would have as its exclusive mandate processes of consent, user interface design, and communication channels whereby children and their parents would be educated regarding their rights and options in cyberspace. The DCB would offer model consent forms, age-related visual aids, and multilingual support platforms to facilitate informed choices. Second, the regulatory agencies must establish a pediatric technology ethics registry where new haptic pediatric medical devices or software intended for children would be registered and undergo premarket ethics review. This will introduce transparency and accountability in the pipeline for innovation and prevent dangerous or noncompliant products from reaching the market.

18.8.3 Child-centric governance model for medical haptics

The center of any regulatory system would be a child-centered model of governance that would put children in their rightful position as rights-holders and not merely as recipients of care. The top priority for inclusivity, accessibility, agency, and long-term best interest would be child-centered model thinking and its application in developing haptic systems. A child-centered model of governance would have to operate on three concentric circles. At a micro level, user interaction needs to be child- and caregiver-friendly, with easy-to-use interfaces enabling them to easily manage data permissions, therapy modes, and device

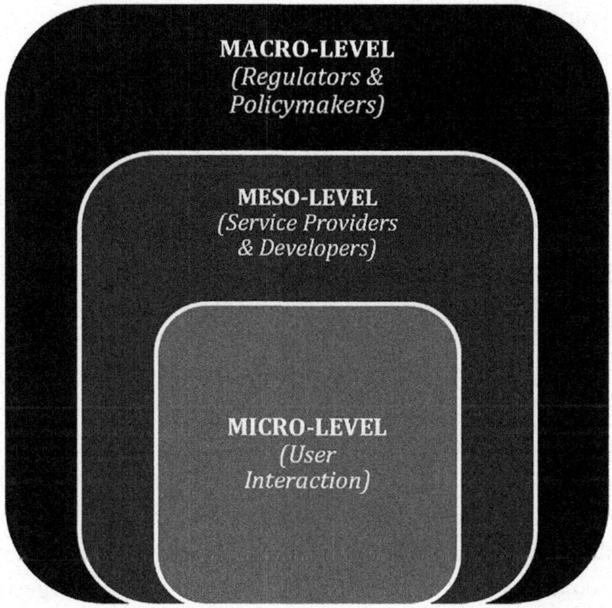

Figure 18.5 Three-tier child-centric governance model for haptic medical systems.

settings (Coombes et al., 2022). Digital literacy assistance and onboarding need to involve user training, and feedback loops built into the system must enable users to signal distress, errors, or discomfort via the system itself. At the meso-level, suppliers and developers like technology firms and medical workers must be accountable for compliance with ethical and legal procedures. These include establishing license procedures that outline ethics training, providing audit trails, and transparency through controlled report systems. At the macro-level, policymakers and regulators must implement general legislation to define pediatric haptic medical systems, their permissible use cases, and associated legal protection (Vogel-Walcutt, 2018). Also, government and standardization organizations have to create an open innovation context through networking with civil society actors, academia, and international institutions. Above all, this framework of regulation should be dynamic by nature. As technology evolves, regulatory sandboxes must be established to create space for experimentation by future innovations under a controlled environment with scrutiny to accommodate iterative learning, real-time risk management, and policy adjustment (Fig. 18.5).

18.8.4 Integration of disability rights in tech regulation

Mainstream technological regulation often fails to consider the specific legal rights of individuals with disabilities. Consequently, disability rights must be integrated into all regulatory drafting and enforcement stages. It starts by mandating universal design standards under which all the haptic systems fall under predetermined levels of accessibility through specifications like the Web Content Accessibility Guidelines (WCAG) and the Bureau of Indian Standards (BIS) guidelines on assistive devices (Swanepoel, 2020). Further, national AI and tech law must implement the United Nations Convention on the Rights of Persons with Disabilities (UNCRPD) into law, which will obligate developers and regulators to comply with Articles 4 (general obligations), 9 (accessibility), and 25 (healthcare) in digital health technology decision-making. To catalyze systemic change, governments need to subsidize inclusive innovation as tax credits and grants to projects achieving disability inclusion levels, and also encourage public-private partnerships to scale up haptic technologies for accessibility for disadvantaged groups (Bratan et al., 2020). Of similar concern is the encouragement of Disability Impact Assessments (DIAs), as with environmental or social impact assessments, which would need to be required of technology developers to consider the impact of their products on disabled consumers, i.e., along the axes of autonomy, usability, accessibility, and harm (Hodson et al., 2023). Integrating disability rights into the regulatory system satisfies international legal requirements and guarantees that technological progress for the largest and most diverse conceivable audience is made, allowing for progress without exclusion or injury.

18.8.5 Actionable strategies for policy and stakeholder integration

An integrated haptic medical systems regulation system is not feasible without all stakeholders' collective efforts, including legislators, technologists, practitioners, researchers, and civil society organization members. This chapter proposes pragmatic policies of harmonizing policy goals, legislative reforms, technology ethics, and rollouts at the systemic levels so that haptic technologies for disabled children are inclusive, safe, and rights-based. These policies are targeted toward the problem of the necessity of harmonization across sectors and the building of long-term ecosystems.

18.9 Role of legislators and law reform bodies

Neither can the role of legislators be overstated, however, in re-shaping existing law which falls short in this respect with amendments and implementing transparent policy which can regulate new technologies in the arena of childhood that are unfolding. Parliamentarians and law commissions have also been advised to give paramount importance to amending existing disability legislation, e.g., India's Rights of Persons with Disabilities (RPwD) Act, 2016, referring to AI, robots, and digital assistive technology (Brownsword, 2019). Equally important is incorporating provisions concerning children in laws like the DPDPA or GDPR-style law in a back-foot manner in response to such pressing issues as consent, profiling, and storage of data from the child's rights perspective.

Secondly, sectoral codes relating to AI use of healthcare equipment and digital medicine have to specifically address pediatric use to safeguard the interests of disabled children. Policy boards can even go a step further with regulation by establishing permanent technology assessment boards with the sole duty of examining pending bills, such as the ethical implications of new technology, and tracking follow-up on actual enactment of such systems in real-life environments (Shivadass et al., 2025). Such institutions would promote compliance with the law, risk avoidance, and realization on a rights basis. Governments can eliminate regulatory lag at once through visionary legislating, building a future-proof legal framework based on innovation as much as equity, security, and belonging.

18.9.1 Technological best Practices for developers and startups

Startups and developers in the haptic and med-tech domain must prioritize ethics in design as a driving force in their product development process, particularly while designing tools for disabled children. This is preceded by inclusive codesign, a process that starts with engaging clinicians, disabled children, and caregivers from the initial stages of development so that interfaces become real-world aware, accessible, and reflective. Bias reduction controls should also be a top priority (Johri et al., 2024). They should involve subjecting AI systems to systematic testing against a wide range of datasets to avoid discriminatory output, broken feedback loops, or the suppression of outlier user profiles. Security and privacy must be designed with technical characteristics such as end-to-end encrypted data processing and edge computing to minimize vulnerability to attacks on centralized storage and dynamic privacy controls so that guardians can specify data visibility, permissioning, and data retention parameters. Of similar significance are the applications of transparent AI technologies to facilitate transparency and demonstrate to end-users how therapeutic suggestions or haptic information is created, attributing credibility and responsibility to automated systems. In addition to these standards, developers must strive to meet internationally accepted safety and quality standards such as ISO 82304-1 for healthcare software and IEC 62304 for the medical device software lifecycle. The accomplishment of these demands establishes regulatory compliance and increases the credibility and ethical standing of the product as part of the pediatric healthcare marketplace.

18.9.2 Healthcare provider training and awareness initiatives

Physicians are on the frontline in applying and organizing haptic technology into children's daily rehabilitation routine. Still, not one has professional training in digital ethics, privacy laws, or a thorough awareness of AI-led systems for Children. The CME training courses for programs must be supplemented with specialist modules on digital consent procedures, risk sensitivity for AI, and ethical utilization of rehab technology to fill in the gaps (Reid & Eaton, 2019). While doing this, rehabilitation hospitals and clinics also need to develop in-house standard operating procedures (SOPs) that encompass the ethical application of haptic devices as part of clinical practice guidelines so that accountability, consistency, and patient safety are inherent. Cross-disciplinary workshop training should also be promoted among child psychologists, physiotherapists, medical doctors, technologists, and lawyers. The training workshops enable two-way learning of patient-centered delivery and design (Sikka, 2021). It helps professionals access the intricate dynamics of technological, ethical, and medical concerns. Each of these construction skills is vital to ensure haptic systems are being used responsibly, to facilitate therapeutic success, and to affirm and not destroy the trusting relationship between children, caregivers, and healthcare workers.

18.9.3 Public-private-academic collaboration models

Innovating and governing haptic medical systems effectively should be governed by collaborating in collaborative governance, whereby the public, private, and academic realms collaborate to marry technological advancement with social and ethical aims. Public-private-academic partnerships (PPPAS) have the potential to be such mass-scale innovation drivers and have the means of co-development, policy experimentation, and joint responsibility. The winning recipe is the establishment of regulatory sandboxes, whereby governments establish contained spaces where regulators, developers, and healthcare organizations get a chance to test new haptic solutions in short-term, targeted policy waivers. Sandboxes enable testing under real conditions with minimal legal and ethical risk. Others are to establish innovation hubs for inclusive tech, wherein universities and government institutions partner with startups to co-create low-cost, context-adapted haptic tools specifically to be employed by marginalized or underserved populations. The hubs might also serve as a means for training, prototyping, and cross-disciplinary research. Multilayered model-based data commons could also be established to support ethical AI development for children's care. Such commons allow anonymized health data to be aggregated and transferred for research and training of algorithms, under ethics board control and to cleared institutions alone. Such governance layers make it easier to protect data and ensure research integrity. Such partnership models cumulatively avert redundant effort, promote good resource sharing, and make technological innovation scalable, successful, and embedded in shared values in society and human rights.

18.9.4 Future research and innovation policy directions

Neuroadaptive interfaces, affective AI, and haptic VR-based systems are on the rise and will revolutionize pediatric medicine at an accelerated pace with immersive, personalized, and interactive disability rehabilitation. As technology advances, it must be imbued with a future ethic. Governments and research institutions must get ahead of the curve in assessing and guiding their application in child healthcare. This would involve providing grants for longitudinal research to measure developmental, emotional, and behavioral impacts of long-term use of such systems, particularly in children with disabilities who are presumably more sensitive or dependent (Goyal et al., 2022). Ethical principles would also have to be developed to guide next-generation human–machine interaction in child-focused settings, so that technology promotes autonomy, emotional well-being, and developmentally appropriate interaction. Moreover, encouraging fellowships in legal technology and interdisciplinary AI governance, digital health ethics, and disability justice PhDs will generate a next generation of leaders who can negotiate this new reality (Nguyen et al., 2023). Policy think tanks must complement such endeavors by providing national blueprints for digital children's health and AI governance and contributions from silenced global South societies. These progressive, inclusive policies will ensure rights-based, responsive policy-making and technology diffusion in child health are responsible, equitable, and sustainable. As pragmatically convenient, note policies define a harmonized approach to ensuring safety, dignity, and inclusiveness in the expanding haptic world of healthcare systems (Feldstein, 2023). Through its facilitation of coordination among legal, technical, and health interests, all stakeholders can cooperatively work toward shaping a digital future with disabled children's welfare and rights at the center stage.

18.10 Conclusion

The application of haptic medical systems in the rehabilitation of children is both a previously unimaginable possibility and a far more deeply intimidating one. Through potentially transforming the provision of healthcare for disabled children through the facilitation of responsive, participatory, and information-based therapy, these systems hold great promise. Nevertheless, as illustrated in this research, this technology provokes directly associated legal and ethical challenges that are inadequately managed under existing regulatory regimes (Pacheco-Barrios et al., 2024). The underlying issues include the lack of robust child-oriented privacy protection, uncertainty over data ownership and consent, algorithmic bias, transparency of AI decision-making, and insufficient robust institutional safeguards. These are compounded when the target group is disabled children, a group arguably the most vulnerable user population for digital health innovation.

Under such circumstances, a rights-oriented, child-centered approach is called for, with special consideration for the risks posed by haptic and AI-based healthcare devices. Frameworks like the GDPR, COPPA, and India's DPDPA set broad principles but do not get to grips with the challenges created by interactive, adaptive systems in real-time and within close care settings. National legislation has to be updated to incorporate the UN Convention on the Rights of the Child and the UN Convention on the Rights of Persons with Disabilities commitments, such as consent, accessibility, and inclusive access to technology. The future necessitates creating an inclusive and safe digital health system with access, privacy, and ethics as implicit elements of innovation. That is, implementing privacy-by-design and ethics-by-design approaches, using privacy-enabling technologies including federated learning and blockchain, and child-centric governance models enabled through ethics committees and consent boards. The ecosystem must be able to ensure not only defense against harm but also the enablement of the dignity, autonomy, and inclusion of disabled children. Lastly, bringing such a vision into reality will be a multistakeholder exercise (Sun et al., 2020). Legislatures will need to provide a responsive legal framework, developers will need to adopt the responsible design culture, physicians will need to acquire digital literacy and child rights consciousness, and transnational institutions must provide convergence support. The ethical application of haptic medicine technology is far from a technical issue alone, but a social response to weigh the advancement of technology against equal, balanced, and free medical care for every child.

References

Team C., Guide to India's Digital Personal Data Protection Act (DPDP Act). cookieyes. (2024),

Almarri, S., & Aljughaiman, A. (2024). Blockchain technology for IoT security and trust: A comprehensive SLR. *Sustainability, 16*(23), 10177. doi:10.3390/su162310177.

Bhargava, H., Salomon, C., Suresh, S., Chang, A., Kilian, R., van Stijn, D., Oriol, A., Low, D., Knebel, A., & Taraman, S. (2024). Promises, pitfalls, and clinical applications of artificial intelligence in pediatrics. *Journal of Medical Internet Research, 26*(1), e49022. doi:10.2196/49022.

Bratan, T., Fischer, P., Maia, M., & Aschmann, V. (2020). Implementation of the UN convention on the rights of persons with disabilities: A comparison of four european countries with regards to assistive technologies. *Societies, 10*(4), 74. doi:10.3390/soc10040074.

Brownsword, R. (2019). Law disrupted, law re-imagined, law re-invented. *Technology and Regulation, 2019*, 10–30. doi:10.71265/ggx02j62.

Bulger, R. E., Bobby, E. M., & Fineberg, H. VInstitute of Medicine (US) Committee on the Social and Ethical Impacts of Developments in Biomedicine. (1995). Institutional ethics committees: Local perspectives on ethical issues in medicine. *Society's Choices: Social and Ethical Decision Making in Biomedicine*. National Academies Press.

Chawki, M., Basu, S., & Choi, K. S. (2024). Redefining boundaries in the metaverse: Navigating the challenges of virtual harm and user safety. *Laws, 13*(3). doi:10.3390/laws13030033.

Coombes, L., Braybrook, D., Roach, A., Scott, H., Harðardóttir, D., Bristowe, K., Ellis-Smith, C., Bluebond-Langner, M., Fraser, L. K., Downing, J., Farsides, B., Murtagh, F. E. M., & Harding, R. (2022). Achieving child-centred care for children and young people with life-limiting and life-threatening conditions—A qualitative interview study. *European Journal of Pediatrics, 181*(10), 3739–3752. doi:10.1007/s00431-022-04566-w.

Cooray, N., Sun, S., Ho, C., Adams, S., Keay, L., Nassar, N., & Brown, J. (2021). Toward a behavior theory–informed and user-centered mobile app for parents to prevent infant falls: Development and usability study. *JMIR Pediatrics and Parenting, 4*(4), e29731. doi:10.2196/29731.

Crandall, R., & Karadoğan, E. (2021). Designing pedagogically effective haptic systems for learning: A review. *Applied Sciences, 11*(14), 6245. doi:10.3390/app11146245.

Cummings, M. L. (2006). Integrating ethics in design through the value-sensitive design approach. *Science and Engineering Ethics, 12*, 701–715. doi:10.1007/s11948-006-0065-0.

Dankar, F. K., Gergely, M., & Dankar, S. K. (2019). Informed consent in biomedical research. *Computational and Structural Biotechnology Journal, 17*, 463–474. doi:10.1016/j.csbj.2019.03.010.

De Fazio, R., Mastronardi, V. M., De Vittorio, M., & Visconti, P. (2023). Wearable sensors and smart devices to monitor rehabilitation parameters and sports performance: An overview. *Sensors, 23*(4), 1856. doi:10.3390/s23041856.

Duval, J. S., Playful health technology: A participatory, research through design approach to applications for wellness. ProQuest. (2022).

Feldstein, S., New digital dilemmas: Resisting autocrats, navigating geopolitics, confronting platforms. Carnegie Endowment for International Peace. (2023),

Felin, T., & Holweg, M. (2024). Theory is all you need: AI, Human cognition, and causal reasoning. *Strategy Science, 9*(4), 346–371. doi:10.1287/stsc.2024.0189.

Giansanti, D. (2023). *Assistive technologies, robotics, and automated machines in the health domain*. MDPI. doi:10.3390/books978-3-0365-7688-6.

Goyal, C., Vardhan, V., & Naqvi, W. M. (2022). Haptic feedback-based virtual reality intervention for a child with infantile hemiplegia: A case report. *Cureus, 14*(3). doi:10.7759/cureus.23489.

Habbal, A., Hamouda, H., Alnajim, A. M., Khan, S., & Alrifaie, M. F. (2024). Privacy as a Lifestyle: Empowering assistive technologies for people with disabilities, challenges and future directions. *Journal of King Saud University - Computer and Information Sciences, 36*(4). doi:10.1016/j.jksuci.2024.102039.

Haleem, A., Javaid, M., Singh, R. P., & Suman, R. (2022). Medical 4.0 technologies for healthcare: Features, capabilities, and applications. *Internet of Things and Cyber-Physical Systems, 2*, 12–30. doi:10.1016/j.iotcps.2022.04.001.

Hine, E., Neroni Rezende, I., Roberts, H., Wong, D., Taddeo, M., & Floridi, L. (2024). Safety and privacy in immersive extended reality: An analysis and policy Recommendations. *Digital Society, 3*(2). doi:10.1007/s44206-024-00114-1.

Hodson, E., Vainio, T., Sayún, M. N., Tomitsch, M., Jones, A., Jalonen, M., Börütecene, A., Hasan, M. T., Paraschivoiu, I., Wolff, A., Yavo-Ayalon, S., Yli-Kauhaluoma, S., & Young, G. W. (2023). Evaluating social impact of smart city technologies and services: Methods, challenges, future directions. *Multimodal Technologies and Interaction*, 7(3). www.mdpi.com/journal/mti. doi:10.3390/mti7030033.

Hoofnagle, C. J., Sloot, B. v. d., & Borgesius, F. Z. (2019). The European Union general data protection regulation: What it is and what it means. In *Information and Communications Technology Law: 28* (pp. 65–98). doi:10.1080/13600834.2019.1573501.

Innovation News Network, Children and GDPR: Protecting children's data online. (2019),

Jafari, N., Adams, K. D., & Tavakoli, M. (2016). Haptics to improve task performance in people with disabilities: A review of previous studies and a guide to future research with children with disabilities. *Journal of Rehabilitation and Assistive Technologies Engineering, 3*. doi:10.1177/2055668316668147.

Johri, A., Sayal, A., Chaithra, N., Jha, J., Aggarwal, N., Pawar, D., Gupta, V., & Gupta, A. (2024). Crafting the techno-functional blocks for Metaverse - A review and research agenda. *International Journal of Information Management Data Insights, 4*(1), 100213. doi:10.1016/j.jjimei.2024.100213.

Kapoor, S., Arora, P., Kapoor, V., Jayachandran, M., & Tiwari, M. (2014). Haptics - Touchfeedback technology widening the horizon of medicine. *Journal of Clinical and Diagnostic Research, 8*(3), 294–299. doi:10.7860/JCDR/2014/7814.4191.

Khalid, U., Naeem, M., Stasolla, F., Syed, M., Abbas, M., & Coronato, A. (2024). Impact of AI-powered solutions in rehabilitation process: Recent improvements and future trends. *International Journal of General Medicine, 17*, 943–969. doi:10.2147/ijgm.s453903.

Kourtesis, P. (2024). A comprehensive review of multimodal XR applications, risks, and ethical challenges in the metaverse. *Multimodal Technologies and Interaction, 8*(11), 98. doi:10.3390/mti8110098.

Lawson, A., & Beckett, A. E. (2021). The social and human rights models of disability: towards a complementarity thesis. *The International Journal of Human Rights, 25*(2), 348–379. doi:10.1080/13642987.2020.1783533.

Li, Y., & Liu, Q. (2021). A comprehensive review study of cyber-attacks and cyber security; Emerging trends and recent developments. *Energy Reports,* 7, 8176–8186. doi:10.1016/j.egyr.2021.08.126.

Marwick, A., Smith, J., Caplan, R., Wadhawan, M., Child online safety legislation (COSL) - A primer. The Bulletin of Technology & Public Life. (2024), 10.21428/bfcb0bff.de78f444.

McGraw, D., & Mandl, K. D. (2021). Privacy protections to encourage use of health-relevant digital data in a learning health system. *npj Digital Medicine, 4*(1). doi:10.1038/s41746-020-00362-8.

Mladenov, T. (2013). The UN Convention on the rights of persons with disabilities and its interpretation. *Alter,* 7(1), 69–82. doi:10.1016/j.alter.2012.08.010.

Nazir, S., Dickson, D. M., & Akram, M. U. (2023). Survey of explainable artificial intelligence techniques for biomedical imaging with deep neural networks. *Computers in Biology and Medicine, 156*. doi:10.1016/j.compbiomed.2023.106668.

Nguyen, A., Ngo, H. N., Hong, Y., Dang, B., & Nguyen, B. P. T. (2023). Ethical principles for artificial intelligence in education. *Education and Information Technologies, 28*(4), 4221–4241. doi:10.1007/s10639-022-11316-w.

Nissenbaum, H. (2011). A contextual approach to privacy online. *Daedalus, 140*(4), 32–48. doi:10.1162/daed_a_00113.

Pacheco-Barrios, K., Ortega-Márquez, J., & Fregni, F. (2024). Haptic Technology: Exploring its underexplored clinical applications—A systematic review. *Biomedicines, 12*(12), 2802. doi:10.3390/biomedicines12122802.

Pantanowitz, L., Hanna, M., Pantanowitz, J., Lennerz, J., Henricks, W. H., Shen, P., Quinn, B., Bennet, S., & Rashidi, H. H. (2024). Regulatory aspects of artificial intelligence and machine learning. *Modern Pathology, 37*(12). doi:10.1016/j.modpat.2024.100609.

Przegalinska, A. (2019). *Wearable technologies in organizations: Privacy, efficiency and autonomy in work*. Springer International Publishing. doi:10.1007/978-3-030-00907-6.

Pujari, R., India—Digital health laws and regulations 2025. Conventus Law. (2025).

Ramones, A. F., & del-Rio-Guerra, M. S. (2023). Recent developments in haptic devices designed for hearing-impaired people: A literature review. *Sensors, 23*(6), 2968. doi:10.3390/s23062968.

Reid, J. E., & Eaton, E. (2019). Artificial intelligence for pediatric ophthalmology. *Current Opinion in Ophthalmology, 30*(5), 337–346. doi:10.1097/ICU.0000000000000593.

Richards, N. M., Hartzog, W., Privacy's constitutional moment and the limits of data protection. (2020),

Shivadass, P., Gayathri, S., Ananya, K., Indian perspective on protecting children's personal data. Indian Business Law Journal (2025),

Sikka, N. (2021). *A practical guide to emergency telehealth*. Oxford University Press. doi:10.1093/med/9780190066475.001.0001.

Sun, N., Esom, K., Dhaliwal, M., & Amon, J. J. (2020). Human rights and digital health technologies. *Health and Human Rights, 22*(2), 21–32.

Swanepoel, M. (2020). Aspects of the impact of the United Nations Convention on the Rights of Persons with Disabilities on South African Health Law: Section 1. *Forensic Science International: Mind and Law, 1*, 100014. doi:10.1016/j.fsiml.2020.100014.

Taylor, A. L. (2016). *Global Health Law: International law and public health policy International Encyclopedia of Public Health* (pp. 268–281). Elsevier Inc. doi:10.1016/B978-0-12-803678-5.00238-1.

Varkey, B. (2021). Principles of clinical ethics and their application to practice. *Medical Principles and Practice, 30*(1), 17–28. doi:10.1159/000509119.

Vayena, E. (2003). *Current practices and controversies in assisted reproduction: Report of a meeting*. WHO, Department of reproductive health.

Vogel-Walcutt, J. J. (2018). *Modernizing learning: Advanced distributed learning*. United States Government Printing Office.

Xie, Y., Fadahunsi, K. P., Kelleher, C., Tarn, D. M., Grace, A., & O'donoghue, J. (2025). Towards an inclusive digital health ecosystem. *Bulletin of the World Health Organization, 103*(2), 170–173. doi:10.2471/BLT.24.292020.

Zhang, L. u, & Kollnig, K. (2024). Theory and practice: The protection of children's personal information in China. *International Data Privacy Law, 14*(1), 37–52. doi:10.1093/idpl/ipad017.

Index

Page numbers followed by "*f*" and "*t*" indicate, figures and tables respectively.

A

Acoustic time reversal, 69
Active touch, 32
Actuation
  mechanisms, 64
  technology, 184*t*
Actuators, 86, 126, 130
  design criteria, 88*t*
  haptics, 94
  integration, 94
  performance requirements, 88*t*
  technologies, 90*t*
Adaptive Braille screens, 49
Adaptive learning tools, 50
Adaptive systems, 110
Advanced actuation technology, 252
Algorithmic opacity, 362
Anterolateral pathway receptors, 29
Aristotle's illusion, 35
Armeo-based rehabilitation programs, 120
Artificial intelligence
  adaptive actuators, 99
  adaptive feedback systems, 16
  adaptive haptics, 252
  algorithms, 126
  applications, 130
  bias, 362
  enhanced telemedicine, 117
  haptics, 115
  integration, 16
  mediated sensory data, 115
  medical training, 131
  prosthetics, 135
  rehabilitation, 135
  remote surgery, 132
  robotics and, 317
  robotic surgery, 132
  role, 129*f*
  simulation, 131
  telemedicine, 132
  transformations, 128*f*
Artificial intelligence-driven haptic systems, 129*f*
  autonomous medical systems and, 165
  computational optimization, 161, 162*f*
  ethical and data security concerns, 159, 160*f*
  ethical barriers, 158
  technical barriers, 158
Artificial intelligence-operated haptic simulation technology, 177
Artificial intelligence-powered robotic surgery systems
  categorization, 338, 338*t*
  integration, 338, 339*t*
  patient selection, 338
  research design, 337*f*, 337
  statistical analysis, 340, 341*f*
  surgical procedures, 340*f*
  variables and data collection, 339, 340*t*
  workflow, 339, 340*f*
Artificial muscles, 92
Artificial skin technology, 43
Assist-as-needed (AAN) therapy, 246
Assistive devices, 193
Atypical somatosensory processing, 38
Augmented reality, 75, 260
  academic implementations, 264
  anatomy, 263
  case studies, 264, 266*t*
  clinical applications, 262, 263*t*
  commercial implementations, 264
  haptic integration hardware, 261
  interventional training, 262
  palpation training, 263
  physical therapy, 263
  procedural training, 262
  rehabilitation, 263
  surgical training applications, 262
Augmented tactile feedback, 233
Automated speech-recognition devices, 49

B

Bilateral control systems, 230

Biomimetic devices, 68
Blockchain-oriented approach, 141
Bone-related minimally invasive procedures, 213
Brain-computer interface, 279*f*, 280*f*
  advances in material sciences, 294
  agency, 290
  autonomy, 290
  bandwidth, 288
  biocompatibility, 288
  classification by invasiveness, 273, 275*f*
  constraints, 288
  electronics miniaturization, 294
  equitable access, 290
  haptic feedback and, 271
  with haptics, 271
  latency, 288
  mental privacy, 290
  multidimensional optimization, 288
  multimodal integration, 293
  portability, 289
  privacy, 290
  resolution, 288
  safety, 288
  scientific challenges, 292
  sensory roles, 273
  system design considerations, 288
  technical challenges, 292
  unprecedented ethical challenges, 290
Brainstem processing, 29
Brushless direct current (BLDC), 87

C

Calibration methods, 14
Capacitive sensing technologies, 225
Capacitive sensors, 225
Catheter-based minimally invasive procedures, 214
Cerebral palsy, 327
Cervical collars, 74
Children's Online Privacy Protection Act (COPPA), 353
Classical haptics control systems, 150
Closed-loop bidirectional mapping, 283
Closed-loop brain–computer interface
  computational backbone, 284
  decoding algorithms, 285
  error suppression, 284
  machine learning, 284
  processing of intracortical signals, 285
  processing signals, 284
  signal processing, 284
  system, 278*f*
Cloud-based tele-rehabilitation, 254
Cochlear implant surgery, 330, 331*f*, 332*f*
Cognitive impairment, 51
Cognitive-motor abilities, 50
Cognitive rehabilitation systems, 48
Communication brain–computer interfaces, 287
Complex surgery, 326
Computer-aided surgery platforms, 3
Computer haptics, 59
Computer vision, 138
Conjugated polymer actuators, 93
Control algorithms, 5
Control systems microcontrollers, 64
Conventional actuator technology, 87
Cortical microstimulation, 282
Cross-validated dual force sensors, 111
C-tactile afferents, 36
Cutaneous feedback, 280
Cutaneous rabbit illusion, 35
Cutaneous receptors
  anatomy, 24
  physiology, 24
Cutaneous somatosensation, 23
Cutting-edge technology, 143
Cyberattacks, 139

D

Data
  privacy, 115, 142
  security, 142
Data protection impact assessments (DPIAs), 142
da Vinci Robotic Surgery System, 47
da Vinci Surgical System, 134, 169, 335, 342
Deep learning
  models, 128, 137
  role, 128
Dental simulators, 73
Dermal-epidermal junction, 25
Dermis, 24
Dielectric elastomer actuators (DEAs), 93
Dielectric elastomers, 92
Digital space, 295
Direct-current motors, 7
Disability rights, 369
Doorsal column–medial lemniscal pathway, 29

Down syndrome, 52
DXRacer gaming chair, 75, 76*f*

E

Eccentric rotating mass (ERM), 94
Elastomer-based systems, 91
Elastomeric soft pneumatic actuators, 85
Electric motors, 87
Electroactive polymer, 85
  actuators, 93
  artificial muscles, 92
Electrocorticography, 276
Electroencephalography systems, 276
Electrophysiology, 22
Electrotactile feedback, 68
Electro-tactile systems, 77
Encryption, 142
Endoscopy, 262
Epidermis, 24
Ethical haptic deployment, 366
Ethics-by-design standards, 366
Exoskeleton devices, 48
Exoskeletons, 228, 262
Exteroception, 23
Eye procedures, 336

F

Feedback
  force, 105, 107*t*, 126
  kinesthetic, 6
  tactile, 6
  vibrotactile, 6
Fiber Bragg Grating (FBG) sensor, 106, 222*f*, 224
Flexible tactile sense, 17
Flight simulators, 79*f*
Force compliance control, 181
Force feedback, 105, 107*t*, 126, 227
  devices, 262
  joysticks, 7
  systems, 77
Fully invasive brain–computer interfaces, 274

G

Gamification, 47
Gaming controllers, 75
Gastrectomy, 335
Gaussian process regression, 109
General Data Protection Regulation (GDPR), 141, 364
Gesture-based haptic systems, 46

H

Handwriting techniques, 50
Haptic feedback, 61, 74, 78, 126, 139, 259
  actuators, 106
  advantages, 12
  algorithm, 64
  applications, 9
  artificial palpation, 213
  case studies, 212
  challenges, 211
  combined methods, 279
  controlled robotic therapy systems, 12
  conventional laparoscopic surgery, 206
  delicate tissue dissection, 213
  design approach, 212
  dimensions, 211
  function, 61, 62*f*
  functional benefits, 286
  fundamentals, 4
  harmonious integration, 64
  integrated systems, 135
  knot tying, 213
  laparoscopic suturing, 213
  limitations, 14
  mechanisms, 182
  minimally invasive surgery, 203, 209
  muscle sense, 279
  performance evaluation and metrics, 212
  practical applications, 212
  rehabilitation robotics, 241
  robot-assisted surgery and, 309
  robotic-assisted minimally invasive surgery, 207
  skin sensations, 279
  stability, 15
  strategies, 212
  surgical training simulators, 207, 214
  systems, 139
  vs. tactile, 65*t*
  technology, 47
  types, 104, 241
  vessel handling, 213
  vs. vibrations, 65*t*
Haptic feedback implications, 215
  clinical adoption, 215
  clinical practice, 215
  ethical, 216
  healthcare systems and policy, 217
  human–machine interaction, 216

social, 216
surgical training and skill development, 215
system design and technology development, 216
Haptic interface, 8
community, 117
rehabilitation robotics, 239
Haptic medical devices
actuation technologies, 189*f*
adaptive control, 189
admittance control, 188
case studies, 192
challenges, 195
control architecture, 183
control strategies for energy efficiency, 188
electronics, 191
energy consumption, 185, 187*f*
energy efficiency, 181
event-based control, 189
fundamentals, 182
impedance control, 188
low-power actuator design, 186
management, 191
model predictive control, 190
open research directions, 195
optimal control, 190
power constraints in medical applications, 185
sensing architecture, 183
sources of energy loss, 185
system-level optimization, 191
Haptic medical robotics
challenges in experimental evaluation, 311
clinical standards validation, 303, 306
control of prostheses and artificial limbs, 309
ethical concerns, 313
evaluation metrics and, 312
latency, 304
legal concerns, 313
medical education and simulation, 309
multisensory feedback, 305
reproducibility and, 312
safety standards, 303, 310
safety validation, 306
sensory complexity, 304
sensory inputs, 305
tele-surgery and, 308
therapeutic intervention and rehabilitation, 308
use cases, 307
Haptic medical robots, 169, 171
advantages, 172
applications, 172, 173
artificial intelligence, 176
benefits, 173
components, 171
constraints, 173
defined, 171
limitations, 172
machine learning, 176
types, 171
working principle, 171
Haptic medical systems, 354, 355*f*
applications in surgery, 149, 150*f*
artificial intelligence audits and, 366
awareness initiatives, 371
bioethics and pediatric digital health, 356, 357*f*
blockchain and, 366
comparative legal matrix, 359, 359*t*
complex medical interventions, 153
components, 148*f*
diagnosis, 149
for disabled children, 353
dynamic medical interventions, 153
ethics committees and digital consent boards, 367
federated learning and, 366
feedback loop, 152*f*
force accuracy constraints, 150, 151*f*
force feedback mechanisms, 147
fundamentals, 147
healthcare provider training, 371
human rights and disability models, 355
interdisciplinary framework, 357
international legal instruments, 358
rehabilitation, 149
response latency constraints, 152
tactile perception mechanisms, 147
theories of data protection and consent, 356
Haptic rehabilitation, 239
clinical applications, 250
predictive modeling, 248
reinforcement learning, 248
Haptic rehabilitation robotics
adaptive machine learning approach, 248
control strategies, 246
force control methods, 246
human-in-the-loop control considerations, 249
interaction dynamics, 246
safety considerations, 249

stability considerations, 249
Haptic rehabilitation systems
challenges, 254
clinical outcomes, 253*t*
performance metrics, 253*t*
Haptics, 42, 73
actuation fundamentals, 86
actuators, 7
brain–computer interfaces with, 271
devices, 65, 125
driven telesurgery systems, 113
equipped training tools, 78
gloves, 63*f*, 228, 261
interaction fundamentals, 240
medical systems, 125
medicine, 46
pen, 50
perception, 86, 104
physics, 261
role, 108
sensation, 261
sensors, 7
simulators, 11
styluses, 261
Haptic systems, 127*f*, 128*f*
components, 5
control architectures, 242
hardware components, 106
Haptic technology, 41, 42, 44*f*, 59, 61, 74, 80, 125, 130*f*
enabled gamification, 47
fundamentals, 104, 126
issues, 81
security practices, 81
Healthcare robotic technology, 3
Health Insurance Portability and Accountability Act (HIPAA), 140
High-end robotic surgery systems, 134
High-resolution encoder-driven servo motors, 87
High-tech full-body haptic feedback suit, 75
Human–computer interaction (HCI), 65
Human haptic perception, 108, 240
Human haptic system, 59
Human–robot interaction control frameworks, 230
Hybrid feedback system, 284*f*
Hydraulic actuators, 89

I

Inner speech decoding, 294
Innovative learning systems, 41
Intelligent haptic control systems, 154
adaptation, 156
anomaly detection, 158
architectural design, 154, 155*f*
artificial intelligence frameworks, 147, 154
closed-loop architecture, 156
computational efficiency and, 157
edge artificial intelligence and, 157
fault prediction, 158
forecasting tactile interactions, 157
human-in-the-loop systems, 157
learning mechanism, 155
machine learning integration, 156*f*, 156
performance optimization, 158
predictive analytics for performance maximization, 157
real-time feedback loops integration, 156*f*, 156
sensory processing, 155
Intracortical brain–computer interface neuro prostheses, 287
Intracortical microelectrode arrays, 277
Invasive surgery laparoscopic instruments, 73*f*
Ionic polymer-metal composites, 92

J

Joystick, 67*f*

K

Kinesthetic feedback, 71*f*, 71, 85, 105, 261, 281
Kinesthetic system, 86

L

Laparoscopic surgery, 231
Laparoscopy, 262, 326
Latency, 14
Lateral spinothalamic tract, 29
Lightweight wearable controllers, 229
Linear resonant actuator (LRA), 62, 94
Lokomat gait rehabilitation robot, 96

M

Machine haptics, 59
Machine learning
adaptive feedback, 109
algorithms, 128, 133
implementation, 109
methods, 16
models, 126

role, 128
Magnetic resonance imaging-compatible robotics, 327, 329*f*
Magnetorheological actuators, 93
McKibben pneumatic artificial muscles, 92
Mechanoreceptor, 24
  density, 26
  distribution, 26
  types, 27*t*
Medical education, 259
Medical exoskeletons, 8
Medical haptics
  actuation, 99
  child-centric governance model, 368, 369*f*
  interface development, 113*t*
Medical robotics
  advanced sensor technologies, 321
  advances, 319, 320*f*, 321*f*
  artificial intelligence and, 320
  autonomous systems, 322
  clinical implications, 323*t*
  integration with emerging technologies, 322
  machine learning and, 320
  semiautonomous systems, 322
  systems, 106, 115
Medical training
  augmented reality, 259
  virtual reality, 259
Meissner corpuscles, 25
Merkel and Meissner receptors, 33
Merkel cells, 25
Metaverse, 80*f*, 80
Microelectromechanical systems (MEMS)
  actuators, 85
  fabrication techniques, 93
  haptics, 93
  sensors, 7
Microsurgery, 231
Minimally invasive brain–computer interfaces, 274
Minimally invasive procedures, 214, 324, 325*f*
Minimally invasive surgery, 312
  conceptual framework, 204
  early studies on sensory deprivation, 206
  fidelity in haptic interaction, 205
  force estimation, 208
  fundamentals, 204
  haptic feedback, 203, 209
  human factor, 206, 208
  passivity theory, 206
  perceptual considerations, 206
  sensing estimation, 208
  sensorimotor control theory, 205
  surgeon-centered evaluations, 208
  surgical interaction, 205
  tactile feedback approach, 207
  theoretical framework, 204
  transparency, 205
  vibrotactile feedback approach, 207
Mixed reality, 261
Motion-capture technology, 46
Motor deficits, 12
Motor systems, 32
Multimodal feedback, 106, 107*t*
Multimodal haptic medical robots, 169, 172*f*, 172, 176
  cloud and data infrastructure, 172
  development and integration, 175
  medical applications, 176
Multimodal sensing systems, 226
Multimodal sensory fusion, 117
Multimodal systems, 45

N

Naturalistic sensation evoking, 293
Natural language processing (NLP), 138
Natural orifice transluminal endoscopic surgery (NOTES), 96
Needle placement, 326
Neural channels, 281
Neural coding principles, 30
Neural networks, 16
Neurodegenerative disease therapy, 250
Neuroimaging, 33
Neurosurgery, 334
Next-generation artificial intelligence models, 163, 164*f*
Nickel-titanium alloys, 92
Nitinol, 92
Nociception, 23
Nonbiomimetic devices, 68
Noninvasive brain–computer interface, 273, 292
Noninvasive minimally invasive procedures, 325
Novel neural stimulation, 294

O

Optogenetics, 277

Orthopedics
minimally invasive procedures, 213
rehabilitation, 251
robot surgery, 134
surgery, 331, 332*f*
trauma simulation, 264
Orthosis, 73
Oxytocin, 36

P

Pacinian corpuscles, 26
Passive touch, 32, 188
Passivity observer/passivity controller (PO/PC) model, 249
Perception
human haptic, 21, 108
smartphones, 61*f*
Performance evaluation metrics, 194, 195*t*
Peripheral channels, 281
Peripheral stimulation encoding, 282
Personal assistant, 48
Phantom limb sensation, 35
Physical touchable haptic technology, 68
Piezoelectric actuators, 7
Piezoelectric devices, 183
Piezoelectric sensors, 44, 225
Piezoresistive sensors, 225
Pin-array matrices, 49
Pneumatic actuators, 7
Pneumatic artificial muscles (PAMs), 92
Pneumatic systems, 7
Posterior parietal cortex, 31
Precision surgeries, 326
Predictive algorithms, 126
Pressure sensors, 7
Primary afferent neurons, 28
Primary somatosensory cortex, 30, 31
Privacy-by-design, 366, 367*f*
Privacy impact assessment (PIA) cycle, 368*f*
Proprioception, 23, 70
Prosthetic control, 286
Prosthetic limbs, 135
Pseudo-haptics, 45
Public-private-academic collaboration models, 372

Q

Quantum computing, 143

R

Rehabilitation, 48
artificial intelligence, 135
augmented reality, 263
devices, 193
orthopedics, 251
stroke, 250
virtual reality, 263
Rehabilitation haptic devices, 247*t*
actuation methods, 244, 245*f*
anatomical classification, 245
architectural differences, 246
classification, 245
design principles of interfaces, 243
mechanical design considerations, 243
sensor technologies, 244
Rehabilitation robotics, 12
haptic interfaces in, 239
for patients with cerebral palsy, 327, 329*f*, 330*f*
Reinforcement learning, 139
Remote surgery, 327
Resistive sensing technology, 225
Retinal vein occlusion (RVO), 333, 336
Rights of Persons with Disabilities (RPwD) Act, 370
Robot-assisted minimally invasive surgery, 96
Robotic-assisted prostatectomy
with artificial intelligence enhancement, 345*f*, 345
clinical consequences, 346
intraoperative artificial intelligence integration, 346, 347*t*
preoperative assessment, 345
Robotic-assisted surgery, 113, 262, 307, 322
challenges, 342, 343*t*
obstacles, 341
procedure, 324*f*
Robotics, 78
exoskeleton joints, 7
neurosurgery, 334, 335*f*
palpation systems, 137
platforms, 3
procedures, 47
Robotic surgery
advantages, 134
Kinesthetic feedback, 86
systems, 120
Robot operating system (ROS), 134

Ruffini endings, 26

S

Secondary somatosensory cortex, 31
Security, 115
Semiactive haptics, 188
Sensor–actuator coordination, 95
Sensors, 126, 130, 135
Sensory deprivation, 49
Sensory feedback system, 45, 62*f*
Sensory processing, 60*f*
Sensory redundancy, 111
Sensory substitution, 282
Series elastic actuator (SEA), 98
Shape memory alloys (SMAs), 85, 92
Silicone elastomers, 91
Simulators, 78
Skin, 59
  cutaneous mechanoreceptors, 25*f*
  hairy, 24
  layers, 24
  shear feedback, 227
  stretch feedback, 227
Smart haptic sensors, 16
Smartphones, 74
Soft pneumatic actuators (SPAs), 91
Soft robotics
  actuators, 91
  approach, 17
  wearable, 252
Software integration, 64
Somatosensory pathways, 30
Somatosensory system, 23, 59, 60*f*
Spinal cord, 29
Spring-mass-damper system, 94
Stereotactic electroencephalography, 276
Stimulation technologies, 277
Stroke rehabilitation, 250
Surgeon wearable haptic devices, 18
Surgical robotics, 3
Surgical robots, 192, 328*t*
Surgical simulator digital twins, 117
Surgical training simulators, 11

T

Tablets, 74
Tactile, 45
  illusions, 35
  sensing technology, 16
  sensor arrays, 16
Tactile feedback systems, 85, 261
  brain–computer interfaces, 234
  device, 77
  machine learning, 233
  for telesurgery, 221
  training system integration, 232
  validation studies, 232*t*
  virtual reality integration, 232
Telemedicine, 97
Teleoperated surgical systems, 10
Teleoperation control strategies, 230
Teleoperators, 78
Tele-rehabilitation, 251
Tele-surgery, 327
Telesurgical tactile feedback, 221
  challenges, 223, 224
  mechanisms, 226
  technologies, 224
  wearable haptic interfaces, 228
Tesla suit, 75, 76*f*
Thalamic neurons, 30
Thermal feedback, 69, 70*f*, 228
Thermal grill illusion, 35
Thermal haptic feedback, 106, 107*t*
Thermosreception, 23
Total knee arthroplasty, 333
Touch
  affective dimensions, 36
  based feedback systems, 48
  emotional and social functions, 36
  gentle, 36
  perceptual phenomena, 34
  psychophysics, 34
  receptors, 24, 28
  role, 37
  social dimensions, 36
  temporal acuity, 35
Touchable haptic technology, 66
Touch Mechanoreceptors, 280
Touch perception
  definitions, 21
  scope, 21
Touchscreen haptic systems, 45
Touch technology, 279,
    see also Haptics
Traditional haptic systems, 152*f*, 152

Training simulators, 97
Tumor localization, 213
Twisted and coiled polymer (TCP) actuators, 92

U

Ultrasonic tactile feedback, 69*f*, 69
Unencapsulated mechanoreceptors, 28
United Nations Convention on the Rights of Persons with Disabilities (UNCRPD), 369
Upper-limb prosthetics, 313
Urological procedures, 231
Urology, 336

V

Ventral posterolateral, 30
Vibration method, 70
Vibrotactile actuators, 96
Vibrotactile feedback, 105, 107*t*, 226
Video games, 75
Virbotactile, 63*f*
Virtual environment, 47*f*
Virtual reality, 75, 260
  academic implementations, 264
  anatomy training, 263
  case studies, 264, 266*t*
  clinical applications, 262, 263*t*
  commercial implementations, 264
  environments, 126
  glove, 75
  haptic integration hardware, 261
  interventional training, 262
  medicine, 97
  palpation training, 263
  physical therapy, 263
  procedural training, 262
  rehabilitation, 263
  surgical training applications, 262
Virtual simulators, 65
Visual haptics, 45
Visual impairment, 49

W

Wearable devices, 71, 72*f*, 262
Wearable haptic gloves, 8
Wearable haptic technology, 66, 68
Wearable medical devices, 94
Web Content Accessibility Guidelines (WCAG), 369
Wireless communication, 82

Printed and bound by CPI Group (UK) Ltd, Croydon, CR0 4YY
14/09/2026
02216409-0012